Advanced Brake Technology

Edited by

Bert Breuer

and

Uwe Dausend

SAE International™

Warrendale, Pa.

Other SAE titles of interest:

Brake Design and Safety (Second Edition)
By Rudolf Limpert
(Order No. R-198)

Electronic Braking, Traction, and Stability Controls
Edited by Ronald K. Jurgen
(Order No. PT-76)

ABS–TCS–VDC: Where Will the Technology Lead Us?
Edited by Josef Mack
(Order No. PT-57)

For more information or to order a book, contact SAE at 400 Commonwealth Drive, Warrendale, PA 15096-0001; phone 724-776-4970; fax 724-776-0790; e-mail CustomerService@sae.org; website http://store.sae.org.

* * * * *

* * * * *

SAE International
400 Commonwealth Drive
Warrendale, PA 15096-0001 USA
E-mail: CustomerService@sae.org
Phone: 877-606-7323 (inside USA and Canada)
724-776-4970 (outside USA)
Fax: 724-776-1615

ISBN 0-7680-1247-3
Library of Congress Catalog Card Number: 2003110330
SAE Order No. R-352
Printed in the United States of America.

Preface

At the invitation of TMD Friction (TEXTAR) some 250 top brake engineers meet at Bad Neuenahr, Germany, annually to hear and discuss expert presentations on the latest developments in road vehicle brakes and brake systems and to exchange technical information and experience in the international bilingual μ Symposium. Many of the attendees are members of the prestigious μ Club, which was founded 25 years ago and gave rise to the μ Symposium.

The editors are grateful to SAE for disseminating outstanding German and European brake engineering knowledge to the world's brake experts through this book. It contains the μ Symposium contributions of the past 4 years covering important areas and aspects of modern brakes, their components and systems and their potentials and interactions with other basic disciplines (e. g. new materials) and automotive systems (e. g. intelligent chassis).

We thereby hope to encourage many automotive engineers and engineering students to turn their attention to the subjects and suggestions of this book and to carry on with their further development.

B. Breuer

U. Dausend

Dedicated to

Georg Naoúm

– founder of the μ Club and creator of the μ Symposia

Contents

1 Introduction

Prof. Dr. B. Breuer, Dipl.-Ing. U. Dausend, Darmstadt University of Technology

Darmstadt, Germany

The technical purpose of brakes is to decelerate moving vehicles through the tire/road contact, to adapt their speed to traffic conditions, and to prevent their involuntary movement once stopped. They transform kinetic energy into thermal energy for storage in the brakes, followed by dissipation to the surrounding air, and must be able to exert sizable torques to the wheels even at standstill. Peak power levels in emergency braking far exceed maximum engine power. Appropriate brake force distribution to the wheels in keeping with road surface conditions (e.g. icy/dry) and braking stability (left/right and front/rear), adapted to the driving state of the vehicle (e.g. gradual speed change versus emergency braking), are vital for accident prevention.

Customers are particularly interested in short stopping distances, ease of operation, prevention of fading and skidding, high braking comfort (actuation, noise and vibration) and long service intervals.

Engineers must meet customer demands as well as strict regulatory and OEM specifications respectively for brake system safety and reliability, dimensions and packaging, comfort, maintenance, recycling and cost.

Nineteenth century automobiles, essentially carriages without a drawbar, had correspondingly very simple brakes. With the advancement of automotive technology and increasing top speeds in the early 20th century, engineers moved from braking the rear axle only to four-wheel systems and replaced mechanical brake actuation by hydraulic systems. The next steps – with cars almost always ahead of commercial vehicles and motorcycles – were disc brakes replacing drum brakes, the addition of a second brake circuit, and antilock brake systems (ABS) – first introduced in Mercedes-Benz cars in 1978. ABS allowed simultaneous braking and steering without skidding, making good automatic use of available tire/road friction, and represented a giant step forward in motor vehicle safety.

Today, driver assistance systems already on the market, such as Adaptive Cruise Control (ACC) and Electronic Stability Program (ESP) cannot function without an advanced brake system which can automatically apply all or individual wheel brakes without a driver's brake pedal actuation, something which was only recently unthinkable. Although we still have good reasons to rely on the physical principle of friction for the core brake functions described above and will continue to do so, new materials, new actuation systems, new development and testing methods, new system approaches and stronger cooperation between different disciplines are finding their way into automotive braking systems, and will enable formidable improvements in subsystems (e.g. chassis) and the vehicle as a whole within the combined human/vehicle/environment system.

The contents of this book demonstrate that brakes are still far from technical saturation, and contain promising potential for future enhancements to safety, economy, ecology and comfort in the transport of people and goods on roads.

2 TMD Friction – Company Profile

TMD Friction Group, Corporate Communication

Leverkusen, Germany

2.1 Changing Markets

Globalization continues to be a key characteristic of the automobile industry. Today in Europe and in the US, manufacturers are moving production abroad, including other continents, often overnight. TMD Friction has already adapted itself to this changing market situation and sees it as the major priority for the next stage of the companies development.

Today the company is amongst the world's leading manufacturers of brake pads and liners for original equipment and the spare parts market. We supply automobile and brake manufacturers as well as the international spare parts market. A decisive step in the company's global approach to world markets was taken with the HSBC Private Equity investment in 2000 – renamed as Montagu Private Equity in March 2003. For the first time in its long history, the company has become primarily focussed on its core business of producing brake pads and had a major source of investment funding for state of the art technical and production facilities. Looking to the future, as well as maintaining its strong position in Europe the company will be concentrating on international markets in NAFTA - MERCOSUR areas as well as on South East Asia and JAPAN.

Local production facilities and strategic alliances will provide the crucial foundation for systematic, successful exploitation of potential growth in these markets, enabling us to improve efficiency in key operations and thereby constantly increase both customer satisfaction and the company's earning capacity, **Fig. 2–1**.

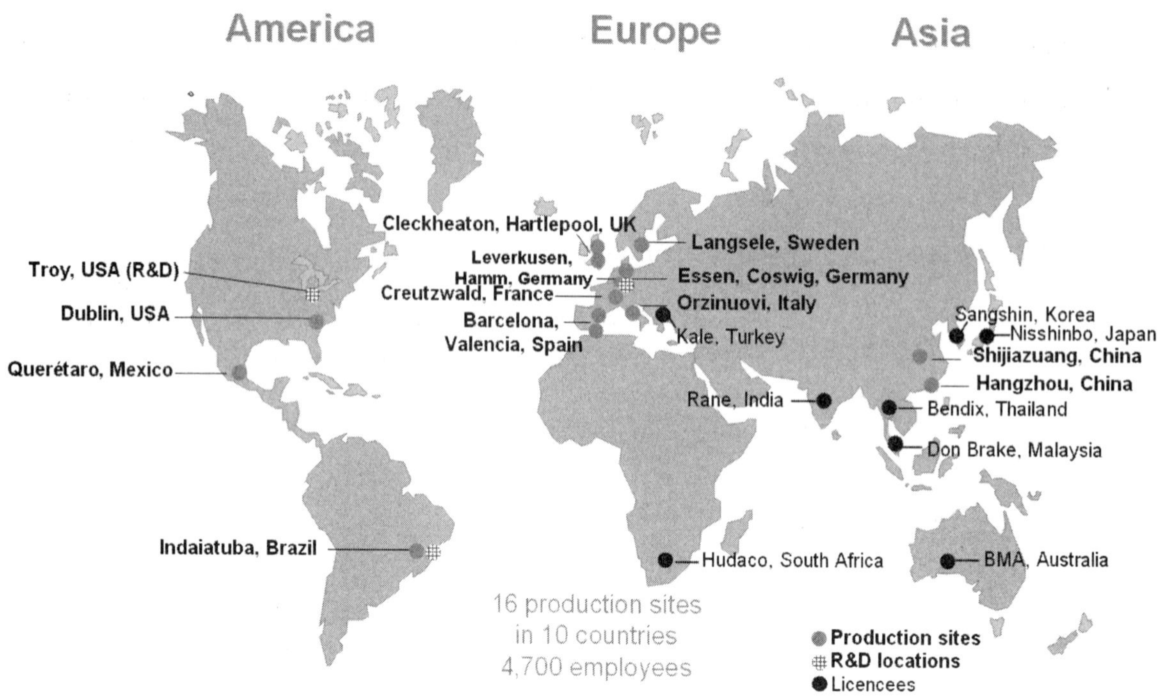

Fig. 2–1: Locations

This is where successful development activities and innovative technologies play a key role.

However, efficient cost management is an equally important factor. Worldwide procurement operations, reduced part variety, standardization of technologies and high flexibility must become a part of our everyday activities if we are to be able to maintain our competitive position on international markets.

The stringent technical challenges and advances in the automobile industry require us to focus upon innovative concepts involving active and passive safety.

This also applies to the supplying of international spare parts markets and their wide range of brake liner requirements.

TMD Friction has already made a significant contribution to improving safety and comfort on the worldwide automobile market.

Systems such as ABS or BAS, which are standard features almost everywhere nowadays, would not have been possible without the use of friction coefficient technology.

Also, new generations of automotive safety systems, such as Collision Avoidance (CA) or Electromechanical Braking (EMB), which will make a key contribution to mobility safety, represent a challenge and a duty at the same time, **Fig. 2–2**.

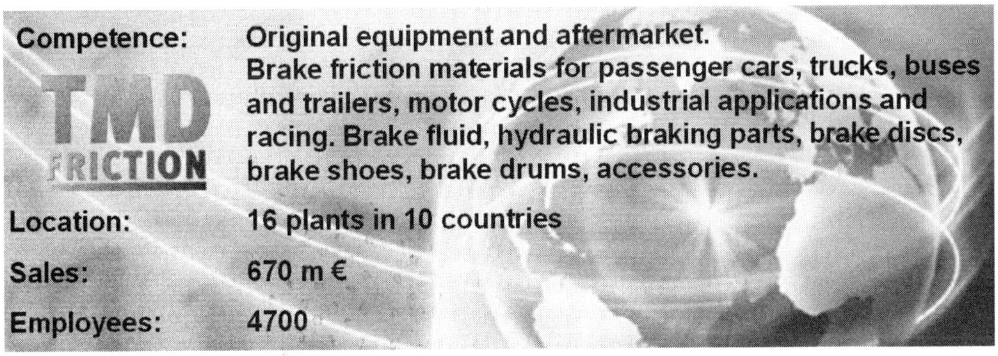

Fig. 2–2: Facts and Figures 2002

2.2 World Market Automotive Brake Pads and Liners

The world market for brake pads is primarily governed by the locations of the major automobile producers and the number of vehicles in these countries. These are currently Europe, NAFTA and JAPAN. Around 80 % of brake pads are produced and used in these markets, **Fig. 2–3**, **Fig. 2–4**.

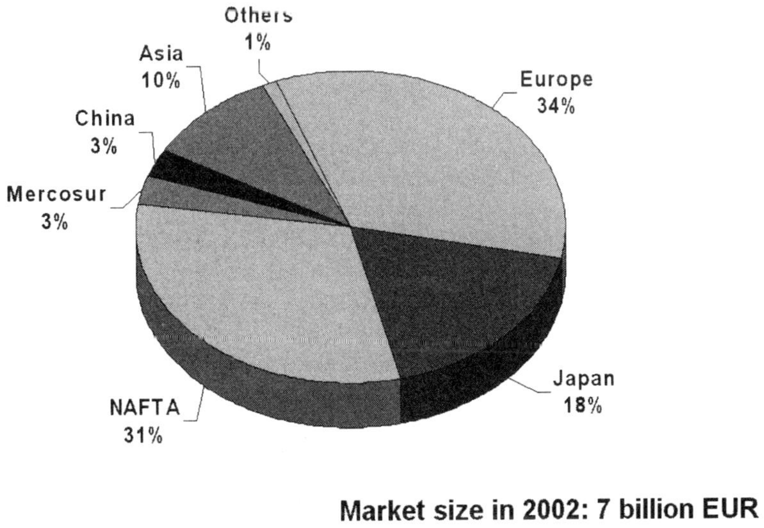

Fig. 2–3: World Market Automotive Brake Liners

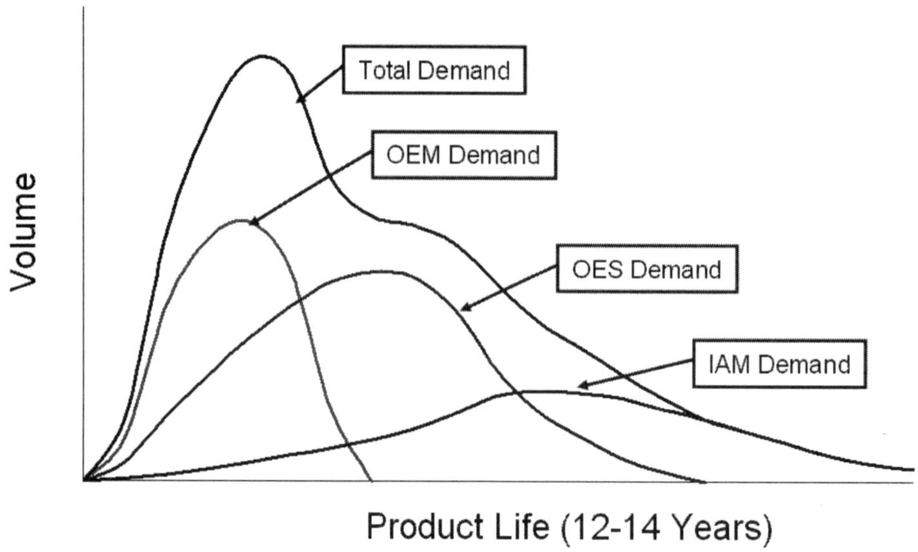

Fig. 2–4: Friction Materials - Demand over Product Life Cycle

However, globalization of the automobile industry and a strong tendency towards emerging markets will bring about changes from this position . Examples include the German producers who now make more than 4 million vehicles a year at overseas production facilities. Here we should also mention the Chinese market which has grown enormously over the past few years.

Against this background the structure of the market for brake pads and liners will undergo major changes in the future.

In the major markets of to-day original equipment volumes will, in the short term, drop or only show marginal growth.

At the same time the demand for spare parts will stagnate due to traffic regulations, such as speed limits and traffic-restricted zones close to cities and residential areas. Furthermore the brake and brake pad product innovations will ultimately lead to a longer service life.

However substantial increases in the demand for OE and spare parts are expected in the future growth areas of MERCOSUR and CHINA, in expanded South East Asian markets and in EASTERN EUROPE.

2.3 Competition Worldwide

TMD Friction commands 29% of the market making it the market leader in Europe and the world's 3rd largest producer, **Fig. 2–5**, **Fig. 2–6**.

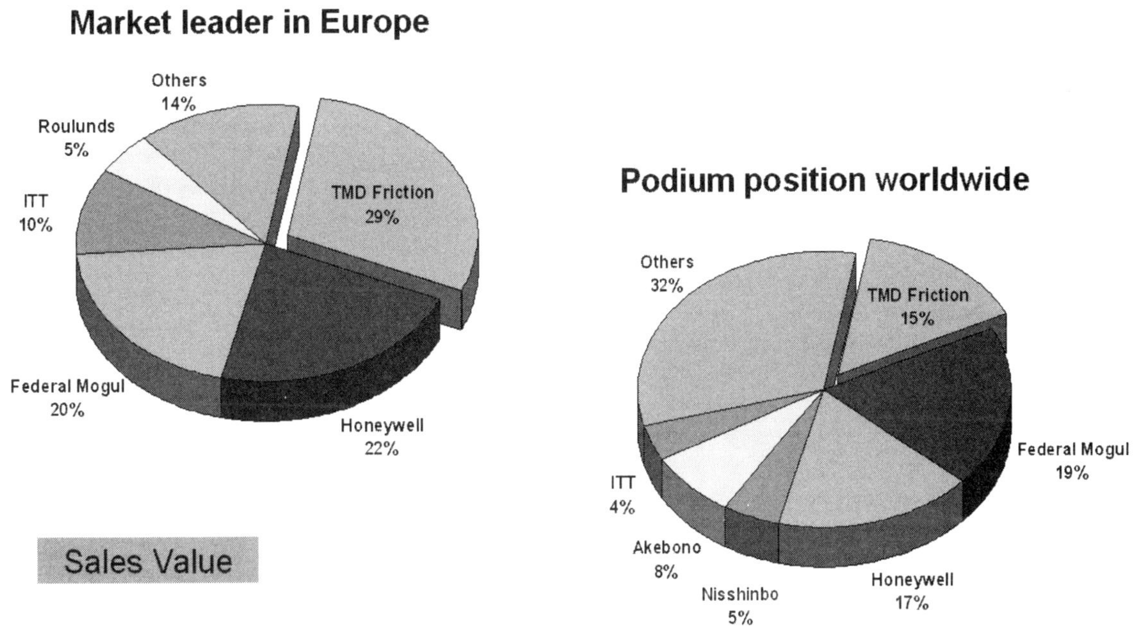

Fig. 2–5: Competition Worldwide

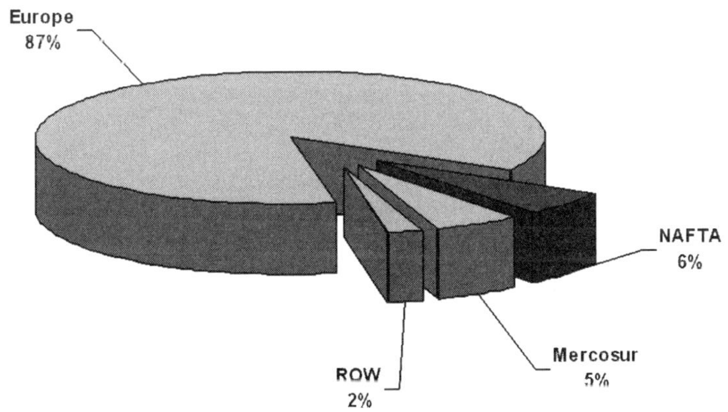

Fig. 2–6: Sales by Geographical Regions

As our strong position in Europe accounts for almost 90% of our sales (**Fig. 2–7**), our strategy will include substantially improving and expanding our position beyond the European continent.

This will be implemented by using TMD Friction production facilities and alliances located in all growth areas of the worlds as well as our European facilities. Turnover includes products supplied via the brake industry.

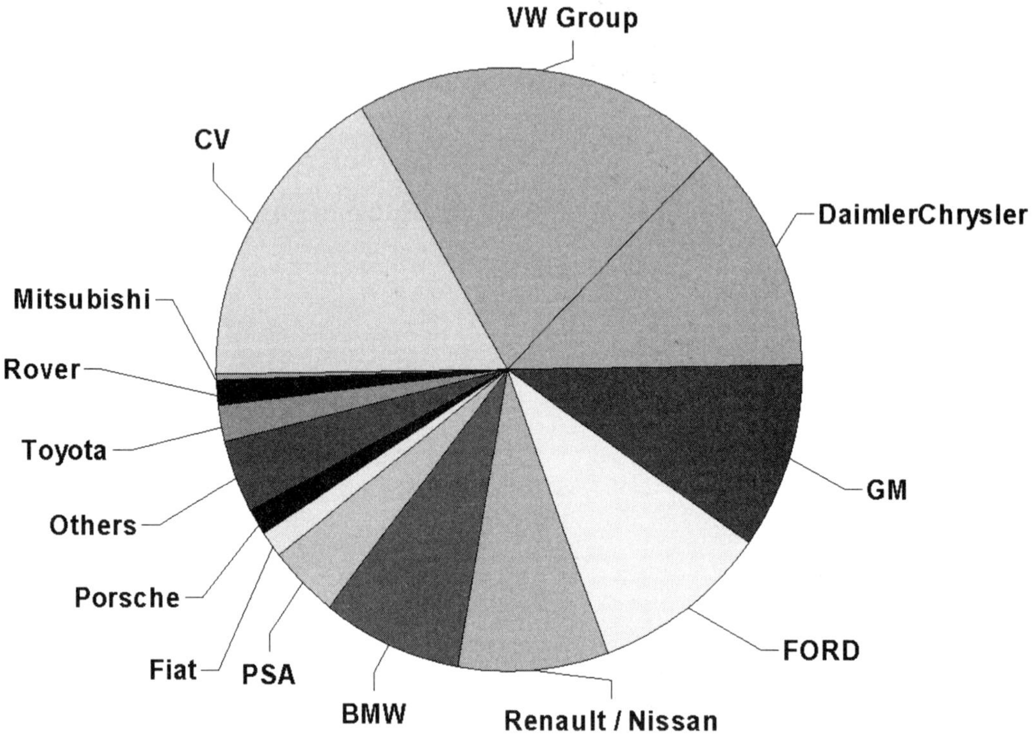

Fig. 2–7: TMD OEM/OES Sales

2.4 Brands and Products

The high market share for original parts and alternative brands in the spare parts markets is to a large extent the result of TMD Friction's strong position on the original equipment market.

However, when it comes to competing for customers and ensuring customer satisfaction, nowadays it is not simply a question of supplying the markets with high-tech original spare parts.

Therefore TMD Friction has developed a multiple brand strategy towards operating on international markets, **Fig. 2–8**, **Fig. 2–9**.

Here, TEXTAR and PAGID OE brands are given top priority in passenger cars. However, segmentation is necessary, on the one hand to take into account aspects like loyalty, reliability and customer service, and on the other hand it is created intentionally for the purpose of improved market penetration, market segmentation and hence maximum customer satisfaction.

	PCP	PCL	CVL	CVP	Industry
TEXTAR	●	●	●	●	
PAGID	●	●	●	●	
MINTEX	●	●			
Don			●	●	
COBREQ	●	●	●	●	
OSID					●

Fig. 2–8: TMD Brands and Products

This strategy is goal-oriented and geared towards potential demand on the independent global aftermarket.

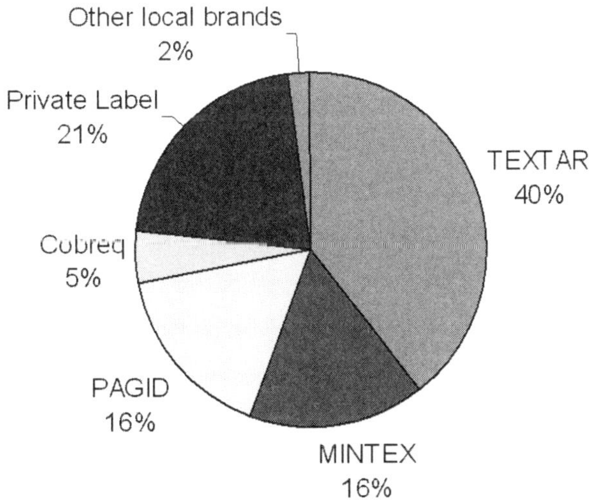

Fig. 2–9: AM Passenger Car Pads - Turnover by Brands 2002

2.5 TMD Friction - Development Process

2.5.1 Organization of R & D and OE Sales

In 2000 TMD Friction centralized research and development at Leverkusen and based it upon the following principles (**Fig. 2–10**, **Fig. 2–11**). However, local support facilities were also established in Detroit USA and Brazil and these will be points of growth over the next few years.

- **Component specialist not a commodity supplier**
- **OE-based business (R&D driven)**
- **Climbing up the value chain**
- **Strong direct access to the system supplier and vehicle manufacturer due to project development**
- **AM products technically approved by OE-Sales/R&D**

Fig. 2–10: OE Sales and R & D Objectives

- One face to the customer - **one TMD responsible** for the customer project from the project acquisition to the delivery contract (Technical Sales Manager)
- Centralized **Research** and **Product-** and **Process-Development** in Germany in 2 locations (Leverkusen and Essen)
- Decentralized market and project specific **Material Development (Customizing), Application Engineering and Testing.**

Fig. 2–11: TMD Group - OE Sales and R & D Organizational Principles

This organization is systematically implemented in the following functions and leads to the TMD Friction Market-to-Customer Process, **Fig. 2–12**, **Fig. 2–13**, **Fig. 2–14**.

Function \ Location	Europe Leverkusen / Essen Germany	Mercosur Indaiatuba Brazil	NAFTA Troy USA
Material Development			
PCP, CVP,	X		
CVL, PCL	X		
Material Development Customizing			
PCP	X	X	X
PCL	X		
CVP, CVL	X	X	X
Applic. Engineering All Products	X	X	X
Testing	X	X	X
Research	X		

Fig. 2–12: TMD Group - R & D Functions

Product Management **PM**	Responsible for introducing the TMD product strategy, convert OEM's and brakemaker's product strategy into the TMD product development.
Product Development **PD**	Material and process development, predevelopment new technologies.
Technical Sales **TS**	Customer project management responsible from the acquisition to the delivery contract (application engineering, pricing, sales).
Product Customizing **PCu**	Project specific material development incl. adaption of process parameters, sampling, support transfer function to production.

Fig. 2–13: Key OE Sales-/R & D Functions

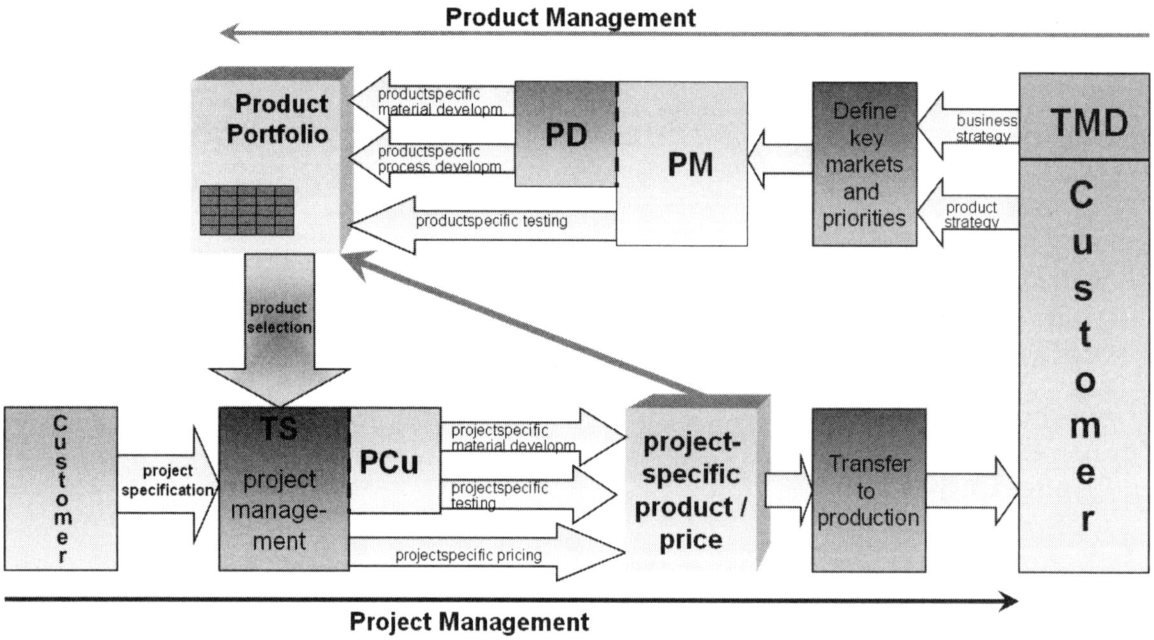

Fig. 2–14: TMD Market to Customer Process

The wide range of tasks of the centralized R & D at TMD Friction can be seen from the following diagrams (**Fig. 2–15** to **Fig. 2–19**).

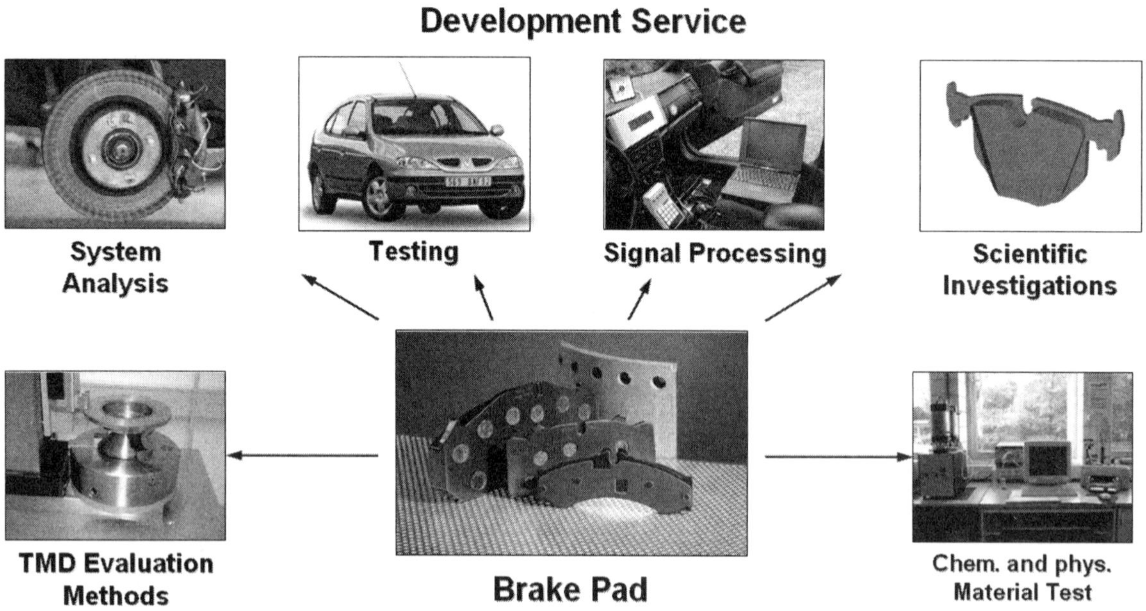

Fig. 2–15: Product Definition

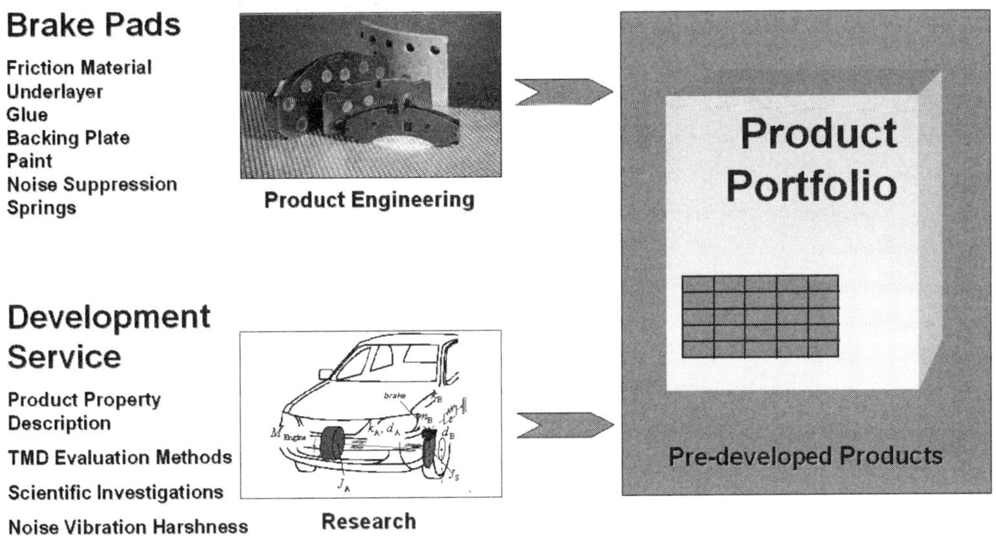

Fig. 2–16: Product Management

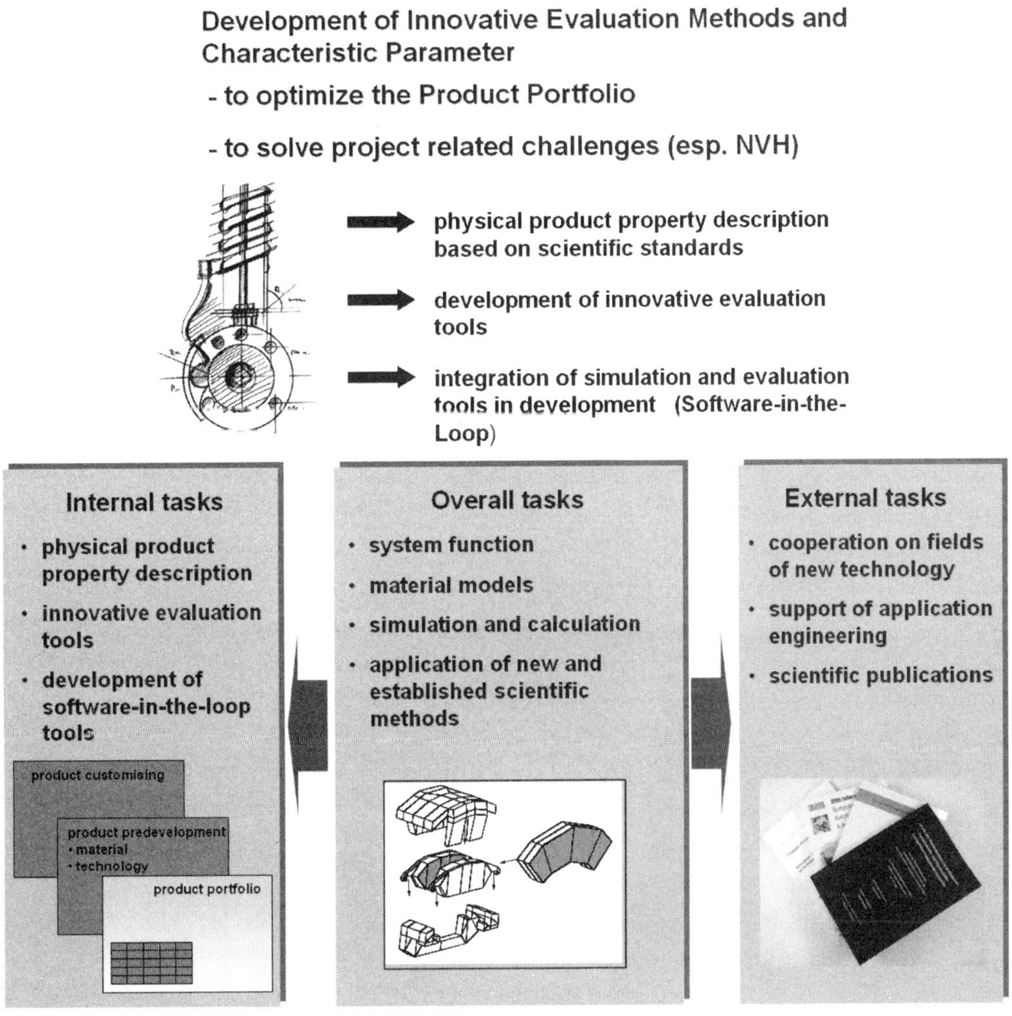

Fig. 2–17: Tasks of Research Group

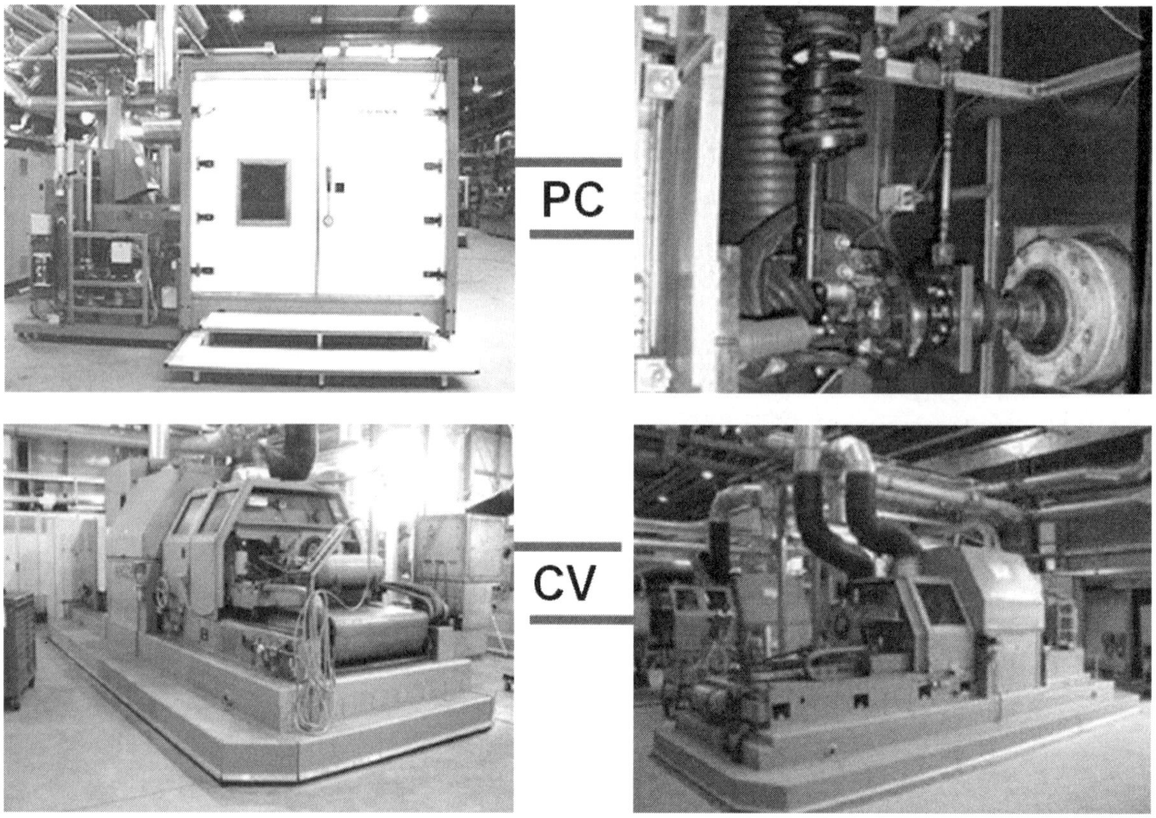

Fig. 2–18: R&D: Dynamometers

Area of vehicle testing building: 3170 m²

Depending on projects:
approx. 80 test vehicles,
approx. 15 test commercial vehicles/ trailers

Test vehicles are fitted with data-recorders.
Measurement and calculation of:

- **Temperature**
- **Pressure**
- **Speed**
- **Deceleration**
- **Pedal travel/Pedal force**
- **Stopping distance**

Fig. 2–19: R&D: Vehicle Testing

These methods are used to achieve our goal of supplying global markets with new products and innovations. This includes Europe, the USA, Japan and Asia (**Fig. 2–20**, **Fig. 2–21**).

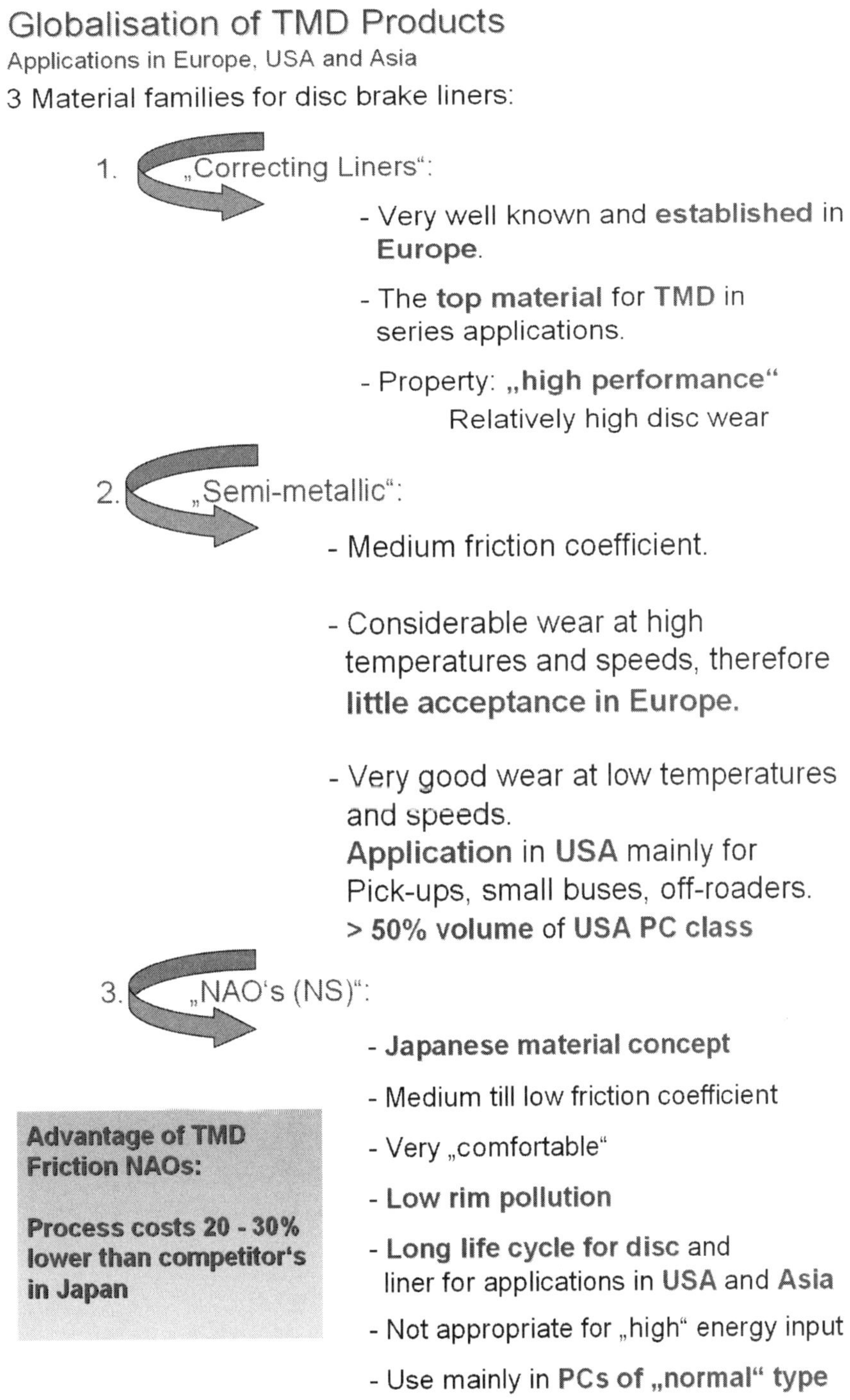

Fig. 2–20: Products for the Global Market

Fig. 2–21: Products for the Global Market – Summary

2.5.2 TMD Friction - Operations

R&D - OE Sales and Technology are closely linked at TMD Friction today, providing optimum conditions for ultimate product efficiency and commercial success.

As suppliers to the international automobile industry all production facilities used by TMD Friction must be certified by the OE customers. This requires a high level of standardization and flexibility to be able to react swiftly to changes in global markets.

This can be achieved at short notice with the internationally recognized Hercules production system, **Fig. 2–22**, **Fig. 2–23**.

The system comprises:

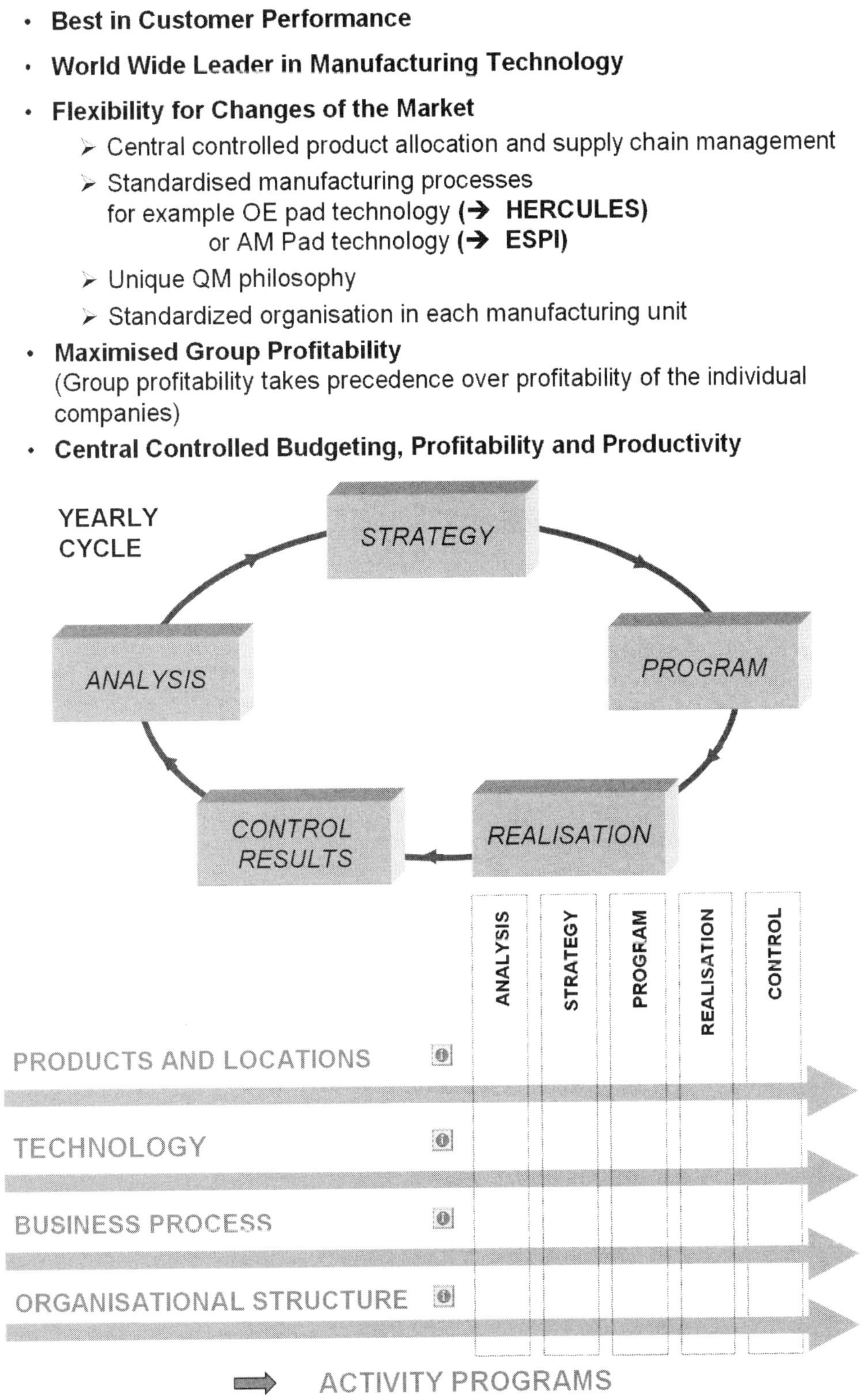

Fig. 2–22: Strategic Alignment Operations - Strategy Process

- **Leadership in production and process technology**
- **Hercules technology as a world wide standard for OE manufacturing units**
    - **Customer oriented line or segment concepts for the full manufacturing process (from pressing to the finished product)**
    - **Cold, hot and warm press technoloy with dry and wet mixes (water and alcohol)**
    - **Batch size oriented single mould press technology**
    - **Linked machinery for finishing**
- **Flexible segments and ESPI technology for all AM products**
- **Standardised machinery and equipment for all manufacturing areas**

Fig. 2–23: Manufacturing Processes and Technology

The following simplified diagram (**Fig. 2–24**) shows the features which make up the individual production stages.

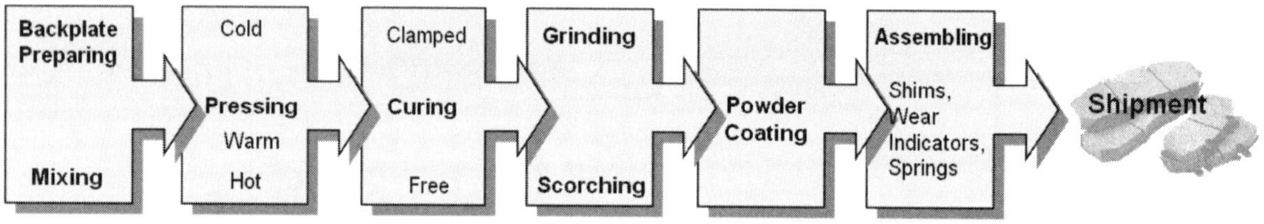

Fig. 2–24: Major Process Steps

Within the TMD Friction Group this technology is used in production facilities in Germany, UK, France, Spain, and the USA and in China. The aim is to apply this standard to all global OE production facilities in order to be able to ensure flexibility in adapting to market requirements, **Fig. 2–25**, **Fig. 2–26**.

Hercules system approved by customer:

- Process standard
- Technology standard
- Quality standard
- Organisation standard

Transfer to other sites following market and customer

Fig. 2–25: Hercules Standard Worldwide

- Centre of Excellence for each product group
- Design of new technology together with preferred supplier
- Installation of new technology
- Homologation and standardisation of new technology
- Transfer technology to other sites

Fig. 2–26: Centre of Excellence

Transferring these technologies enables TMD Friction to create worldwide competence centers and access new global markets.

2.6 TMD Friction - The Future

Over the last few years we have endeavored to reinforce our leading position in Europe as a foundation for the next stage of a focus on world markets.

We have standardized our production technologies and implemented Hercules, and therefore have been able to facilitate transfers to global markets. We now have locations in all developing areas in the world. Heads of R & D and OE Sales are in contact with the automobile industry on a worldwide scale.

Audits, constant rationalization and subsequent continuous development are a part of our everyday work.

We have invested almost € 100 million.

Therefore we have created an optimum basis which will be developed further. This will involve making use of our own resources as well as our successful joint ventures to provide ever more active and passive vehicle safety.

It is not only a question of achieving our ultimate aim of operating in global original equipment markets but also of being able to supply spare parts to worldwide markets in which we have until now been very much underrepresented.

The Montagu Private Equity investment has provided us with solid financial support.

The details mentioned on the previous pages refer to 2002 and can therefore be regarded as facts. However, any reorientation is plan and prediction.

Nevertheless, our priorities for our future business development are clear (**Fig. 2–27**):

- **In Europe we will reinforce our current strong position on the original equipment market.**
- **We will continue to use our strength on European OE to build our market share the European spare parts market**
- **In the world market beyond Europe we are clearly underrepresented in both OE and AM trading.**

- **We believe that our technical capability will give us numerous opportunities to contribute to innovations in the automobile and brake industry.**
- **We will make use of these opportunities to operate on a global scale through our local R&D and manufacturing facilities as well as our JV's and other alliances.**
- **We will maintain our worldwide multiple brand strategy which is both market-oriented and based upon continuity in market segments and with business partners.**

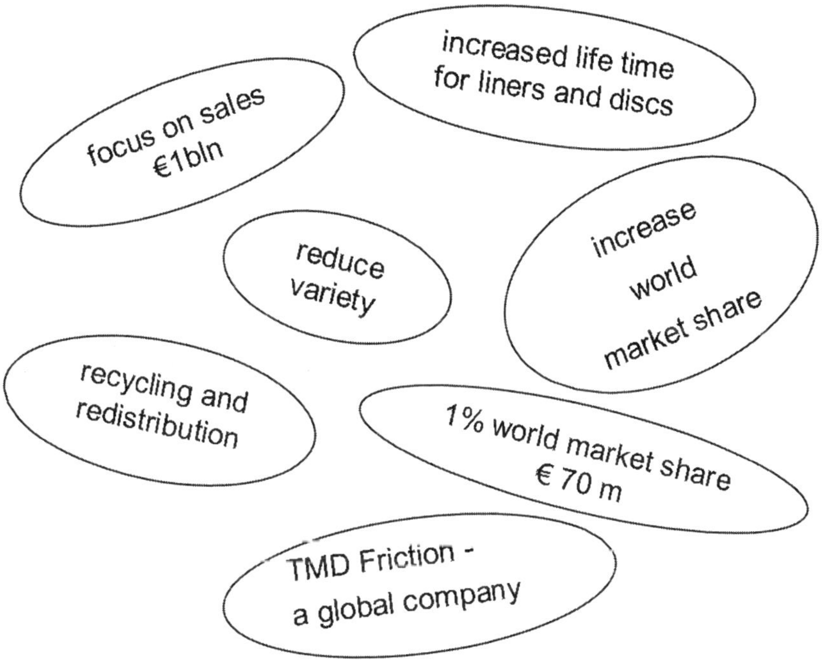

Fig. 2–27: Priorities for our Future Business Development

In conclusion, we shall considerably build on our position in Europe enabling us to make a decisive step towards increased growth and profitability in the wider global markets.

Part I

Basics, Materials and Methods

3 Braking from the Behavioural Scientist's Viewpoint

Prof. Dr. B. Spiegel, Spiegel Institute Mannheim

Mannheim, Germany

Summary

Gaps in the interface between man and machine can generally be compensated by both sides. While it is true that the brakes have not yet completely fulfilled their adaptation tasks, they have done so for the greater part, and are indeed capable of bridging deficits (e.g. by means of ABS) that cannot be be fully closed by the human side of the equation. The adaptational success on the human part, on the other hand, is still less than satisfactory, since this success is impaired not so much by a lack of plasticity (adaptability), but rather by his inadequate training in this area. In the meantime, the technological sector has started to make attempts – e.g. by means of solutions such as the brake assistant – to close those gaps that in principle could also be closed by the driver himself. The *brake by wire* principle offers a wholly new range of possibilities by which also demanding tasks of adaptation can be solved.

In the early 1960s, at a NASA press conference someone asked Fred Sänger why space flight was to be carried out using humans – his humorous answer to this question was as follows: "The human being is the only feedback-stabilized and above all *multifunctional* control and servomechanism that can be mass-produced by unskilled workers to an adequately cheap degree."

And it is this control mechanism that is the subject of the following. As vehicle engineers, your focus is – at least by preference – directed at the interval between the operating element and the contact patch of the tires – and rightly so, too – but this control mechanism nevertheless forms part of the overall system.

Let us continue to consider the driver as a control mechanism for a moment. Seen from this viewpoint he must undoubtedly by classified as a supplier component – one, however, that you as engineers would look upon with extreme scepticism. True, he does possess highly positive characteristics, ones that in some cases are nigh on incredible. He has, for example, an unsurpassed plasticity, one that would be the delight of a constructor of automatic handling devices if he could only gain control of it.

On the other hand, he is also characterized by a number of deficiencies that make him a pronounced *bottleneck component with a series of weak points.* The feature most deserving of complaint is the completely inadequate *replication stability.* The intraindividual scatter is enormous, even in the case of the highly experienced driver, not to mention the interindividual scatter.

There are any number of specific bad habits in his repertoire, and I will be dealing with these very soon. But beforehand a few comments on this remarkable "supplier component". It is precisely here where the vehicle manufacturer has no opportunity whatsoever to run quality-control checks, and naturally he does not have any chance at all to reject this component in favour of one he is developed himself. After all, the driver is the component who foots the bill for the whole shamboodle, he does not want to be merely a passenger on the journey, but instead even derives great pleasure from playing an active role in the process, too.

What, then, are the driver's specific bad braking habits? On the basis of systematic behavioural observations we know the following: He brakes *too weakly, too shallowly, too late, too strongly, too long, too often* – which does not necessarily mean that he does not have any other failures besides. Fig. 3.1 to Fig. 3.6 provide a better impression of what is meant here.

Let us first take a look at braking *too weakly*, **Fig. 3.1**. This is the realm of the *panicky braker*, as the corresponding subject in the testing routines is somewhat exaggeratedly termed. He is not called so because he brakes out of fear, but rather because he is in an state of panic when and while he is braking. In reality he does not need to be panicky, and he is not necessarily panicky by nature – in any case he is *brake-shy*. It is common knowledge that this type of driver is very widespread, and he is found among motorcyclists even more frequently than among car drivers, this for the simple reason that the motorcyclist is subject to an additional dilemma. On the one hand, the deceleration is for by far the greater part enacted by means of the front wheel, while on the other he cannot afford to block the front wheel under any circumstances – or at most for a fraction of a second – since this effect would inevitably result in his involuntarily dismounting his steed at high speed.

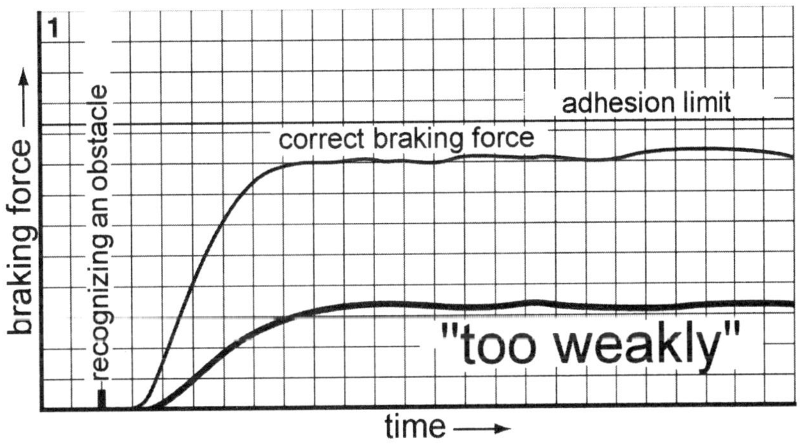

Fig. 3.1: Brake force characteristic at braking too weakly

The situation regarding braking *too shallowly* (**Fig. 3.2**), is similar, and in most cases is closely connected with braking too weakly. Such a driver amplifies the braking power only hesitantly. Only very few drivers are at all aware of the fact that a short braking distance is covered predominantly at the start of the braking process, at a high speed, i.e. in a phase when a considerable part of the braking distance can be forfeited most quickly.

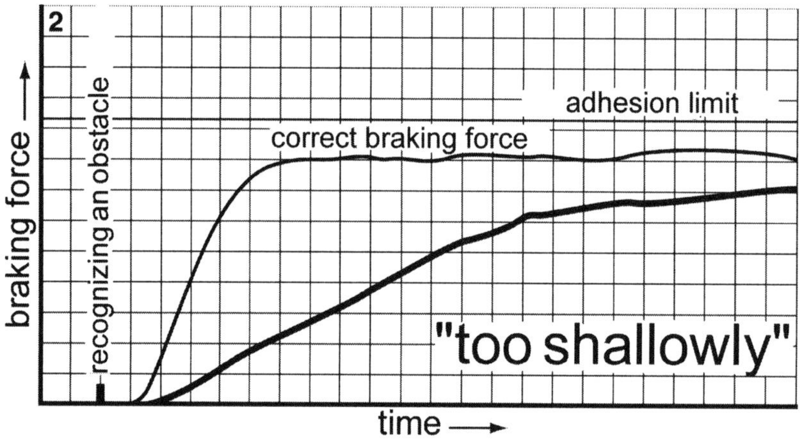

Fig. 3.2: Brake force characteristic at braking too shallowly

The next bad habit – that of braking *too late* (**Fig. 3.3**) – is a peculiar one. This is not, as one might expect, a delay in the reaction time. Instead, the driver – even in an extremely tight situation – takes his foot from the accelerator pedal, but then hesitates a moment, as if paralyzed, as if to see whether the situation might change for the better again. This is an override of the braking inhibition that takes its time – most pronouncedly in the motorcyclist.

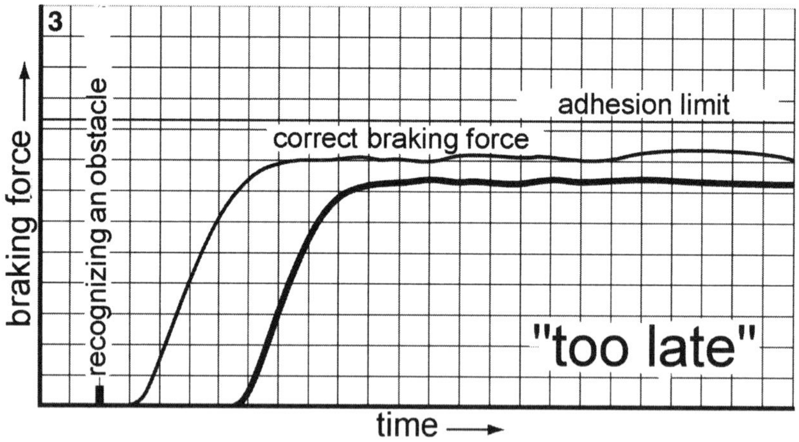

Fig. 3.3: Brake force characteristic at braking too late

The counterpart to this is the *scared braker*, who brakes *too strongly* (**Fig. 3.4**) and applies too much braking power, frequently over and beyond the adhesion limit. This is again a habit that is observed among motorcylists in particular. If the rider takes his hand off the brake lever again and the direction of travel is straight ahead, then nothing can really go wrong. The scary moment often occurs, however, when negotiating a curve, namely when the driver can see into a new section of road and there espies a threatening obstacle, but when he himself is still in a inclination position. In many such cases, it is not necessary to apply the brakes in the first place, but the perceived obstacle automatically makes the rider do so. This aspect has led to the development of a number of brake-design approaches, ones that I will be coming back to in a moment.

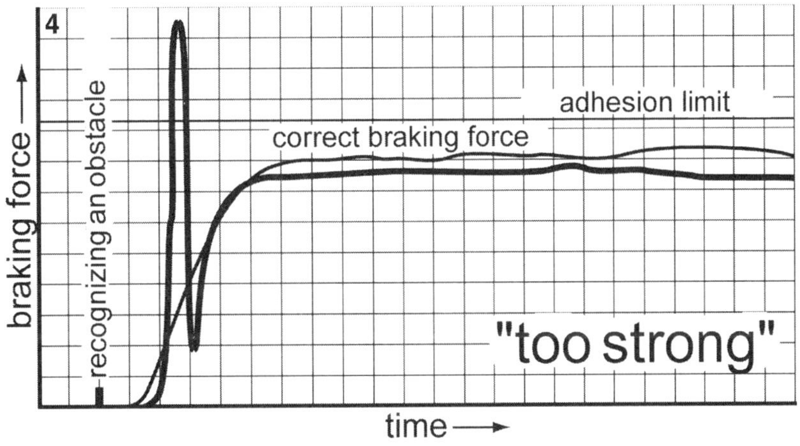

Fig. 3.4: Brake force characteristic at braking too strongly

The next two bad habits are not ones that occur when applying full brakes. Some drivers from the large family of the panicky brakers, to whom we have already been introduced, somehow manage to make serious mistakes even when applying brakes lightly, types of behaviour that researchers do not encounter at all in routine tests, unless, of course, they have been specially arranged. For example the constant braker, who activates the brakes much *too long* (**Fig. 3.5**) on downhill gradients, e.g. for the entire downhill ride after crossing a mountain pass, yet only applying a slight braking pressure – particularly sparing on the brakes, he thinks, but in truth exerting a destructive local overheating effect.

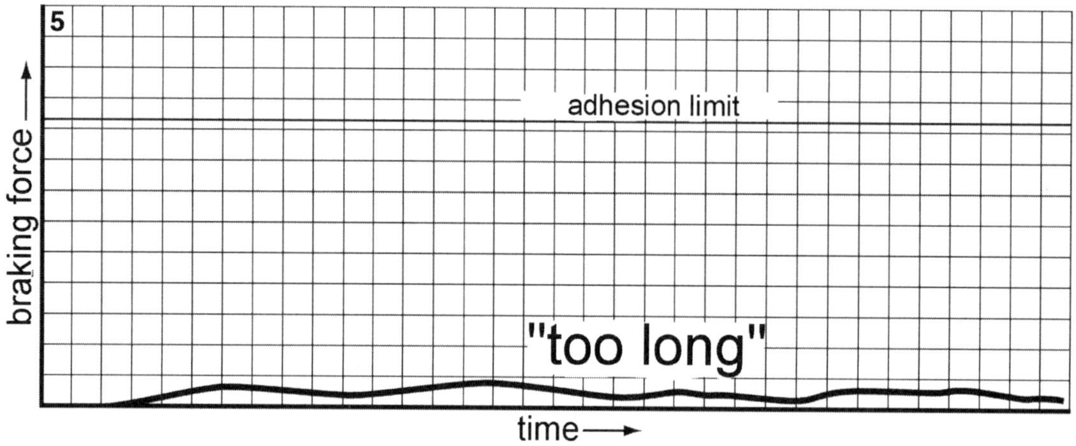

Fig. 3.5: Brake force characteristic at braking too long

Not quite as bad, but far more frequently encountered is the last type in the series, namely the driver who brakes gently, but much *too often* (**Fig. 3.6**), at least once before every curve or bend. This type is closely related to the panicky driver and here is obviously practicing soothing relaxation by braking. I jest not – this form of braking indeed exerts a psychically relieving effect on the driver.

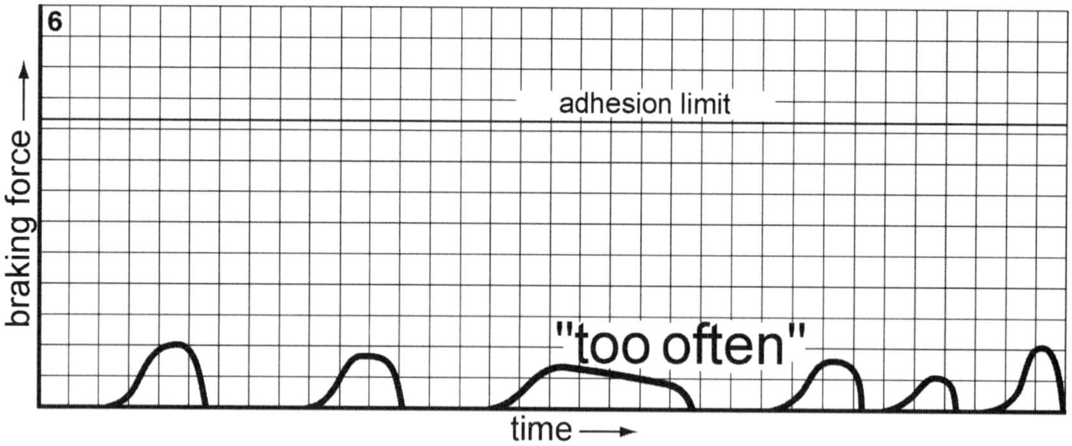

Fig. 3.6: Brake force characteristic at braking too often

That just about rounds off the types of erroneous braking habits. I need not, least of all in this expert circle, describe the consequences. But how do we extract ourselves from this situation, one that applies the whole world over?

The tasks that are assigned to us can be fully and completely split up into ones for man and machine, into software and hardware, namely into assignments for the technological world on the one hand and ones for the training of the drivers themselves on the other.

Technology has already accomplished a major part of its task with the development and implementation of ABS. An aspect that still deserves strong reproach, on the other hand, is the obstinacy with which most of the manufacturers hold on to the separate operation of the brakes on the front and back wheels of motorcycles – numerous illustrous examples of the contrary notwithstanding. Separate brakes simply *cannot* be activated simultaneously along an optimal curve [3-1] – this is the consequence of the lacking channeling capacity of actions that demand awareness –, meaning that the motorbike rider can do nothing other than concentrate on his front wheel. The conventional braking system of a motorcycle is, to make a slight exaggeration, a construction for a control element that does not even exist. [3-2].

The desirable single-lever activation should be connected with an automatic brake-power proportioning system similar to that used in cars, wherever possible still retaining its function even after the ABS becomes defunct, just like *Peraves* already practices on a small scale with his *Ecomobil*, a cabin motorcycle.

A last major item on the behavioural scientist's list of Christmas presents he'd like from the technological sector, and one that affects not only the motorcycle, but indeed everything on wheels, would be a greater uniformity in the characteristics of the brakes. The wide variety of options open to the user is a burden and a hazard risk at the same time. With his differentiated adaptation to a specific automobile, a person switching the brand or type suddenly no longer matches the car that he now has to drive, and his integration into the new vehicle is impaired. From the behavioural scientist's viewpoint, it is imperative that the manufacturers reach agreement on a few select characteristics, each of which is at an anthropotechnical optimum. This is an aspect that is currently of topical relevance in connection with the introduction of the *brake by wire* technology.

All the same, no matter what demands are still being made of the technological side: on the other side of the coin, the human side, the tuition and training of the driver, the picture is far more unfavorable.

Going back to the image of the control device for a minute, then better teaching and better training can be translated into meaning that the *programming* of the controller must be improved – and that is far more than just an image! These are indeed real programmes [3-3].

While it is true that man has more or less totally freed himself from the rule of rigid behavioural programmes – genetically mediated programmes as they still hold sway in the animal realm –, he still possesses programmes, namely those he has acquired himself. Here he is relatively free to choose, he can acquire them or not, he can refine them by training, or he can reprogramme them. An acquired programme, once it has been fully assimilated, is by and large similar to a genetically mediated one, a natural "inborn" programme. Once you've learned how to ride a bike or how to swim, you will never forget again – just as if the programme had been ingrained in you from your very inception.

The programmes do not contain merely the chain of the individual actions, but also the consequences to be anticipated, the reactions of the system. We're thus dealing with both motor as well as sensory sequences that are closely intermeshed with each other – they are *sensomotor programmes.* Just one example that stands for a thousand other situations: a running person who suddenly brakes abruptly, as can be seen, for example, in a soccer match, has to adopt a considerable backward leaning stance. But this does not happen as a result of continuous minute adjustments – he would not make it in time – but instead, so to say, as a result of anticipated commands to lean back in the individual moments that require hardly any retrospective fine corrections – the more differentiated his programme – i.e. the better he is trained - the fewer the number of corrections that have to be made. This sensomotor programme corresponds very closely with the *motion draft*, an extremely useful term for motor behaviour. No motion complex without a previously made motion draft, which, however, does not have to be thematized for the person involved. The action is oriented towards the motion draft. In a certain way, it is – or course not verbally coded – the action specification, meaning that firstly a useful motion draft can be obtained, and secondly to transform this into action as exactly as possible – two entirely different tasks [3-4].

For our topic this means the following: The driver so to say puts his motion draft for the braking process and thus also for the full stop "on constant hold" and need only activate it as the circumstances dictate. Of course, if he is never before slammed on the brakes at the speed he is currently driving, then this is going to be just a vague motion draft, while for the experienced or even the highly skilled driver the motion draft is already highly precise and detailed, and one that coincides very well with what actually follows.

And thus we arrive at the crux of the matter. If today the accident researchers state that in the majority of accidents that occur people do not brake enough, in other words the strived-for full braking does not actually come about, if numerous accidents happen that could be avoided if the people had better mastership over their brakes, then this is due to the lacking skills on the part of the driver, to his inadequate teaching and a lack of training. Or to revert to the image of the control device: to his wanting and incomplete programming in terms of braking.

Were you aware, for instance,, that slamming on the brakes – just even once - does not form part of the teaching programme for learner drivers, and is not part of the driver's test either? I was not. The driver's test for motorcyclists demands one single full braking – from a starting speed of 50 kilometres an hour! Sure, there is a lot of regulation and over-regulation, and driver's licence category I is turned into A-unrestricted and category III into B, but no-one feels responsible for the most important aspect, namely how the car driver or motorcyclist manages to bring his mount to a standstill safely and in good time. Good driving instructors already teach far more than is required in the driver's test, but this impairs their competitiveness versus the "cheap operators". For all driving pupils – especially those learning how to ride a motorbike a highly important aspect would be to practice how to brake with the longest possible delay, time and time again, at different starting speeds, since only then do the correct sensomotor programmes establish themselves properly and the correct motion drafts imprint themselves.

But even this is not yet enough. These programmes need nurturing, and nurturing means that they must be used. Artur *Rubinstein* is reported to have made the following bonmot: "If I do not practice for just one day, then I notice it. If I do not practice for three days, then my wife notices. And if I do not practice one whole month, then it is the audience that notices." – Programme disintegration through non-use! But when are the programmes actually put to use? It can take months, maybe even years before the situation arises in which the driver must slam on the brakes. A well-trained driving pupil will thus deliver his best results in terms of full braking soon after having taken the driver's test, after which his performance will generally become poorer. After all – who voluntarily practices how to apply full braking? At least the motorbike rider enjoys fully opening the throttle far more than full braking. And if he is in a mood for practicing, then it is bound to be how to negotiate bends and curves, and not how to brake. The only way out of this dilemma – and it is an extremely narrow one, since it is taken by only very few people – takes the form of safety-training or perfection-polishing courses. These are, of course, fully optional, meaning that one finds oneself asking whether it might not be appropriate to consider making follow-up safety-training courses compulsory – a demand that would naturally never be heard to come from the lips of the driving instructors or their associations, but that instead would be the domain of the politicians responsible for traffic affairs.

The problem is thus very much a human one, in fact it lies in the teaching of a person how to drive. Technology has made its essential contribution. The last major step it took to close a gap – one which could never have been perfectly closed even by the best training – was the development of ABS. Maybe technology can still be relied on here and there to help bridge certain instructional and training deficits – as it has succeeded, for example, with the brake assistant – but its next major challenge is still on its way: the topic of *brake by wire* is now on the agenda and will ultimately lead to a rearrangement of the relation between man and machine.

And so, ladies and gentlemen, I think the next time that engineers and behavioural scientists shall have to get together again will be when the *brake by wire* technology is approaching the hot phase.

3.1 References

[3-1] Eberspächer, H.: Psychological Consideration on Brake Use Patterns of Motorcyclists. Safety Environment, Future: Proceedings of the 1991 International Motorcycle Conference, ed. by Institut für Zweiradsicherheit, Forschungshefte Zweiradsicherheit, Nr.7, Bochum

[3-2] Spiegel, B.: Die obere Hälfte des Motorrads. Vom Gebrauch der Werkzeuge als künstliche Organe. 2nd ed., Stuttgart 1999, p.63f.

[3-3] ibid., p.34ff.

[3-4] ibid., p.70f.

4 New Material Technologies for Brakes

Dr. T. Haug, Dr. K. Rebstock, DaimlerChrysler AG

Ulm, Germany

Abstract

Brake discs made of grey cast iron are produced today in large quantities in seconds clock pulse. In spite of less tolerance and greater quality demands, extremely small unit costs are realised. On the one hand, the reason for this may be found in the low prices of the raw materials, refined production technology and large production numbers. On the other hand, the technical potential of grey cast-iron has essentially been exhausted.

For further technical innovations in the area of brakes, new materials such as Al-MMC and CMC play an important role. The potential has already been demonstrated. In part an assembly installation has already been successful (see rail traffic) or will be shortly (see SLR at the IAA 99). The main hindrance to further assembly installation of a large number of items is the high installation costs. Therefore, in the future, along with the reduction of raw material costs (for example, C-Fibre) the main direction that will need to be taken to further reduce production costs, will have to be in the realisation of a production technology with reliable processes suitable for large volumes. Otherwise, alternative brake disc material will only have a reduced significance.

4.1 Introduction

Due to higher braking efficiency and a more even braking moment, the disc brake is currently being used increasingly in cars, trucks and rail vehicles. Improved comfort and increased safety are the consequences. Due to the balanced characteristics (strength, heat retention capacity, heat conductivity) and the lower production costs, iron-based materials (grey-cast iron, graphite cast iron, steel) have assumed a dominant position in brake disc production.

Whilst steel is used in high-efficiency brakes in high-speed vehicles, grey cast iron is the standard disc material for trucks. Light build is playing an increasingly important role in vehicles. Aluminium has largely replaced iron alloys in engine production, car bodies and chassis. The weight of a grey cast iron brake disc as a rotating, un-sprung mass, is disproportionately high in aluminium chassis. This leads to increasing problems (e.g. screeching). The relatively high thermal expansion coefficient leads to "hot spots", the consequence being hot juddering, whilst the wearing creates thickness errors and leads to cold juddering. Fading is also a common problem, which occurs with the known disc materials above 500^0. In the car, corrosion at low running times determines the time of replacement. In each case, the grey cast iron disc is not the durable disc in today's cars. These are all reasons for thoroughly researching new brake materials.

Various particle-reinforced aluminium alloys have been examined over several years, on the basis of its light-build potential [4-1, 4-2]. Carbon-fibre reinforced carbon (C/C) is used in aircraft due to its low weight, despite considerable problems with comfort and wearing. CMC = Ceramic Matrix Composite, which is still a relatively new material, is currently being examined in respect of its suitability for high-efficiency brakes [4-3]. The main issue is its low weight and high wear-resistance. As described in the following section, DaimlerChrysler is using this new material intensively and extensive knowledge has also been collated on CMC.

4.2 New brake materials

In addition to the conventional brake materials already described, which are based on grey cast iron, the following materials have, with varying degrees, been tested or are already being used in brake discs:

- Particle-reinforced Al (Al-MMC)
- Carbon fire reinforced carbon (C/C)

– Fibre reinforced ceramics or Ceramic Matrix Composite (CMC).

Steel, as a material used in rail vehicles, will not be discussed any further below.

4.2.1 Particle-reinforced aluminium (Al-MMC)

Due to its low density of 2.6 g/cm^3, aluminium has a vast weight advantage over grey cast iron. Adding ceramic particles with very fine granular fractions (∅ approx. 4...12 µm), a so-called dispersion resistance is achieved, which creates a significantly improved high-temperature strength. Preferably, 20...30 % by volume of ceramic particles are used (SiC or Al_2O_3) [4-4, 4-5]. Higher proportions improve wear resistance, but also make it more difficult to work the material, which consequently leads to increased production costs. The essential stage of producing Al-MMC is realising a homogenous distribution of the fine ceramic particles in the Al matrix. The ability of the ceramic particles to bond with the Al melt plays a significant role. Normally, the ceramic particles are stirred into the melt by a special process and cast into bars, which are then delivered to the brake disc manufacturer. Typical material key values are shown in **Table 4.1**. In the case of Al-MMC materials, the low maximum application temperature 450...535°C is the limiting factor for the use of brake disc materials. Thermal overload must be ruled out, since irreversible damage occurs on the brake disc which can lead to spontaneous failure. The high thermal conductivity of up to 180...190 W/mK assists heat extraction into the pot or hub and increases aeration efficiency, so that overall, the maximum temperatures fall to a lower level. However, use of Al-MMC is currently only possible for small, light cars (e. g. Lotus Elise), since in the case of heavier vehicles, a thermal overload in extreme situations cannot safely be ruled out.

The situation with rail vehicles is different. In this case, ceramic-containing Al brake discs are suitable not only for niche applications, but offer a potential alternative to conventional brake materials in all important areas. Thus, goods trains with up to 25 t axle weight can for the first time cross the Alps with only two brake discs per axle [4-6]. The weight saving in the bogie amounts to up to 50%. The weight saving is particularly interesting in local traffic, since considerable energy can be saved due to the large number of stops and thus braking actions. Even at high-speeds, Al-

MMC brake discs are a real alternative. As a result of the very efficient heat conductivity, the low tendency to the so-called hotspots (heating surface) is taken into consideration, so that the thermal tension in the friction ring is reduced. Aluminium brake discs are being tested intensively on goods trains, the underground and Deutsche Bahn (German Federal Railways) high-speed trains [4-7].

4.2.2 Carbon fibre reinforced carbon (C/C)

This material was originally developed for thermally stressed structural components used in air and space travel. In addition to the temperature resistance, the low density of approx. 1.8 g/cm³ is an important criterion for the entire range of application. The potential of C/C as a friction material was recognised in the 60's. It was first used as a lamella aircraft brake on Concord. Today, C/C is used as a brake material on almost all Boeing and Airbus aircraft. A special case is its use in racing. C/C is now used in approx. 90% of brake systems world-wide.

C/C materials are today produced by the expensive CVI method (Chemical Vapour Infiltration) or the more cost-effective resin impregnation and pyrolysis methods. Efficient brake materials can unfortunately only be produced using CVI. In the initial phase, PAN fibres (poly acrylo nitrile) are processed and bound by means of a C-precursor. The subsequent carbonisation process leads to a highly porous structure rich in C fibres, which are compressed during the expensive CVI process. With this method, the C-matrix is separated from the gas phase, usually hydrocarbon-rich gas, over several hundreds of hours [4-9]. The reaction dynamics must be kept low enough to enable a homogenous separation of the C-matrix – otherwise, the porous channels grow quickly on the outside and only one layer of the structure to be impregnated is available.

Table 4.1: Characteristics of new brake materials compared to GG26 Cr

Material		GG26Cr	Carbon/Carbon		Al/MMC		CMC	
			Hitco	Carbone Industrie	30% Al_2O_3 Lanxide	20% SiC Alcan	C/C – SiC DLR	C/SiC DC
Density	[g/cm³]	7.1	1.8	1.8	3.0	2.8	1.9	2.2
Strength	[MPa]	250…350	140…170	90…150	—	380	160	70…140
Expansion coefficient II ⊥	[$10^{-6}K^{-1}$]	10…14	 0.3…1.14 5…10	 -1…+2 —	15…18	17	 -1…+2.5 2.5…6.5	 3.5 5.4
Thermal conductivity II ⊥	[W/mK]	43…36	 12.5…21.9 4.6…7.5	 — 10…90	100	180	 17…18 7.5…17	 30…45 20..30
Weight specific thermal capacity	[KJ/kgK]	0.5…0.9	0.7…2	0.7…1.7	0.8…1.2	0.8…1.1	0.7…1.5	0.8…1.8
Volume specific thermal capacity	[MJ/m³K]	3.7…6.4	1.2…3.6	1.2…3.1	2.5…3.7	2.3…3.1	1.2…2.7	1.4…3.3
Maximum application temperature	[°C]	720	1600	2000	450	450	1300	1300

C/C materials show extensive pseudo-plastic breaking behaviour. The C-fibres reinforce the brittle carbon matrix. The reason for the resistant breaking behaviour lies in the optimised fibre/matrix link, which is so weak that in the event of an overload, energy decreasing mechanisms such as "fibre-pull-out", micro fracture formation and fracture spread occur, leading to a pseudo-plasticity. The compound material is normally produced in a 2D fibre structure. This leads to anisotropic, thermo-mechanical properties (see **Table 4.1**). In addition to the low density, the strong direction-dependent thermal expansion coefficient and the strong direction-dependent thermal conductivity are emphasised. The temperature stability at approx. 1800...2000°C, is extremely high, but oxidation starts only from approx. 400°C. This is the reason for the lamella configuration being used in aircraft brakes.

Depending on the material composition and micro-structures, C/C-combinations show very varied tribo results. For braking systems, a frictional value of between 0.35 and 0.45 is used [4-10]. Depending on the fibre arrangement, degree of graphitization, and fillers, the frictional value, however, can fluctuate between 0.2 and 0.8 [4-11].

The main problem of braking systems is wearing resistance. Particularly at low temperatures of $<$ approx. 150°C, a shattering takes place which ensures that in the case of aircraft, 75% of the wearing occurs during taxiing on the runway and only 25% on actual landing. With the disc/lining arrangement used for example in Formula 1, oxidation leads to increased wear, which means that the disc thickness can be precisely defined for each race. Another problem centres around the cold frictional value, which leads to the so-called "early morning effect", predominantly at high humidity, as a result of the residual porosity, i.e. during initial braking, an inadequate frictional value is achieved. The main benefit, in addition to lower density, is the excellent fading resistance, even at very high temperatures, and an increasing thermal capacity as temperature increases.

4.2.3 Fibre-reinforced ceramic (CMC)

Fibre-reinforced ceramics based on SiC as the matrix and carbon as the reinforcing fibre limit the brittleness of monolithic ceramics up to the highest possible temperatures. As a manufacturing method, like C/C, the CVI method and the polymer route were tried and tested in the 80's. In the 90's, the liquid siliconisaton method was developed. All three technologies are "near-net-shape" processes, which can be used to produce large-format, complex components. Typical applications are space travel (Reentry Technology) for hot structural components.

In 1993, the first report was published on the potential of liquid siliconised C/C as a brake material [4-12]. This method does not create highly rigid materials, but has the advantage of being able to produce an almost pore-free, fibre reinforced ceramic with high wear resistance through relatively few processing stages, and therefore represents the cost-effective alternative to the manufacture of brake discs. On the basis of current facts, the costs of the C/C brake material could be reduced.

DaimlerChrysler have developed a patented manufacturing process [4-13], which focuses on the following four basic stages:

a) Mixing short carbon fibres, an-organic powder (e.g. carbon) and carbon resin to form a compressible compound
b) Hot pressing to form the required structure
c) Carbonisation of the structure to porous C/C at 900...1100°C
d) Siliconisation at 1500...1650°C

The first stage a) is based on producing a pourable, homogenous compressed compound. The hot pressing process (stage b) serves to compress and moisten the resin. Stage c) converts the resin into carbon. This creates a clear porosity, since during thermal decomposition of the resin, gaseous components escape and the transition between polymer – carbon causes a considerable change in density from approx. 1...1.6 g/cm^3. In the final stage, the siliconisation, the porous structure is impregnated with liquid silicon. Due to its good wettability, the structure becomes fully absorbed, like a sponge within a very short time. The liquid silicon reacts with the carbon to form SiC ceramic. The difficulty lies in the fact that liquid silicon reacts only with the carbon matrix and not with the carbon fibre, which creates a fibre-reinforced ceramic. Thus, in addition to the micro-structure, the siliconisation method is also important. The compound material consists of both carbon fibre, the

SiC matrix and also small quantities of the residual, free silicon, which is non-reacting. The residual porosity at approx. 2% is significantly less than that of the C/C brake material.

DLR in Stuttgart have developed a C/C-SiC material using the siliconisation method, which is reinforced with carbon fibres and differentiated by both its fibrous structure and also the significantly higher percentage of fibres [4-3]. Thus, compared to the short-fibre reinforced C/Si from DaimlerChrysler, a reduced density (1.9 as compared to 2.2 g/cm³) and an higher strength is achieved. However, the costs are increased through the use of tissue and the high percentage of fibre. The strength is heavily dependent upon the choice of fibre and the fibre conditioning, whilst the other material values have only a slight influence. Like with C/C, the expansion coefficient and the thermal conductivity are strongly anisotropic. As a result of the lower fibre content and the higher proportion of ceramic, a significantly improved thermal conductivity is achieved by the DaimlerChrysler C/SiC than with the DLR C/C-SiC.

Tribological tests have demonstrated the high potential of CMC materials. The wear resistance is excellent and exceeds that of all other known brake materials. With characteristic frictional pairing, frictional values of 0.6-0.9 are achieved. Specially modified lining materials are currently under development.

4.3 Potential analysis of brake disc materials

In addition to the material analysis carried out by structure examinations, oxidation tests, strength calculations and measurement of material emissions for an initial screening, the favoured materials were subjected to basic tribological tests. The materials were subsequently produced in brake disc geometry and subjected to alternate test bed examinations on a Krauss frictional value tester and an inertia test bed. Theoretical calculations were carried out at the same time. By means of the Finite Element Method (FEM), the behaviour of the materials was determined in typical load situations and also the contact pressure and temperature distribution. Particularly the energy distribution between brake disc and brake lining, as well as the thermal extraction through convection and radiation, were determined. Stop braking, consequential stop braking and "Großglockner" departures were simulated [4-20].

4.4 Testing

This analysis demonstrates the high potential of the CMC materials, further application-related test runs should therefore be carried out on these materials.

4.4.1 Testing in high-speed trains ICE

As part of the BMBF Joint Project (Project 03N3009A8 and 03N300930), fibre-reinforced ceramic brake discs were developed for use in the high-speed train. In addition to the development of materials and the test-bed runs, the objective of this project was to demonstrate the test status using the train. Thus, shaft and wheel brake discs were fitted into a high-speed test carrier and tested in operation.

During the development work on fibre reinforced ceramics for high-performance brakes for the high-speed train, in the DaimlerChrysler Research Centre in Ulm, reinforced carbon masses were converted to ceramics using the carbon fibre siliconisation method (see section 4.2.3). To develop the C/SiC materials, various matrix systems, fillers and fibrous coatings were used, which siliconised the produced C/C mass and characterised it in respect of structure and strength behaviour. Selected compound masses were tested for their wearing and frictional behaviour at the Company Knorr Bremse, Braking Systems for Rail Vehicles, in Munich. Initially, brake disc rotors of various material qualities, with 280mm diameters were supplied to Knorr Bremse AG in Munich for test bed examinations. After prioritising the various tribo-pairings, shaft brake discs (diameter 680 mm) and wheel brake discs (D=750 mm) were produced to full scale and tested extensively on the inertia test bed at Knorre Bremse AG. The test bed results for the shaft brake discs (WBS) and the wheel brake disc (RBS) were so positive and the brake disc was found to be in such a good condition that they were tested in the final project phase of 1998 in the high-speed train. The train test confirmed the massive potential of the CMC brake disc development.

Fig. 4.1 and **Fig. 4.2** show the CMC-Fullscale-Protoypes produced from fibre reinforced ceramic. The weight saving of 60 – 65% on cast steel is considerable).

Fig. 4.1: CMC shaft brake disc ∅ 640 mm (weight reduction compared with cast steel: 60 %)

Fig. 4.2: CMC brake disc ∅ 750 mm (weight reduction compared with cast steel: 65 %)

4.4.2 Mercedes-Benz SLR

At the Detroit Motor Show 1999 the CMC high-performance brake from DaimlerChrysler was for the first time presented to the public in the Mercedes-Benz SLR (**Fig. 4.3 and Fig. 4.4**). This is a vehicle that combines the attributes of the triumphant McLaren-Mercedes F1 car with the stylistic elements of the 300 SLR from 1954. The turbo-charged V8 accelerates this vehicle to a top speed of 330 km/h. In addition to the vehicle attributes, it has been possible for the first time, to present the properties of the fibre reinforced ceramic used in the high-performance brake:

- Brake disc rotor from fibre-reinforced high-performance ceramic
- Light build through density of 2.2 g/cm^3 in relation to GG26 Cr (7.1 g/cm^3) leading to improved agility in driving behaviour
- High specific braking efficiency
- Corrosion resistance
- Acoustic internal damping of the material higher than GG26 Cr by a factor or 10, leading to a reduced tendency to screech
- Thermal expansion coefficient of only 4 x 10^{-6}/K, leading to reduced hot juddering
- Lower brake disc wearing to achieve full durability and reduce cold juddering
- Constant thermal properties in the temperature application range

Fig. 4.3: LR equipped with CMC brake disc, Detroit Motor Show 1999

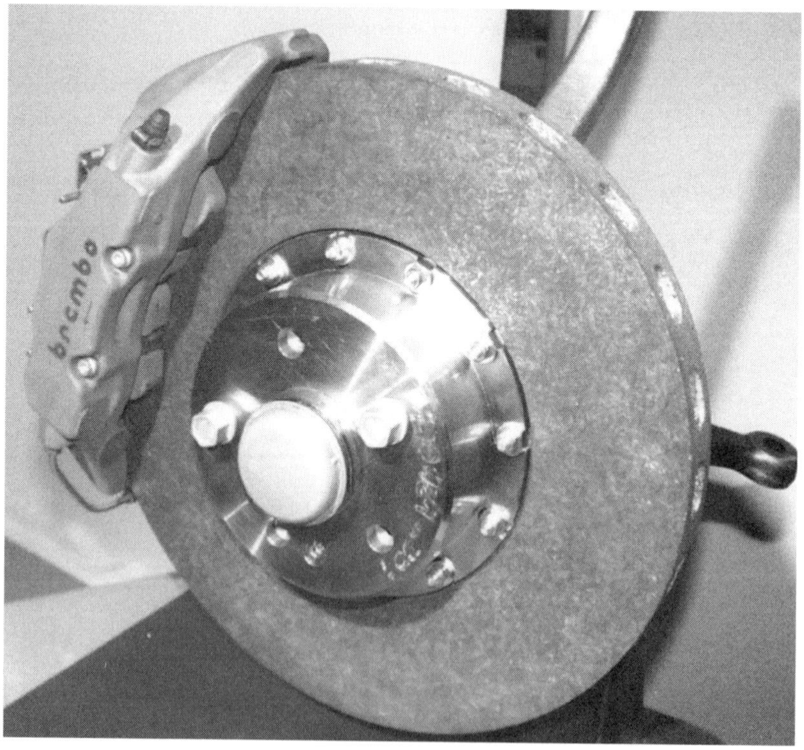

Fig. 4.4: Steering knuckle with CMC high-performance brake, mounting and caliper

4.5 References

[4-1] Fuganti, A.; Lorenzi, L.: Experimental study of MMC brake discs. 29th ISATA Conference, 3. – 6.6.1996, Florence, ISBN 0-947719-79-2

[4-2] Buch, F. H.; Denholm, M. J.; Konzaki O.; Weintrub, M.: A Status Report Aluminium Metal Matrix Composite Disc Brake Rotors. 29th ISATA Conference, 3. – 6.6.1996, Florence, ISBN 0-947719-79-2

[4-3] Krenkel, W.: Ceramic high-efficiency brakes from C/C-SiC. DGM-Training Seminar "Ceramic Composite Materials", 20./21.10.98, Stuttgart

[4-4] Wycliffe, P.: Friction and Wear of Duralcan Reinforced Aluminium Composites. in Automotive Braking Systems, Alcan Int. Ltd., SAE-Paper 930187

[4-5] Dwivedi, R.: Performance of MMC-Rotors in Dynamometer Testing, Lanxide Corp., SAE-Paper 940848

[4-6] Wirth, X; Rychli, W.: New brake discs from aluminium for rail vehicles. Weekly publication NEUE BAHN, 22.09.95

[4-7] Wirth, X.: Rail & commercial vehicle disc brakes in respect of new frictional materials, Company publication KNORR-BREMSE aktuell, Volume 25, December 96

[4-8] Tatarzycki, M.; Webb, R. T.: Friction and Wear of Aircraft Brakes, ASM Handbook, Vol. 18, S. 582-587

[4-9] http://www.messier-bugatti.com/carb-ind/ci-cct.htm

[4-10] Gibson, D. W.; Taccini, G. J.: Carbon/Carbon Friction Materials for Dry and Wet Brake and Clutch Applications, SAE Paper 890950

[4-11] Gkogkidis, A: The frictional behaviour of poly-granular carbon and carbon-fibre reinforced carbon with grey cast iron brake discs, Karlsruhe, Techn. University, Dissertation 1986

[4-12] Schanz, P.; Krenkel, W.: Description of the Mechanical and Thermal Behaviour of Liquid Siliconised C/C Proc. of the HT-CMC, ECCM6, 20 – 24, Sept. 1993, Bordeaux

[4-13] Patent Application DE 197 11 829 C1

[4-14] Burckhardt, M.: Chassis Systems: Brake dynamics and car brake systems, 1st Edition., Würzburg: Vogel, 1991

[4-15] Bühl, H.; Mörgenthaler, K.D.; Näumann, E.: CFC-Components in automobile building, BMFT-Project, Ref 01 ZA 091

[4-16] Company Publication: Lanxide Corporation, USA: Data Sheet Reinforced Metals, 1994

[4-17] Company Publication: Alcan Aluminium Corporation, Duralcan USA: Material Data Sheet, 1996

[4-18] Company Publication: Hitco Technologies Inc. USA: Material Data Sheet, 1996

[4-19] Krenkel, W.: CMC Materials for High Performance Brakes, 27th ISATA Conference, dedicated on: Electric, Hybrid & Alternative Fuel Vehicles and Supercars, Aachen,1994

[4-20] Füller, K.H.: Tribology, mechanical and thermal behaviour of new brake materials in car disc brakes, Dissertation, Univ. Stuttgart, 1998

5 Investigation of the Thermal Behaviour of Vented Disc Brakes by Means of CFD

Dipl.-Ing. C. Repmann, Volkswagen AG

Wolfsburg, Germany

Abstract

A particular disc brake design is used as an example to describe a method for calculating temperatures of disc brakes under time-independent conditions. For the calculations a *CFD* (Computational Fluid Dynamics) code is employed that determines distributions of local temperatures in the flow field as well as the disc brake rotor. Results of the numerical simulations are compared with those of experiments. It is demonstrated that the numerical data are useful for interpreting the experimental findings.

5.1 Introduction

In recent years a trend towards faster vehicles with more powerful engines developed in the automotive industry. At the same time manufacturers strive for improved energy efficiency of their cars. This is to be achieved by restraining vehicle weight and reducing aerodynamic drag. The latter results in an increased covering of the forward areas of vehicles, thus minimizing drag in the underhood and the wheel arch zones. Consequently the air flow towards the brake assemblies is reduced even though increased top speed means higher energy dissipation inside the brakes. The requirements for the least possible weight and size of components and assemblies prevent an increase in the dimensions of the brakes on a par with their higher loadings. Therefore measures for improved efficiency of the brakes have to be taken.

An upper limit on the temperature of the brake assembly is dictated by certain material properties. This means that the dissipated kinetic energy has to be disposed of more efficiently. There are three modes of heat transfer: Convection, radiation and conduction. Possibilities of increasing radiation and conduction are restrained by shape and material properties. Therefore the focus is on convection. This means the heat is transported out of the brake by the surrounding fluid, i.e. the air. Two approaches are possible: Increasing the mass flow towards the brakes and making better use of the mass flow that is available. The latter aspect will be elaborated on in this paper.

The brake disc is responsible for the major share of the convective heat transfer away from the brake assembly. To this end the rotor is frequently equipped with cooling ducts. This ventilation facilitates a significant increase in cooling performance. The flow through the vents is highly complex and cannot easily be influenced. The results presented in the following sections are intended to give an improved understanding of the details of the convective heat transfer inside the cooling ducts. This is meant to pave the way for increasing the heat transfer from the vent passages.

5.2 Vented brake rotor designs

5.2.1 Hat interfaces to the inner and outer rims

The cooling passages of a ventilated brake rotor cause a subdivision of the rotor into two distinct rims, the outer surfaces of which are the rub cheeks. In an assembled state in a car these rims either face inside or outside and are therefore referred to as inner and outer rims. The rims are connected by fins that act as vanes for the air flowing through the cooling ducts. One of the rims of the disc brake is connected to the hat that is attached to the hub. Examples for the hat interfacing either with the inner or the outer rim of the rotor are given in **Fig. 5.1**. As can be seen, the connection of the inner rim to the hat causes a severe blockage of the flow path into the passages. This can only partially be compensated for by holes introduced into the hat area of the rotor. In contrast an interface of the hat with the outer rim allows the air to be sucked into the ducts without subjecting it to such severe turns. This can be expected to lead to higher flow rates through the ducts with the result of a better cooling performance.

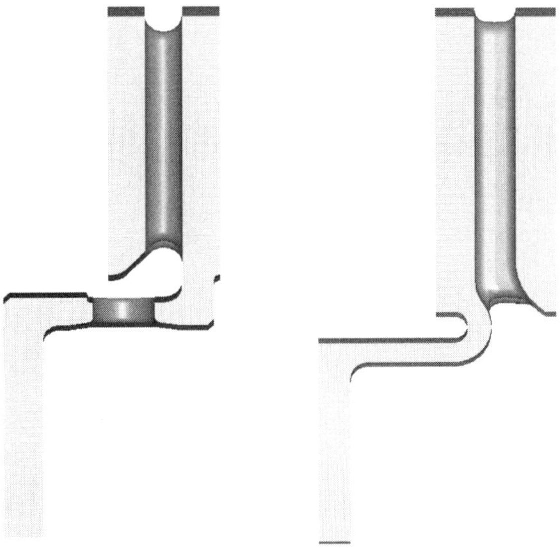

Fig. 5.1: Design variants of vented brake rotors with inboard (left) and outboard (right) air entry

5.2.2 Investigated configuration

The results presented in this paper pertain to a rotor design where the hat area is connected to the outer rim. The angle of the ribs is a speciality of the design. This is illustrated by the sectional view of **Fig. 5.2** displaying the wetted surface of the cooling ducts. The angling of the ducts leads to two different operating conditions depending on the direction of the rotation.

Fig. 5.2: Sketch of a vented brake rotor with straight angled cooling ducts

5.3 Experiments

5.3.1 Setup

The experiments are performed on test rigs with largely isolated disc brakes mostly without assembly parts. No attempts are made to reproduce the actual driving conditions. The reason for this approach is the desire to reduce the efforts needed for the tests. Moreover it is assumed that no severe qualitative changes result from mounting the rotor designs in a car.

The experiments consist of two parts. An isothermal measurement of the flow rate $\dot{V}$ through the cooling ducts and a measurement of the steady-state cooling performance $\dot{Q}$ .The flow through the cooling ducts is caused by the pumping effect of the rotating disc. The experiments result in the flow rate as a function of the rotational speed of the disc. The steady-state cooling performance describes the temperature of the disc as a result of the heat flux into the rotor at selected rotational speeds. The measurement of the cooling performance is conducted by subjecting the rotor at constant rotational speeds to a constant heat flux until the temperature stabilizes. The heat flux into the rotor is generated by friction between the rub cheeks and the pads. In terms of heat flux this setup corresponds to a genuine vehicle brake. A steady state is chosen to avoid influences from the thermal capacity of the rotor. This storage of heat is not relevant to the convective qualities of the disc brake.

For brake rotors of similar designs a largely invariant share of radiation and conduction can be assumed for comparable operating conditions. Therefore the temperatures derived from the measurements of the cooling performance are primarily indicative of the quality of the convective heat transfer from the cooling passages. This conclusion can be drawn because for brake rotors with similar or even identical outer shapes the respective shares of convection from the outer surfaces are similar as well.

Measuring the cooling performances and temperatures is rather expensive. Therefore a simpler means of characterizing the cooling efficiency of brake rotors is desirable. A rather straightforward approach consists of measuring the isothermal flow rate through the cooling ducts due to the radial pumping effect of the rotation. Since this fluid flow serves to dissipate the convective share of the heat flux from the cooling passages of the rotor, a correlation of the flow rate and the cooling performance may be assumed for similar geometries and surface conditions. One of the objectives of the measurements was to check the above hypothesis. The experiments were performed at the FH Wolfenbüttel.

5.3.2 Results

According to **Fig. 5.3** the measurement of the isothermal flow rate through the passages of the rotor shown in **Fig. 5.2** displayed a significant difference between the two possible directions of rotation. As was to be expected, the operating condition with less turning of the flow in the inlet area of the rotor vents led to a higher flow rate. According to the presumed correlation of flow rate and cooling performance a significant temperature difference can be assumed.

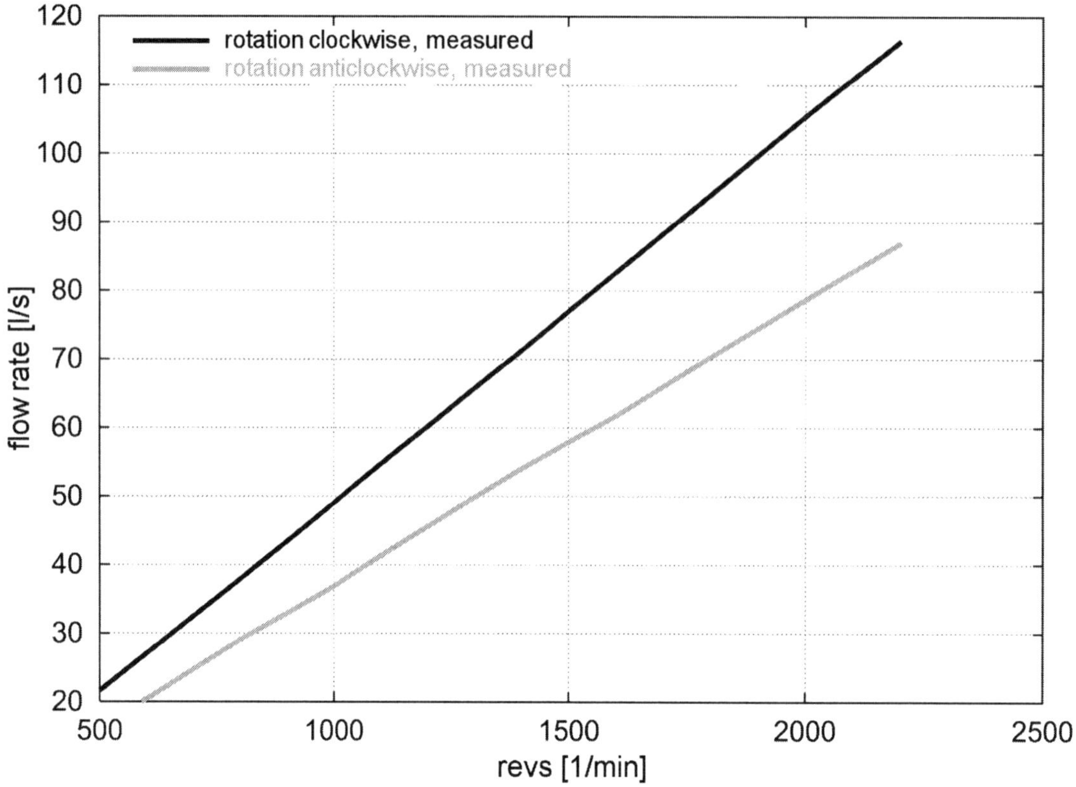

Fig. 5.3: Isothermal flow rate $\dot{V}$ through the cooling ducts of a brake rotor with angled ribs. Data derived from experiments

In contrast to this the plot of the experimental results in **Fig. 5.4** shows only minor temperature differences between the two conditions. This phenomenon had to be investigated more thoroughly. Doing so by experiment would have been very complicated and expensive. The local flow and heat transfer conditions inside of a rotating opaque object are hardly accessible for measurements. Therefore the means and possibilities of a numerical investigation were preferred.

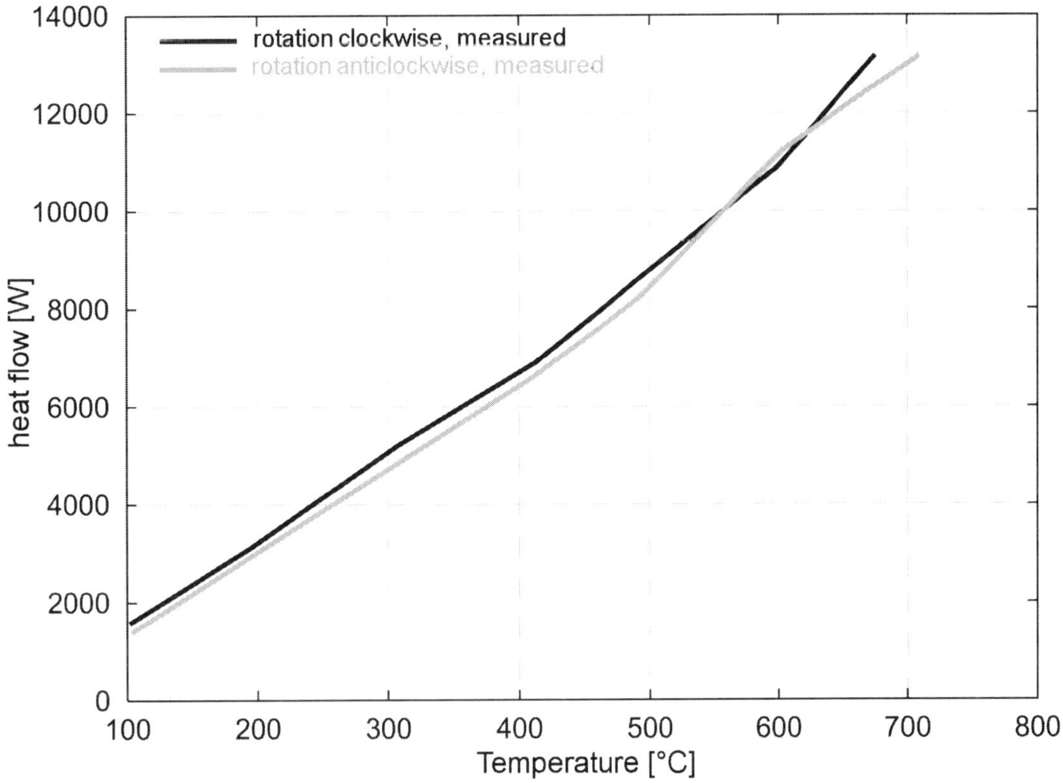

Fig. 5.4: Cooling performance $\dot{Q}$ of a brake rotor with angled ribs rotating at $n = 1013\ \text{min}^{-1}$. Data derived from experiments

5.4 The numerical model

5.4.1 Numerical code

The flow solver employed for the calculations is *Star-CD*, a commercial code. It was used for solving the time-averaged *Navier Stokes* equations [5-1]. They are a coupled system of nonlinear partial differential equations comprising the momentum equations, the continuity equation and the energy equation. The CFD code also solves the energy equation for solids thus calculating a temperature distribution for the flow field as well as for embedded solid objects. In addition to the heat transport phenomena of convection and conduction the code also simulates heat transfer through radiation.

The experiments were conducted for steady-state conditions. These circumstances can be modelled numerically in a very efficient manner by confining the simulated geometry to a segment of the entire disc brake. A further significant simplification can be achieved by modelling turbulence as steady-state. Turbulence [5-2] is a highly transient phenomenon but for most engineering purposes it can be approximated as time-independent. *Star-CD* offers a number of different turbulence models of the above category. A systematic evaluation and ranking of different turbulence models was conducted. As a result the calculations described in this paper were performed with a *two-layer* turbulence model [5-3] consisting of a kerng model [5-4], [5-5] and a model proposed by Wolfshtein [5-6].

Further preliminary investigations dealt with additional numerical aspects e.g. the choice of suitable differencing schemes for the different equations that have to be solved. An alternative to the standard model of thermal radiation was evaluated as well. Thus efficiently running the code on parallel computers in transient mode for potential future needs was made possible.

An essential aspect for the investigations is the possibility of integrating user subroutines into the code. This way special features can be added to the capabilities of the solver. In this fashion temperature dependence of conductivity was introduced for the cast iron of the brake disc as well as the air. The conductivity of the rotor material was found to have a significant impact on the temperature distribution inside the disc even in steady-state conditions. Another user subroutine was used to facilitate the heat flux into the rub cheeks of the rotor by means of heat sources placed at the surfaces. Thus a realistic modelling of heat generation through friction could be achieved.

5.4.2 Spatial discretization

Performing the numerical calculations requires a mesh of the geometric area that is to be investigated. The mesh is a discretized representation of the continuous volume. It consists of separate adjacent three-dimensional volumetric cells described by their respective vertices or nodes. It is this representation that can be processed by the solver. The mesh for the brake rotor is depicted in **Fig. 5.5**.

Fig. 5.5: Computational mesh for the segment of the brake rotor under investigation

The calculation is not performed on the entire geometry but on a segment of it. By taking advantage of the rotational symmetry of the disc all relevant geometrical information are contained in such a segment. This segment of the brake rotor is embedded in an adjacent fluid volume as is shown in **Fig. 5.6**.

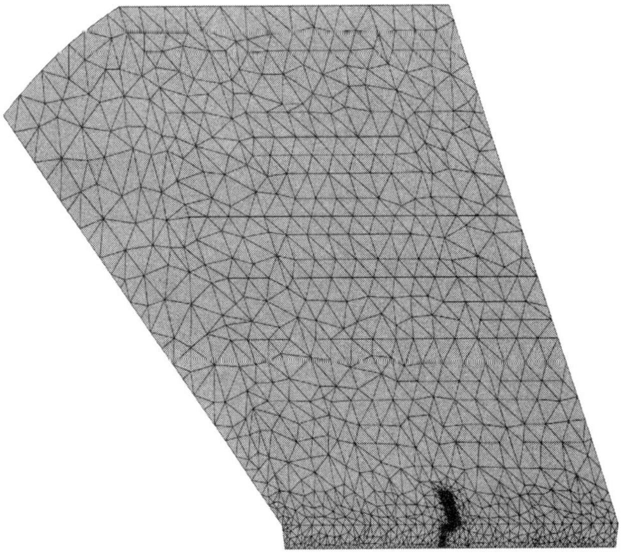

Fig. 5.6: Computational mesh for the segment of the brake rotor under investigation including the surrounding volume of air

5.4.3 Physical aspects of mesh generation

The concept of boundary layers is of great importance for understanding the procedure of the meshing process. The boundary layer is the area of a viscous flow field dominated by the influence of a wall. This zone is essential for momentum and heat transfer. **Fig. 5.7** shows the conditions inside a boundary layer for the example of a straight duct with heated walls. Right next to a wall the fluid has the same velocity as the wall. In a duct with stationary walls this means that the fluid particles have zero velocity – they stick to the wall. Under such circumstances the fluid particles also have the temperature of the wall. Because of the flow through the duct the fluid experiences a temperature rise towards the wall. The gradients of the velocity and temperature field inside the boundary layer are very steep. These gradients of the velocity and temperature profile determine the local momentum and convective heat transfer.

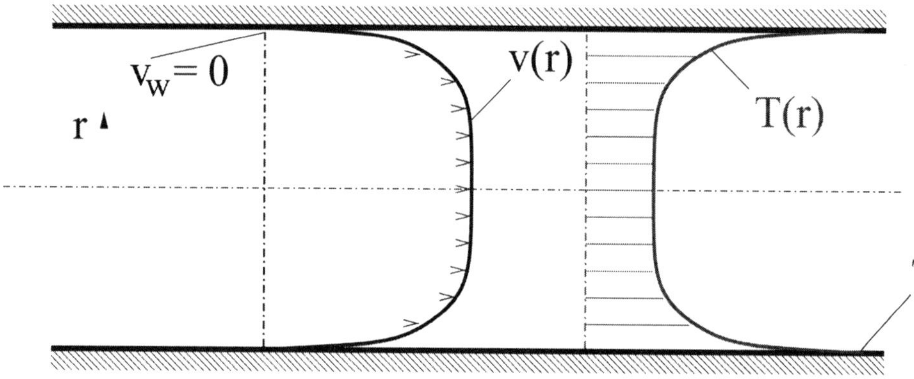

Fig. 5.7: Sketch of a velocity and temperature boundary layer in a duct with heated walls

Correctly simulating the temperature field next to the walls is of paramount importance for a realistic prediction of the heat flux from the walls into the fluid. **Fig. 5.8** shows the mesh for the ducts and their surroundings. It is advisable to discretize the areas inside the boundary layer with smaller elements of particularly suitable shape.

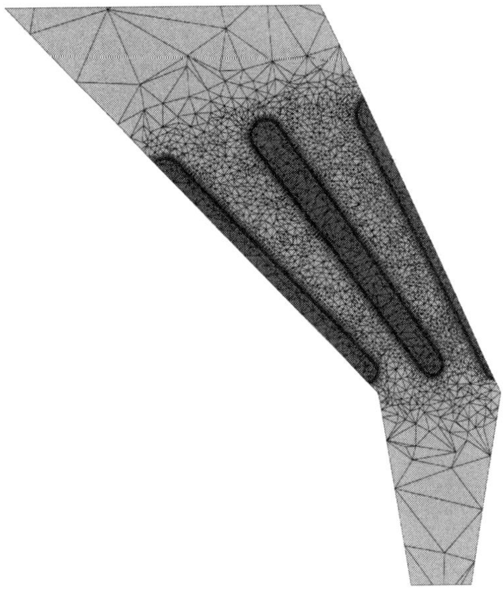

Fig. 5.8: View of the computational mesh in the area of the cooling ducts

An enlarged view of the wall area in **Fig. 5.9** exemplifies that adjacent to the walls semi-structured prismatic cells of decreasing thickness were used instead of the unstructured tetrahedra throughout the rest of the mesh. The prisms facilitate the numerical resolution of the steep gradients inside the boundary layer with an affordable number of elements. At the same time the tetrahedra allow efficiently filling arbitrarily-shaped regions. This hybrid approach towards mesh generation combines the numerical advantages offered by semi-structured prismatic meshes with the flexibility afforded by tetrahedra.

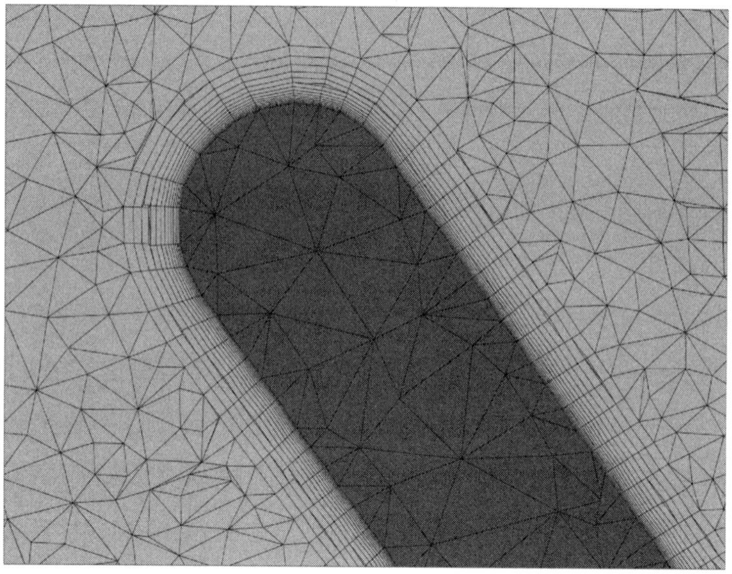

Fig. 5.9: Enlarged view of the computational mesh in the upper area of a rib

5.5 Numerical results

5.5.1 Isothermal flows

In contrast to the experiments the numerical simulation offers in-depth information on the flow field inside the cooling passages of the brake rotor. **Fig. 5.10** exemplifies this by showing the local fluid velocities in a field cut through the rotor. The velocities are depicted relative to the motion of the disc which rotates clockwise. The isothermal flow through the two ducts is relatively homogeneous. Along the trailing faces of the ribs, i.e. the suction sides, the velocity of the flow decreases as the outlet is approached. Above the ribs the flow separates from the ribs and small recirculation regions develop.

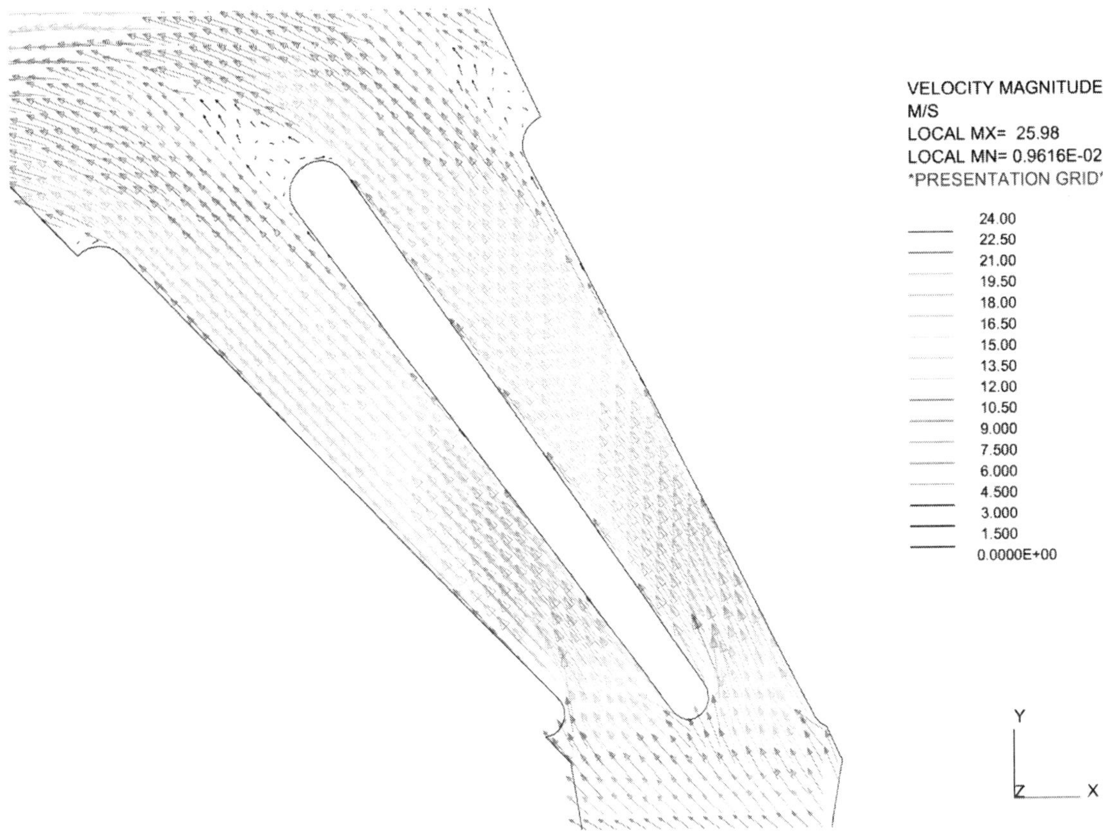

Fig. 5.10: Numerical results for the velocity vectors relative to the brake rotor spinning at $n = 1013\ \text{min}^{-1}$ in a clockwise direction

A change in the rotational direction to counterclockwise results in an entirely different motion of the fluid as can be observed in **Fig. 5.11**. The fluid sucked into the ducts is subjected to a severe change of direction. This causes it to mostly flow along the pressure side of the vanes – the leading faces in terms of rotation – while the suction sides show rather low velocities over their entire length. This distribution of the flow velocities has a negative impact on the amount of air pumped through the rotor. These results coincide with the experimental data of **Fig. 5.3**.

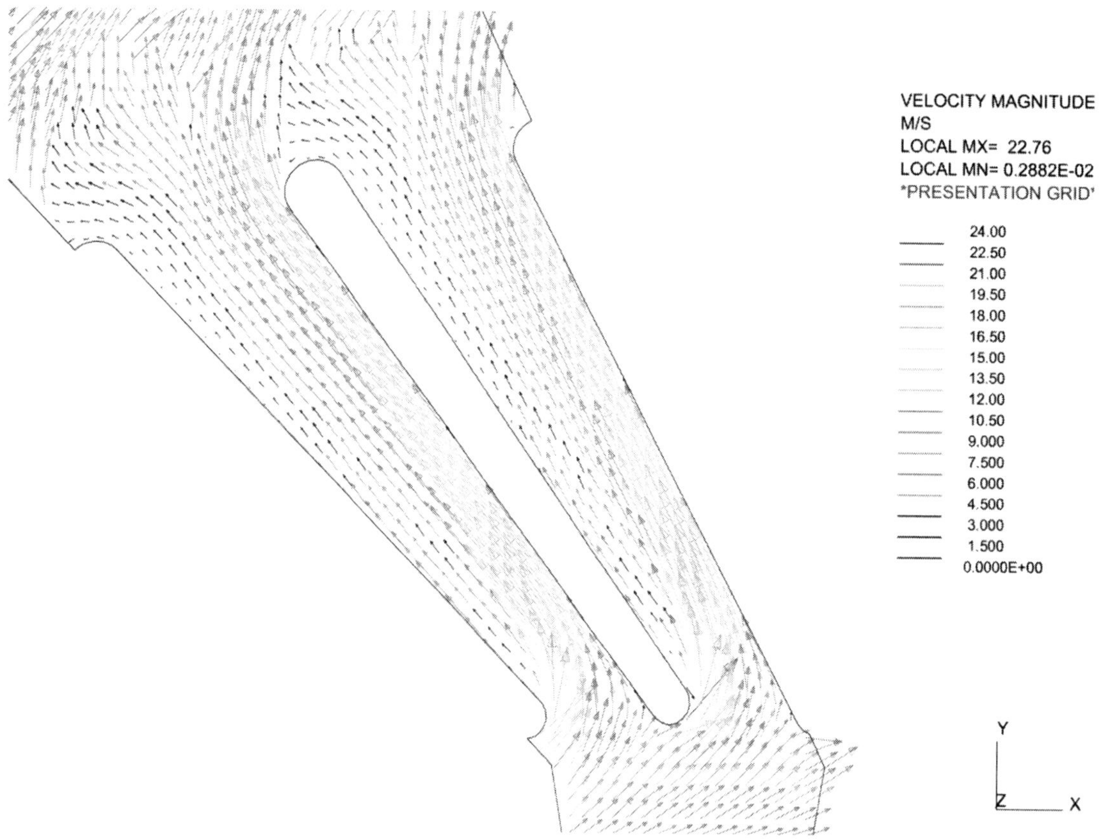

Fig. 5.11: Numerical results for the velocity vectors relative to the brake rotor spinning at $n = 1013$ min^{-1} in a counterclockwise direction

The flow through the passages is strongly influenced by the inlet conditions. **Fig. 5.12** shows an enlarged view of this entry area. At the lower side of the ribs almost in the middle of the curvature a stagnation point (marked P in the figure) can be observed. There the fluid has zero relative velocity. The flow is split along the streamline ending in the stagnation point. From there on the fluid moves attached to either side of the ribs towards the outlet. The relatively thin boundary layers starting at the stagnation point are clearly visible. To the right of the stagnation point the flow around the strongly curved surface of the ribs attains a high velocity. Downstream the boundary layer thickness increases, as can be seen by the growth normal to the wall of the area with low velocities. The thickening of the boundary layer is a characteristic phenomenon correctly predicted by the numerical simulation.

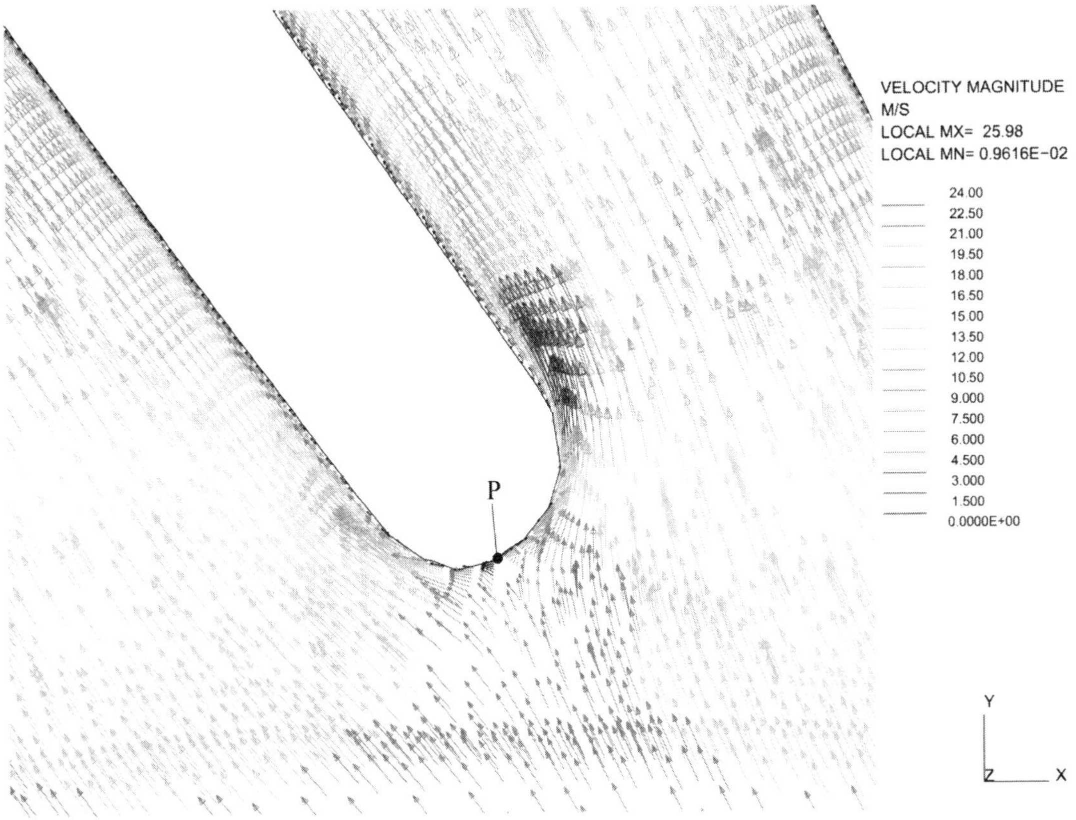

Fig. 5.12: Numerical results for the velocity vectors relative to the brake rotor spinning at $n = 1013\ \text{min}^{-1}$ in a clockwise direction

Reversing the rotational direction of the disc makes the stagnation point (again marked *P*) move to the left hand side of the vanes. **Fig. 5.13** shows fluid originating from the left flowing around the lower curved surface of the ribs where it is significantly accelerated. The subsequent deceleration at the straight wall results in an increase in static pressure. The fluid in the boundary layer already suffered a reduction in energy through viscous effects. It does not have sufficient energy to advance into the region of the adverse pressure gradient. As a consequence the boundary layer separates from the wall. This creates a recirculation region. Due to influences of the three-dimensional flow the fluid behind the separated region is again dragged towards the outlet. However following the separation the average velocity in the vicinity of the suction side of the rib remains significantly reduced. This is the reason for the reduced flow rate if the disc spins counterclockwise.

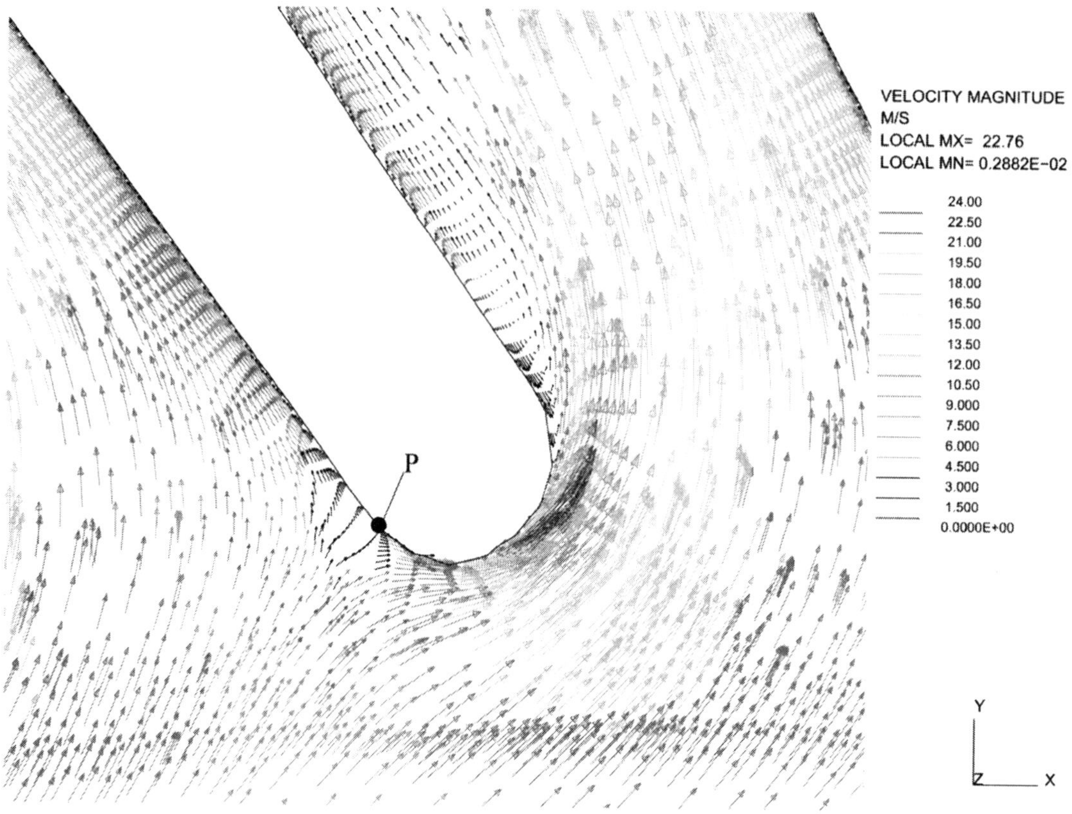

Fig. 5.13: Numerical results for the velocity vectors relative to the brake rotor spinning at $n = 1013$ min^{-1} in a counterclockwise direction

Unfortunately simulating the extent of the recirculation remains a challenge. The turbulence model used for the calculations is not particularly well suited for predicting separations. This is a general flaw of the commonly available turbulence models that are both numerically efficient and robust. Special attention has to be drawn to this topic because the size of the separated region significantly influences the amount of air flowing through the passages of the rotor.

Fig. 5.14 shows the flow rate through the passages derived from experiments and calculations. The values for the clockwise rotation are in very good agreement. The counterclockwise mode of operation displays larger discrepancies that are attributed to the flawed modelling of the size of the recirculation region.

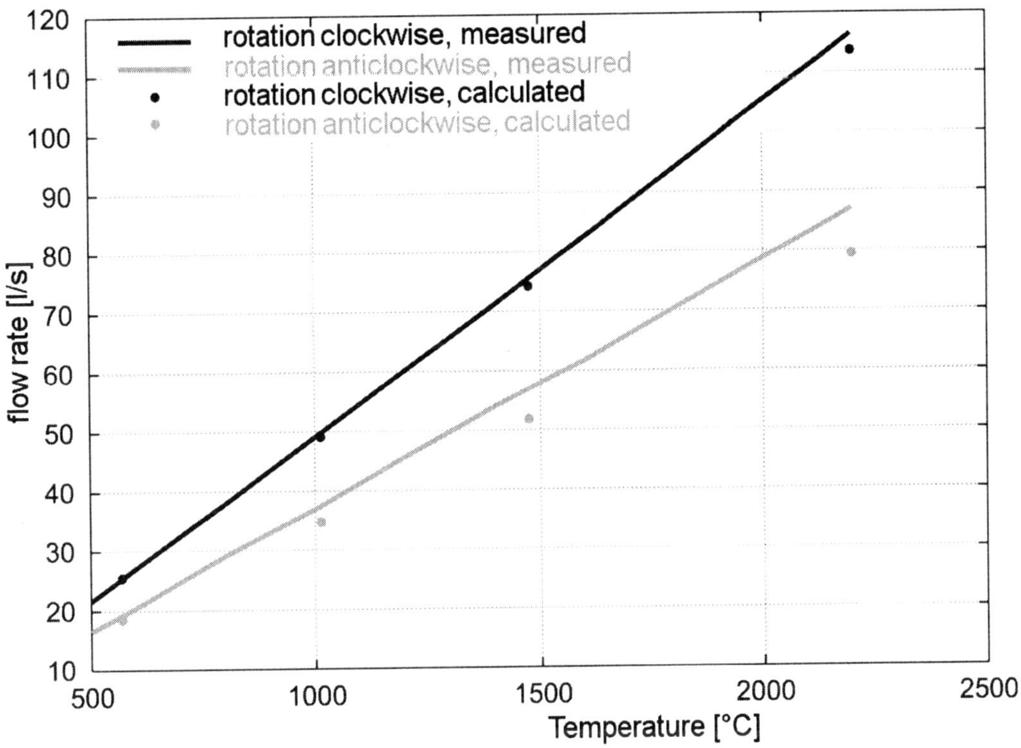

Fig. 5.14: Comparison of experimental and numerical isothermal flow rate $\dot{V}$ through the cooling ducts of a brake rotor with angled ribs

5.5.2 Flows with heat transfer

Fig. 5.15 shows a comparison of averaged temperatures of the rotor brake plates derived from experiments and calculations. The data are qualitatively as well as quantitatively in satisfactory agreement. The numerical results for the cases with a heat load of $\dot{Q}$ = 10 kW are to be considered as an example.

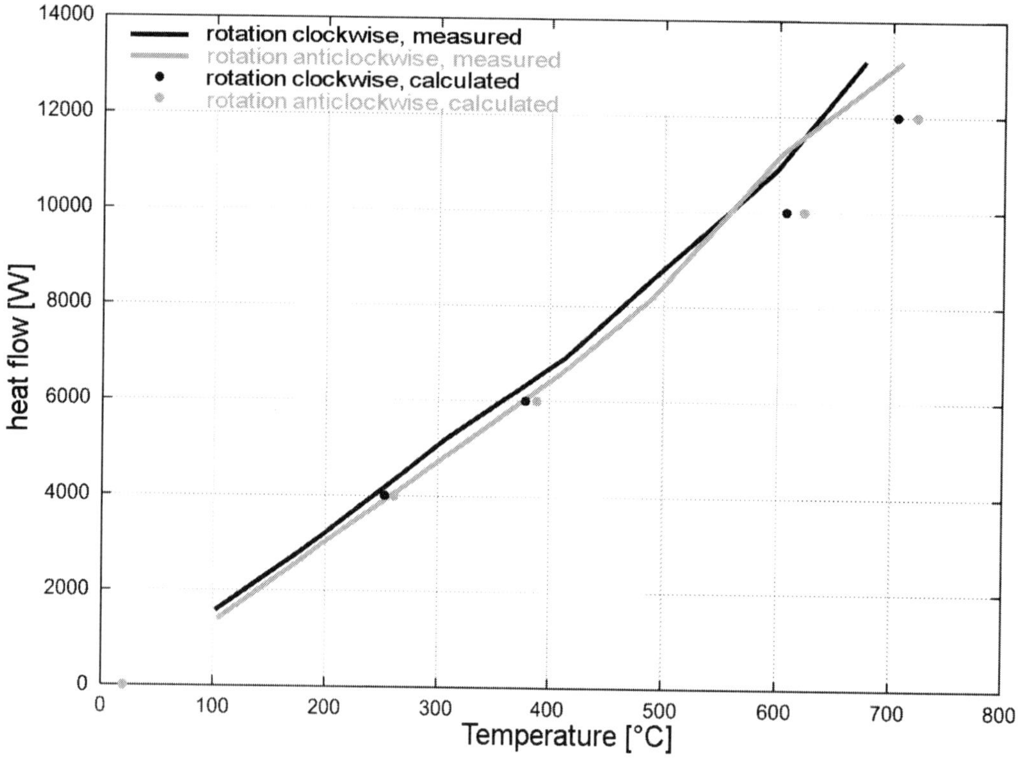

Fig. 5.15: Comparison of experimental and numerical results for the cooling performance $\dot{Q}$ of the brake rotor with angled passages spinning at $n = 1013$ min^{-1}

When the rotor spins clockwise, according to **Fig. 5.16**, there are extensive areas of low temperatures, indicated by the bluish hues. This is a sign of little mixture inside the fluid flowing through the ducts. Therefore the transfer of heat from the surfaces of the passages to the fluid away from the walls is limited. The air exits the passages at an average temperature of $t = 188$°C ($T = 461.15$ K). Taking into account that the temperature of the passages is von $t \approx 620$°C ($T = 893.15$ K) the efficiency of the brake rotor as a heat exchanger is rather low.

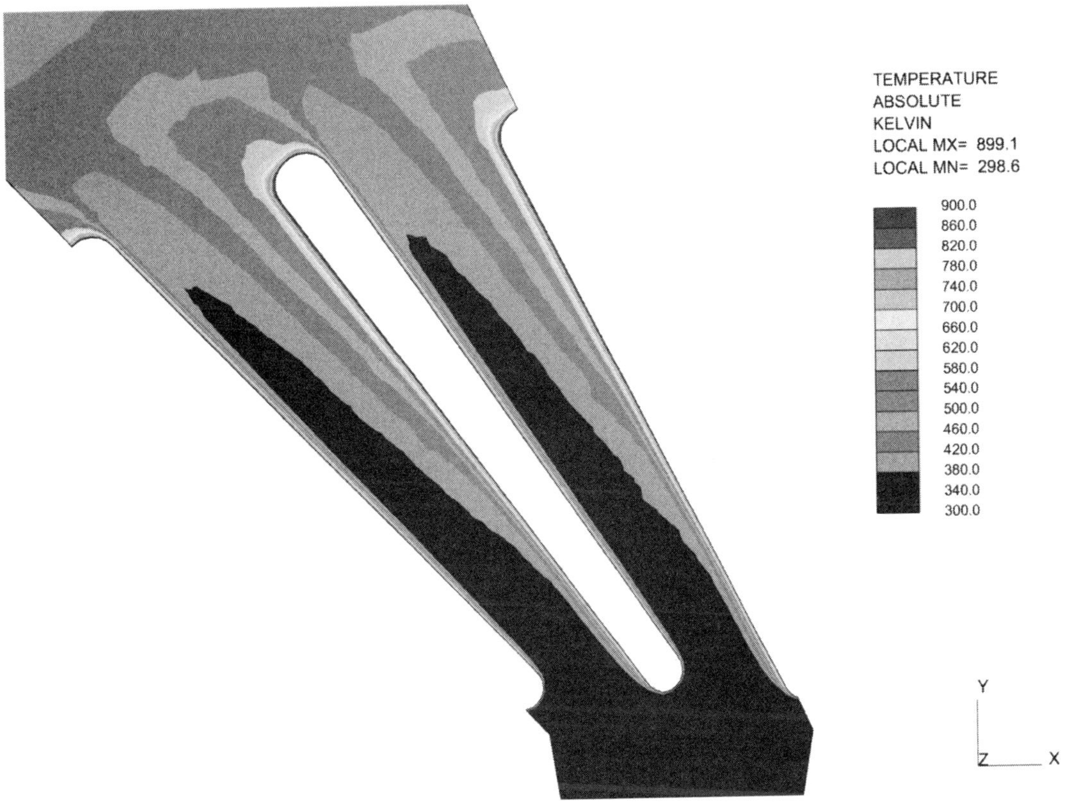

Fig. 5.16: Numerical results for the temperatures of the fluid flowing through the cooling passages of the brake rotor spinning at $n = 1013\ \text{min}^{-1}$ in a clockwise direction. The heat flux for the entire rotor is $\dot{Q} = 10$ kW

Fig. 5.17 shows different conditions if the rotor operates counterclockwise. The averaged temperatures at the exits of the ducts rise to hier $t = 243°\text{C}$ ($T = 516.15$ K). A stronger mixing of the fluid near the wall and at the centre of the ducts is visible through the earlier rise of the temperatures in the fluid downstream.

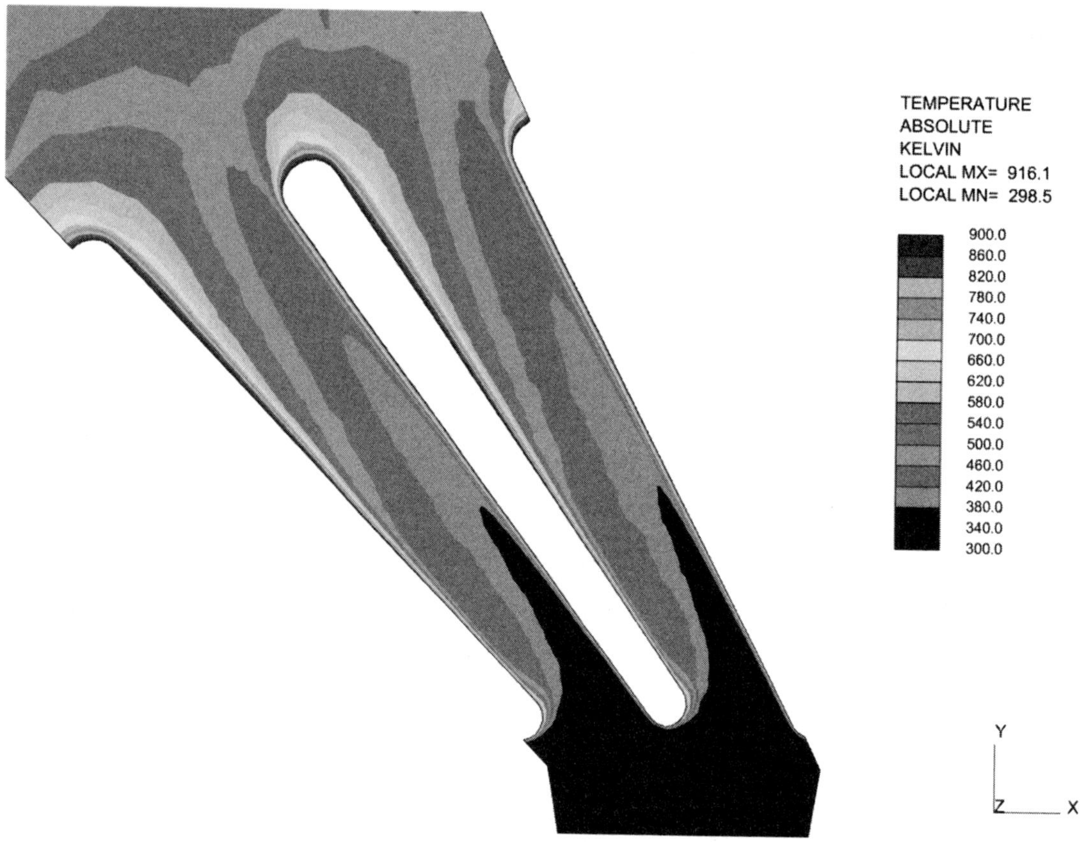

Fig. 5.17: Numerical results for the temperatures of the fluid flowing through the cooling passages of the brake rotor spinning at $n = 1013\ \text{min}^{-1}$ in a counterclockwise direction. The heat flux for the entire rotor is $\dot{Q} = 10$ kW

Fig. 5.18 is intended to shed some light on this phenomenon. The differences of the velocity magnitudes at four field cuts perpendicular to the axis of a duct are displayed. The magnitude of flow velocities for a counterclockwise rotation was subtracted from the values for a clockwise rotation. This means that positive values denote areas with higher velocities when the disc spins in a clockwise direction. Those areas that cause the higher average flow rate in the case of clockwise rotation can be recognized to be located predominantly near the centre of the duct, i.e. away from the walls. The zones decisive for the heat transfer are close to the wall. In these areas positive and negative values seem to be in balance. Herein lies the explanation for the differences in the flow rate for the opposing directions of rotation not being reflected in the temperatures attained by the rotor. Even though the mass flow is significantly higher when spinning the disc clockwise, no adequate use of the additional fluid sucked through the ducts is made. Therefore the temperature is hardly lower than for the other operating condition. So it becomes clear why the isothermal flow rates through the passages of the investigated rotor are no adequate measure of the cooling performance of the brake disc.

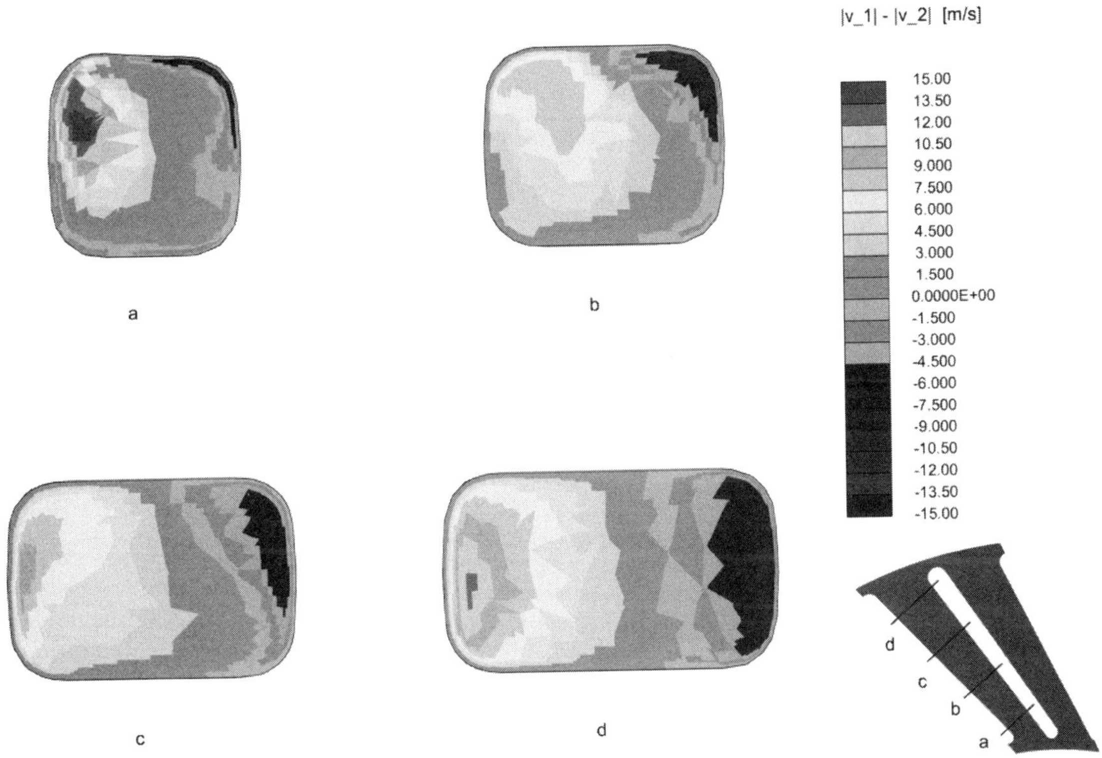

Fig. 5.18: Numerical results for the differences in velocity magnitude between the two directions of rotation of the brake rotor at $n = 1013$ min^{-1}. The values for counterclockwise rotation are subtracted from those for clockwise rotation. The heat flux into the entire rotor is $\dot{Q} = 10$ kW

According to **Fig. 5.15** only minor differences in temperature result from changing the rotational direction. As an example the temperature distribution in the rotor for a clockwise rotation is shown in **Fig. 5.19**. The highest temperatures can be observed around the rub cheeks immediately exposed to the heat flux from the friction process. The entrance area of the ducts benefits from favourable convective conditions leading to lower temperatures. This effect is augmented by thermal conduction into the cooler hat area.

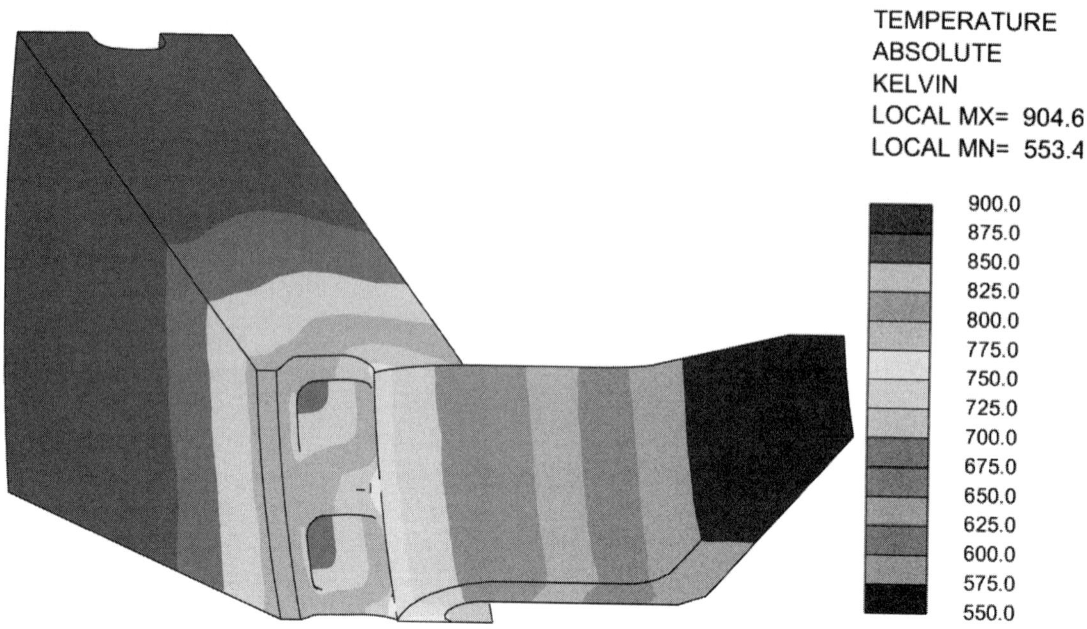

Fig. 5.19: Numerical results for the temperatures of the solid brake rotor spinning at $n = 1013\ \text{min}^{-1}$ in a counterclockwise direction. The heat flux into the entire rotor is $\dot{Q} = 10$ kW

Fig. 5.20 shows the distribution of convective heat flux per unit area on the wetted surface of the ducts for a clockwise rotation. It becomes clear that by far the highest heat transfer occurs at the inlet region of the passages where thin boundary layers, high relative velocities and low fluid temperatures coincide. Closer to the outlets of the ducts the heat transfer is reduced significantly.

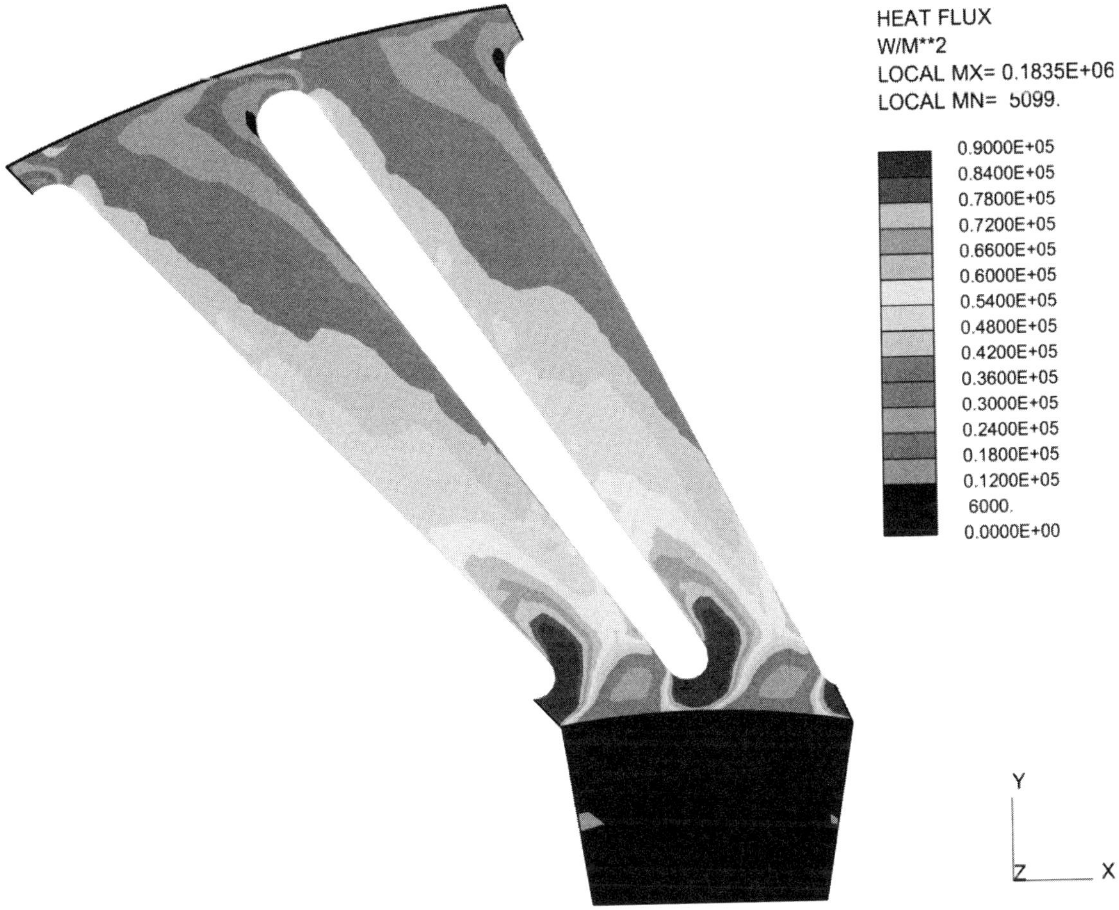

Fig. 5.20: Numerical results for the heat flux per unit area on parts of the wetted surface of the solid brake rotor spinning at $n = 1013\ \text{min}^{-1}$ in a clockwise direction. The heat flux into the entire rotor $\dot{Q} = 10$ kW

A change in direction retains the areas of maximum heat transfer at the entrances of the ducts with a significant decrease downstream (**Fig. 5.21**).

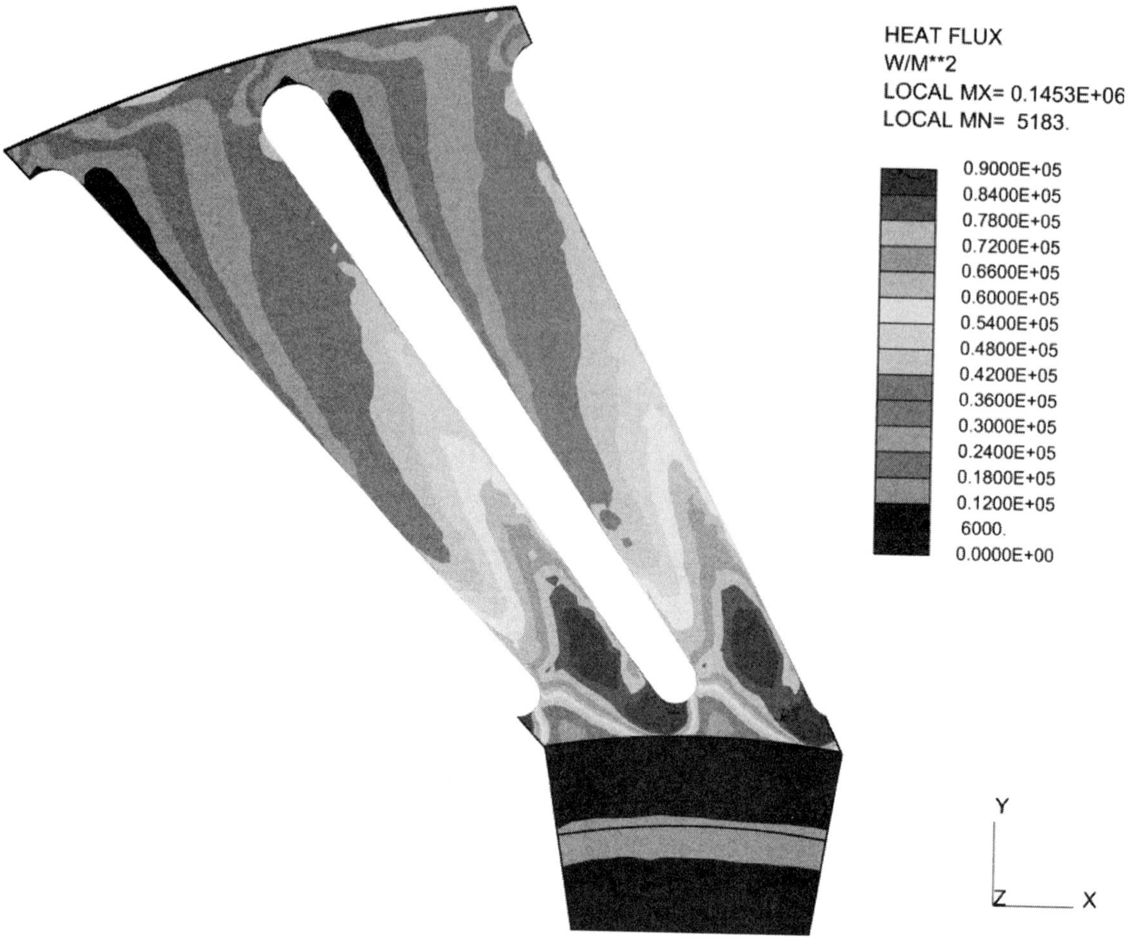

Fig. 5.21: Numerical results for the heat flux per unit area on parts of the wetted surface of the solid brake rotor spinning at $n = 1013\ \text{min}^{-1}$ in a counterclockwise direction. The heat flux into the entire rotor is $\dot{Q} = 10$ kW

The numerical results described above can serve as an indication towards approaches to improving the cooling performance of vented brake rotors. Is is clearly advisable to maximize the areas of high heat transfer at the inlets of the ducts. Additionally the deterioration of the heat flux near the outlets of the ducts should be countered through suitable measures of flow control.

5.6 Conclusions

A brake rotor geometry with straight angled cooling ducts was used as an example for experimental and numerical studies. The results demonstrated that *CFD* is well suited for analyzing the heat transfer behaviour of vented disc brakes. The validation of the presented method for calculating segments of brake rotors serves as the basis for a systematic investigation and evaluation of the convective cooling performance of vented disc brakes. Employing *CFD* for the analysis and development of new rotor designs offering improved convective heat transfer is envisaged. This can form the basis of general guidelines for the future design of vented disc brakes with enhanced cooling performance.

5.7 References

[5-1] Schlichting H, Gersten K: Grenzschicht-Theorie. 9. Auflage. Berlin, Heidelberg: Springer-Verlag; 1997.

[5-2] Pope SB: Turbulent Flows. 1. Auflage. Cambridge Univ. Pr; 2000.

[5-3] Rodi W: Experience with Two-Layer Models Combining the ke Model with a One-Equation Model Near the Wall. AIAA 91-0216; 1991.

[5-4] Launder BE, Spalding DB: The Numerical Computation of Turbulent Flows. Comp Meth in Appl Mech and Eng, 3:269–289; 1974.

[5-5] Yakoth V, Orszag SA, Thangam S, Gatski TB, Speziale CG: Development of turbulence models for shear flows by a double expansion technique. Phys Fluids, A4, 7:1510–1520; 1992.

[5-6] Wolfshtein, M: The velocity and temperature distribution in one-dimensional flow with turbulence augmentation and pressure gradient. Int J Heat Mass Transfer, 12:301–318; 1969.

6 Future Utilization of Stationary Brake Design in Simulation and Analysis of Brake Systems

Dr. U. Sailer, Dr. D. Mahnkopf, Dipl.-Ing. S. Knoll,

Robert Bosch GmbH

Stuttgart, Germany; Detroit, USA

Summary

The future use of the stationary brake design in brake system simulation and analysis is a central question for the development of existing development tools. As a Full-Brake-System-Supplier, Robert Bosch GmbH accompanies the development process of stationary brake design with the software tool CoBrake (Computation of Brake Systems). Not only the increasing use of dynamic brake control systems in modern brake systems, but also reappearing doubts about the accuracy of results obtained by stationary calculation for even the simplest of brake maneuvers, raises the question whether the classic stationary brake design will still be sufficient in the future. At the beginning of this article, concepts and parts of CoBrake's implementation are explained. In addition, some peculiarities of the calculation realized within CoBrake are described in detail. Furthermore the boundary conditions are presented which decisively influence the question of stationary or dynamic brake design. The article is concluded by examining some of those boundary conditions in detail, which leads to first conclusions about what calculation methods are to be used in the future.

6.1 Motivation and Introduction

In the past, different tools were used for the design, simulation and analysis of brake systems or components at the individual development locations. That is, different calculation bases, result representations, and a partly different handling of customer projects existed. The motivation and long-term vision of CoBrake is to standardize these tools on a global basis by means of the Intranet. Realizations, algorithms, rules and data that are of worldwide interest and validity for several project applications in the simulation and analysis of brake systems are made readily available as a company-internal global "single source". CoBrake reproduces – within a reasonable scope – the reality that is later on verified by testing the brake systems. The long-term vision of CoBrake is the computer-aided accompaniment of the entire development process of passenger car brake systems, and not just their calculation.

6.1.1 Conception

CoBrake is built in a 3-tier-architecture and consists of any amount of clients, an application-server and data resources, (ref. **Fig. 6.1**).

The application is loaded over the Intranet into the user's browser when the program is started. The application includes the user interface and the basic functionality of the calculation. The middle layer consists of a web- and application-server. While the web-server offers online-help and additional information, the application-server is used to provide additional calculation resources, for example commercial tools and interfaces to those tools. Currently the resource layer consists of an Oracle database and a file system. The data of the brake and vehicle components are stored in the database. For a more elaborate description of the software and hardware concept of CoBrake refer to [6-1, 6-2].

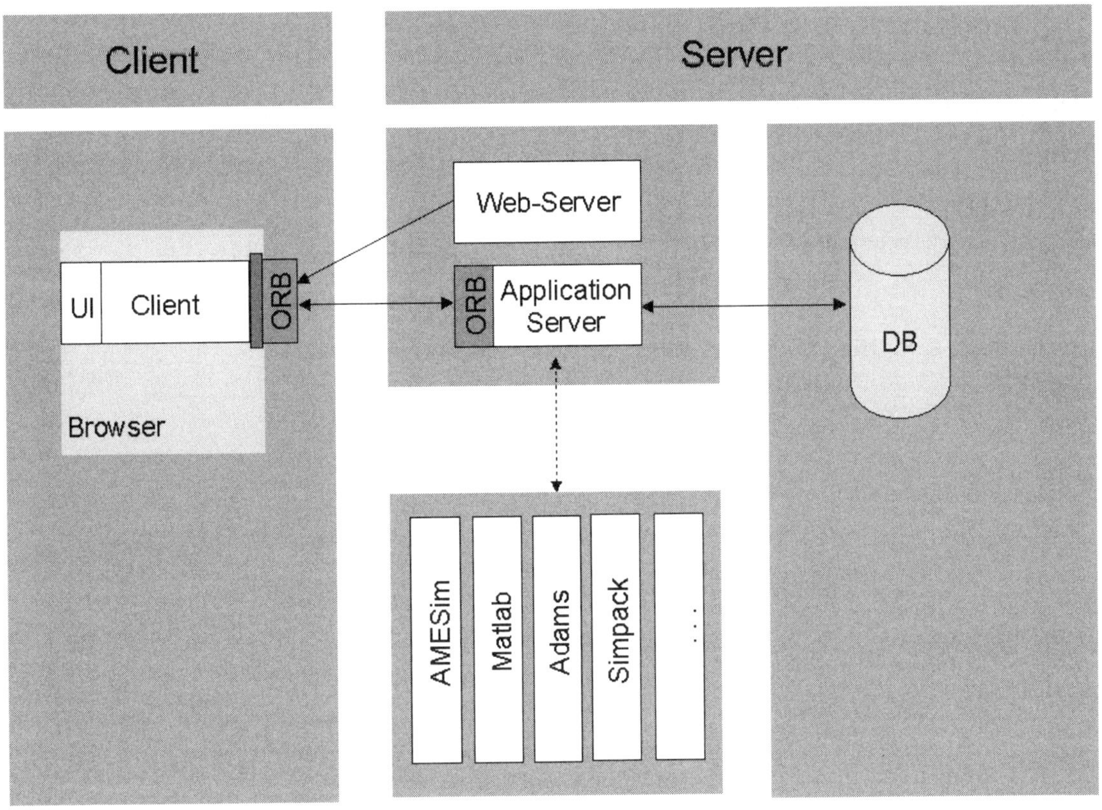

Fig. 6.1: 3-Tier Architecture of CoBrake [6-1, 6-2]

6.1.2 Calculation Scope of the Stationary Brake Design Program

The calculation of the service and parking brake is based on the classic stationary brake design 6-3, 6-4, 6-5]. A passenger car brake system consists of components, which either already form a module (e.g. disk) or can be combined to form such a module (e.g. wheel brake). The modular and hierarchical structure of the brake system was carried over into the program and data architecture. For a given (hierarchical) component, different models can be chosen that differ in their parameters and/or the calculation (see also **Fig. 6.9**). New systems, for example the EHB (Electro-Hydraulic-Brake), were easily included because of the object-oriented, modular implementation as well. For an elaborate description of the modeling of the service brake system refer to [6-6]. The thermal calculations within CoBrake concerning the wheel brakes will be extensively represented in a separate publication.

6.1.3 Automatic Generation of Technical Customer Reports

The result of every brake system design and simulation is a presentation to internal or external customers. Reproducibility and standardization, without giving up certain flexibility, is of enormous significance in this case. The general sequence in CoBrake is: predefined reports (without result data) are available as so-called Templates. CoBrake then reads the selected Template, executes the simulations defined in it and fills in the results. The finished report is saved in a format that allows further editing in word processors (**Fig. 6.2**).

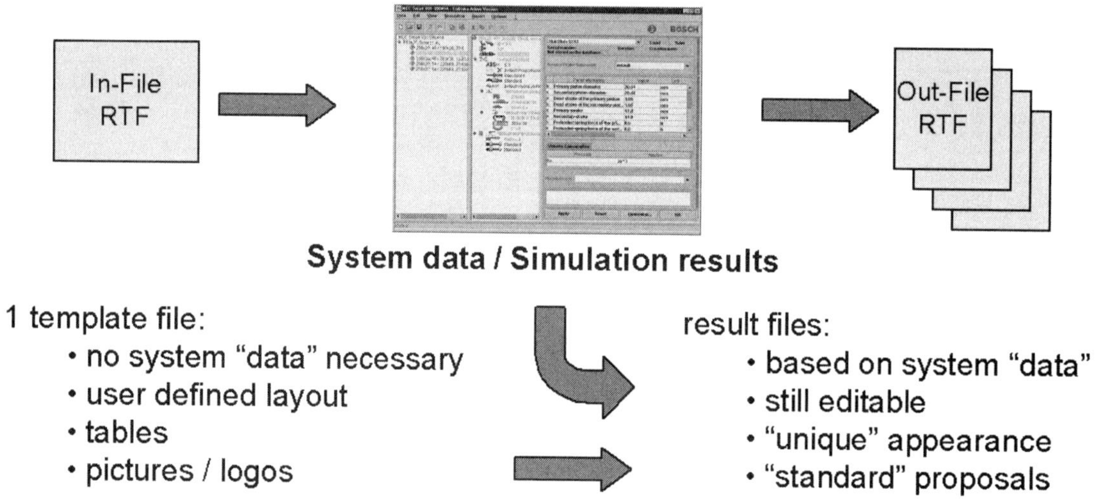

Fig. 6.2: Generation of Customer Reports

The Template itself is generated as a RTF-File using a common word processor, supplemented by additional instructions to define the simulation, the representation, etc and place holders for the results. RTF guarantees tool and platform independence since it is a publicly available standard.

6.2 Particular Approaches for the Stationary Brake Design with CoBrake

This chapter describes some particular aspects of the brake design and their solution in CoBrake.

6.2.1 Treatment of Points of Discontinuity

When stationary modeling brake systems, ambiguities appear with some components between input and output variables. They represent discontinuous step-functions in the mathematical sense (see e.g. **Fig. 6.3**). A variation of the input parameters – normally sufficient to determine the system response – does not guarantee any that the range of output variables is completely grasped. If, for example, the "Jump-In" Fth of the vacuum booster is exceeded during a finitely large calculation step, the output force exhibits a step (F_j) and because of that, a step also occurs at the input force of the master cylinder. The consequence is a discontinuity in the pressure curve.

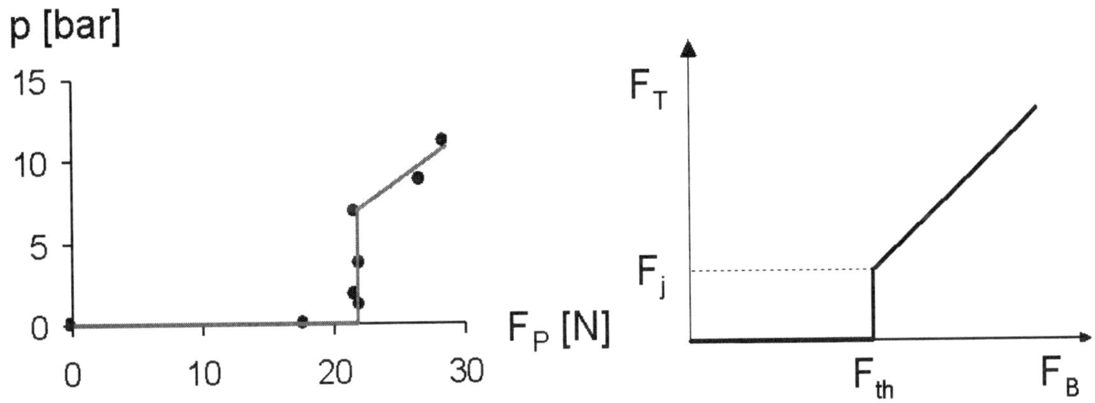

Fig. 6.3: Vacuum Booster-Characteristic Diagram (left: Measurement, right: Model)

As can be seen in **Fig. 6.3**, measurement points absolutely exist in the area of the discontinuity. However they cannot be represented by the stationary calculation. The effect of the ambiguity in the modeling of the vacuum booster is represented in **Fig. 6.4**. A linear interpolation between the solution points before and after this discontinuity would be absurd.

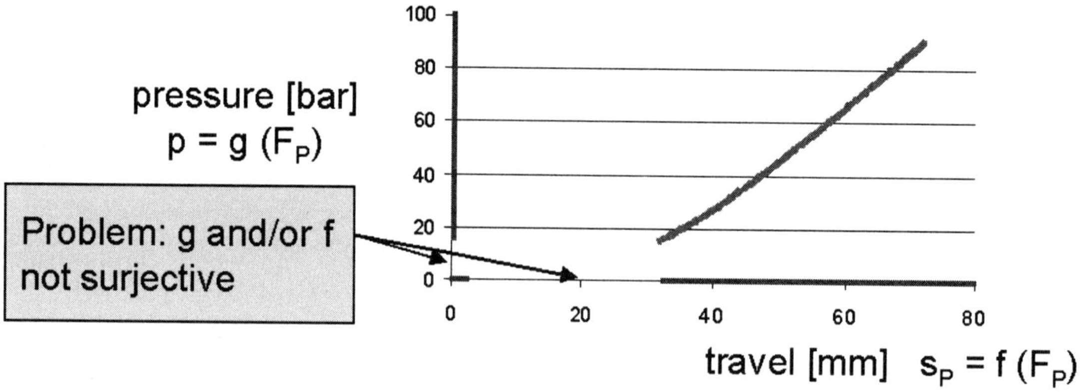

Fig. 6.4: Pedal Travel - Pressure Diagram

An approach to solve the problem by replacing all appearing discontinuities with adequately steep linear curves would falsify the component behavior on one hand, and would force too small incremental widths on the other hand. Therefore that approach is ruled out.

In CoBrake, additional simulation steps are inserted at points of ambiguities. The basic principle is represented in **Fig. 6.5**.

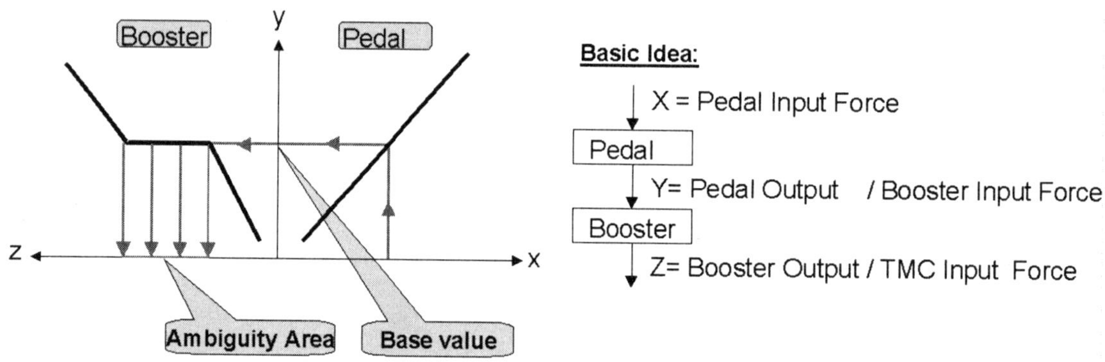

Fig. 6.5: Basic Idea for Treatment of Ambiguities

The start values, i.e. the location of the discontinuities, are determined by an approximation method. Generally more than an ambiguity can be encountered between two regular steps (e.g. the application pressure of the rear wheel brake can fall into in the area of the "Jump-In" of the vacuum booster). The possibility to analyze these discontinuities, to sort and to then remedy them is not pursued, since the Greedy-Algorithms [6-8] also achieves the goal and delivers a closed recursive algorithm [6-7]. If multiple ambiguities are detected during the calculation of a regular step, then only the ambiguity of the first block in the chain is treated. Ambiguities appearing in parallel become inevitably serialized through the implemented sequence of calculation. Another advantage of the procedure is that knee-points (C^0-steady) in characteristic lines can be declared as ambiguities by additional introduction of a point at the knee-point. The procedure can also be applied to them. The knee-points become part of the solution, independent from the underlying simulation step width. **Fig. 6.6** shows a simulation result using the procedure in the comparison to a measurement of a midsize class vehicle. A comparison with **Fig. 6.4** shows that the procedure treats the critical area in a physically meaningful way.

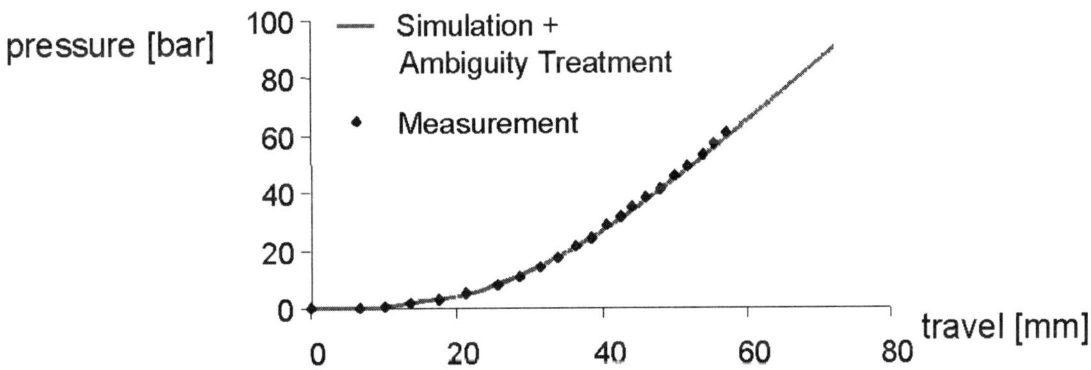

Fig. 6.6: Pedal Travel - Pressure Diagram after Treatment of the Ambiguities [6-7]

6.2.2 Simulation Coupling (Between Service Brake Calculation and Thermal Simulation)

Up to now CoBrake uses separate descriptions for the applications:

- Service Brake Sizing
- Parking Brake Sizing
- Thermal Sizing

However, with respect parameterization, a common database is used for the different hierarchical components.

To determine the amount of energy input, the thermal simulation needs information about the brake force distribution, which is identified through the service brake sizing. In the simplest case, the thermal simulation requires only one calculation run of the service brake and from there obtains a constant brake force distribution for a constant deceleration. This brake force distribution is then used in the simulation for all stops with the same deceleration. However, as soon as the temperature dependency of the friction coefficient of the lining is taken into consideration (fading), the brake force distribution varies and must be calculated again. A closed circle emerges between thermal simulation and service brake sizing. **Fig. 6.7** shows that the influence cannot be neglected. Different simulations are represented for constant and temperature-dependent friction coefficients of a front wheel brake in comparison to test results.

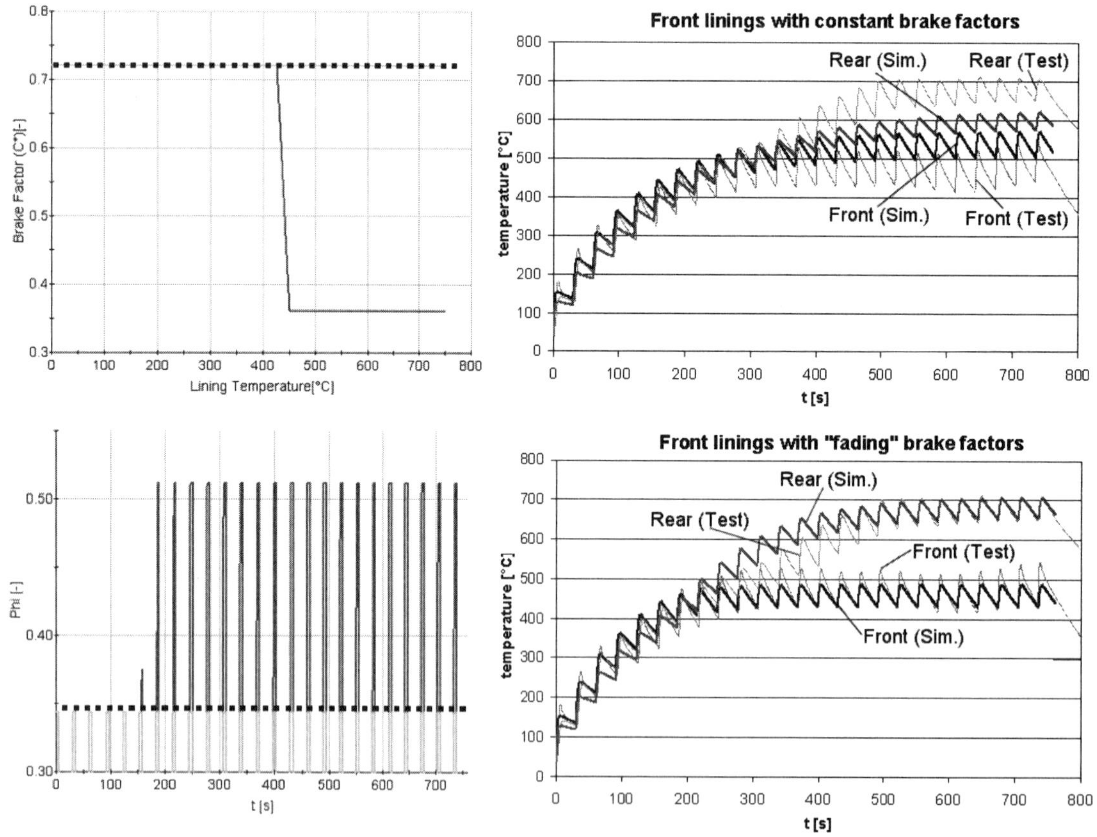

Fig. 6.7: Influence of the Temperature-Dependent Friction Coefficient on the Brake Force Distribution and on the Temperature of the Wheel Brakes

To solve that coupling, following possibilities were examined:

- Execution of the service brake calculation at each time step: that delivers the most exact results, however requires a lot of calculation time and memory space.
- Forecast (Pre-Simulation) of all possible friction coefficient-combinations and storage of a matrix of the brake force distributions $\Phi = f(C^*_{front}, C^*_{rear})$: That solution is easy to implement. Based on the fact that many unnecessary simulations must be executed in that case this solution is also ruled out and is modified to the so-called:
- "Lazy Evaluation Method": The brake force distribution for the currently applicable friction coefficients is calculated and stored. It turns out that the matrix of the actually occuring friction coefficient-combinations amounts to, at most, only 30-40 percent of all that are theoretically possible. In an adequately close proximity of already existing calculated points, the "Lazy Evaluation" uses both these old values as well as linear interpolation. The calculation time and the memory expenditure are improved while simultaneously obtaining good results. Therefore, this method finds its application in CoBrake (**Fig. 6.8**).

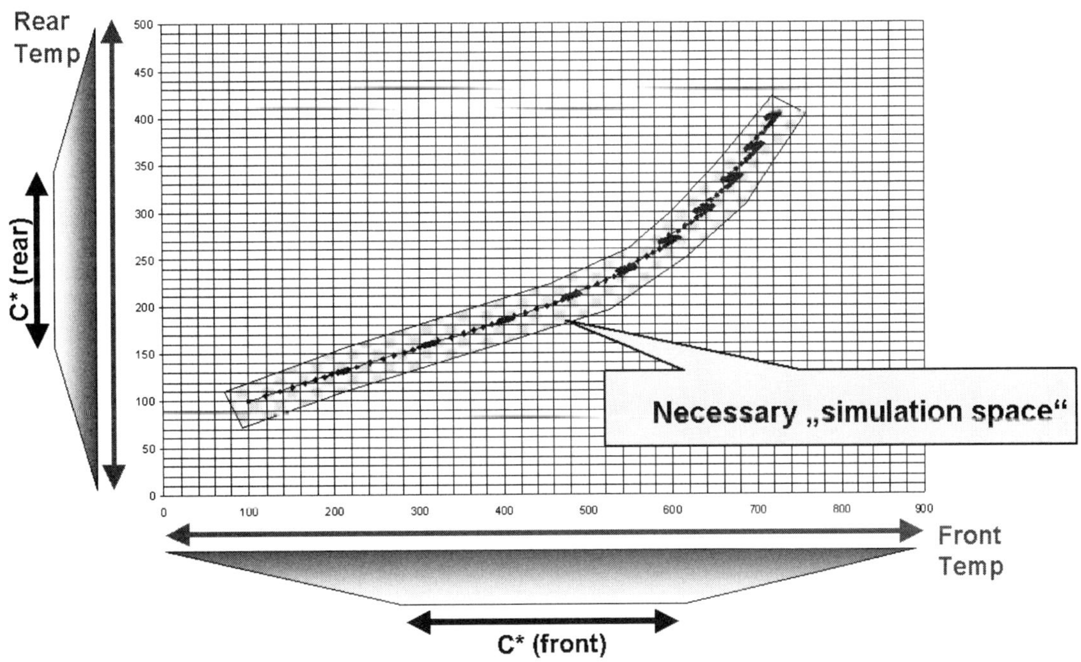

Fig. 6.8: Coupling Service Brake Calculation and Thermal Simulation with "Lazy Evaluation"

This current approach of executing separate partial simulations must be checked again for dynamic simulations which consider the effects of brake control systems.

6.2.3 Submodels for Considering Different Parameter Sets

The quantitative and qualitative availability of parameters changes during development. Calculation programs must take this into account.

In CoBrake, different models (submodels) that differ in their parameters and/or their calculation rules can be defined on every hierarchical level. Models that require only a few simple parameters are used in the beginning phase of development. Later, with increasing knowledge about additional parameters as well as more exact values, accordingly more complex models are available. In **Fig. 6.9**, the brake lining is represented as an example.

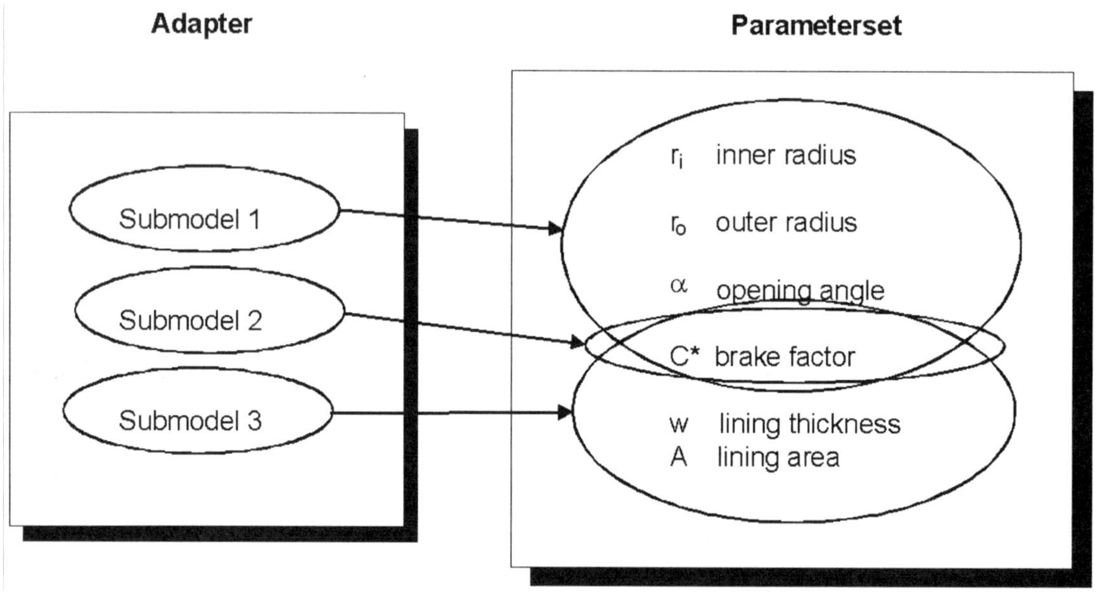

Fig. 6.9: View of the separate component brake lining on its parameters [6-6]

The cool-down characteristic e.g. is determined by thermal simulation in a very early development stage with a model that is based on geometrical data and material parameters. As soon as the first prototype exists, cooling curves can be recorded from which cooling coefficients can be extracted. They serve as parameters for a more powerful model that has better accuracy for certain thermal driving maneuvers.

6.2.4 Stochastic Simulation

Brake design programs work mainly with single nominal values for each parameters. That causes the results to have the same character: they represent particular results. However, the ultimate long-term goal of simulation tools should be the system-prediction, which has the character of requirements.

To achieve that, tolerance bands for the system parameters have to be introduced, which are considered accordingly in the simulation. The results can then be compared to requirements (a $<$ target value $<$ b; target value $>$ min; target value $<$ max), and with that the development responsibilities can be improved internally between components and system development, as well as externally between suppliers and automobile manufacturers.

The stochastic simulation helps to answer the following questions:

- Consideration of parameter fluctuations caused by different operating conditions (for example friction coefficient changes, vacuum level at the vacuum booster).
- Parameter fluctuations based on production tolerances.
- Parameter fluctuations caused by aging or wear of the component.
- Estimation of the influence of unrestricted design parameters.
- Effects of only unclearly known parameters.

CoBrake is completely programmed in the Java language. Starting with Release 11 the commercial simulation tool MATLAB offers the possibility to understand Java code. With that, it was possible to implement an interface connecting CoBrake and MATLAB (CoMa). The CoMa-interface offers the entire CoBrake functionality including database access. Presently, the stochastic simulation is executed using the Statistical-Toolbox of MATLAB and the CoMa-interface.

As an example, a simulation evaluating pedal feel is shown in **Fig. 6.10**. The production tolerances of the vacuum booster were varied. It is obvious that the customer gets a better statement about attainment of his requirements from this simulation and different histograms (e.g. pedal travel at a deceleration of z = 0.5 g between 34 and 45 mm), than by means of a single value or curve.

Using the example of the stochastic simulation by connecting to MATLAB on the middleware (application-server, see. **Fig. 6.1**), the development philosophy of CoBrake can be shown quite well: Globally harmonizable methods and their pre and post processing, which are standardizable and of interest in all projects, are integrated as permanent components and made available via Intranet. Special inquiries, as shown in **Fig. 6.10**, are answered quickly and cost effectively by using simple interfaces to commercial tools. That solves at least partly the seemingly unsolvable conflict between flexibility (project- or customer-specific questions) and rigidity (standardization for general questions). When certain stochastic simulations are increasingly being demanded, and therefore accessible to standardization, they are incorporated into CoBrake as a permanent component. That does not necessarily mean that the code is implemented again. It means that just the simulation control and data management is integrated into CoBrake. Whether the calculation ultimately runs in MATLAB or in Java, whether on the client or on the middleware, is not of interest to the user.

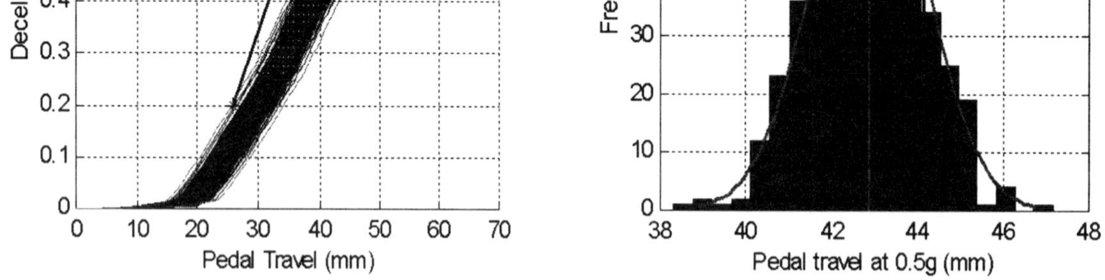

(left: with Boundary Curves of Customer Requirements, right: Histogram with z = 0.5 g)

Fig. 6.10: Stochastic Simulation of a Pedal Travel – Deceleration Curve (500 Variations)

6.2.5 Mixed Stationary and Non-Stationary Simulation

Using the vacuum booster as an example, the limitations of a purely stationary sizing calculation can very easily be shown. Caused by multiple actuation of the vacuum booster, the vacuum level decreases (see **Fig. 6.11**). The customer requirement is that a specified deceleration (z) can be reached using only a specified, limited pedal force during the third stop in a sequence.

Since the vacuum booster is not operated in a stationary condition, but rather the pressure level must be monitored over the time, this question cannot be answered with just a stationary calculation. CoBrake solves this question also using the CoMa-interface. The vacuum booster is modeled in MATLAB/SIMULINK [6-9] and the remaining components are taken from CoBrake. The result is a mixed stationary and non-stationary approach.

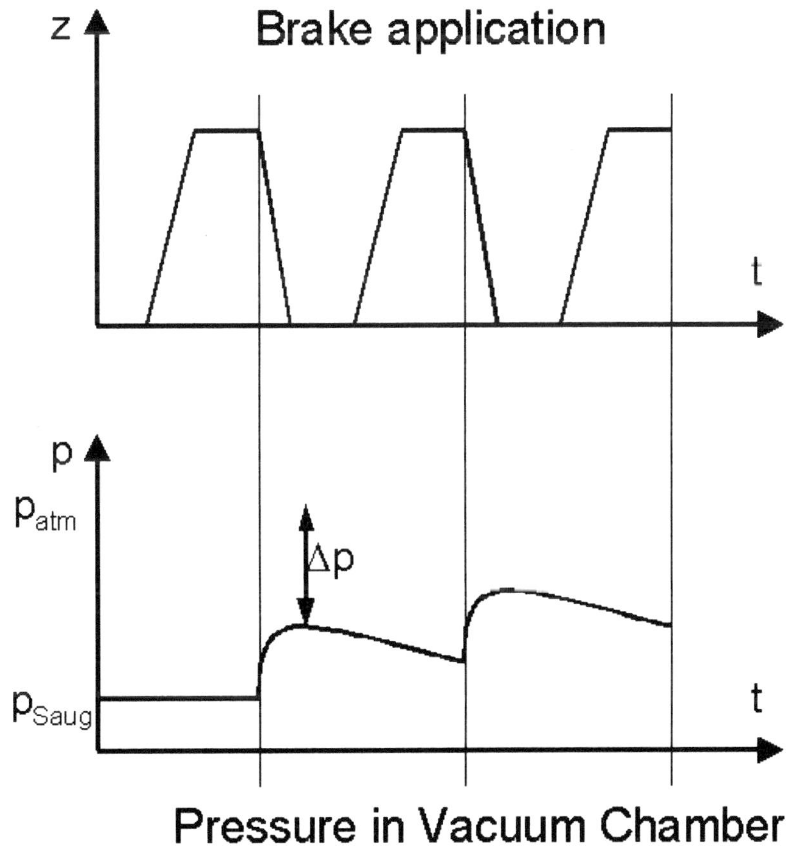

Fig. 6.11: Multiple Actuation of the Vacuum Booster

Using this approach, a simulation environment was created to solve the sizing problem of repeated brake applications described above. The validation of the model was done on the basis of comparisons with measurements. As sizing parameters, the diameter of the reaction disc d_A and the volume of the vacuum chamber V_U are chosen. As a new evaluation criteria, the relationship between the time-variant boost-ratio of the nth stop i_n and the time-invariant (stationary) boost-ratio i_0 is chosen (**Fig. 6.12**).

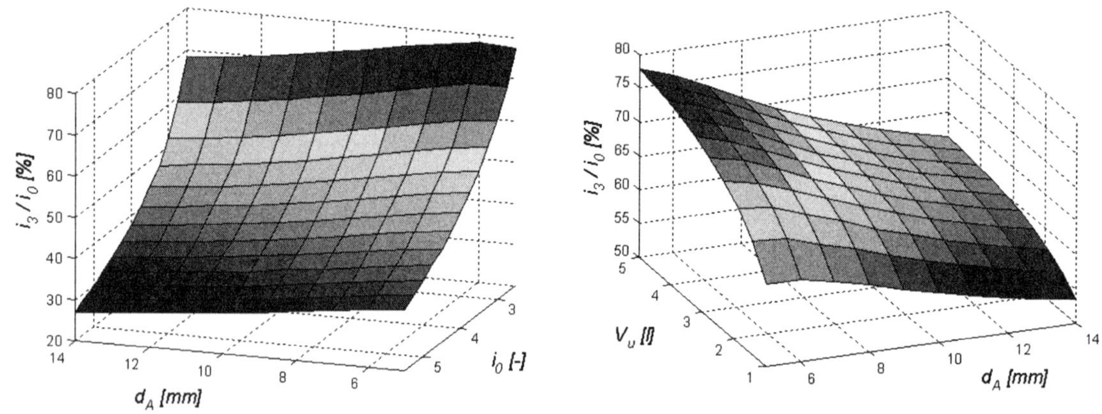

Fig. 6.12: Non-Stationary Sizing of a Vaccum Booster [6-9]

If a vacuum booster does not meet the requirements of repeated brake applications, its performance can be improved by varying the reaction disc geometry within the design limitations. If the requirements still can not be met, the volume of the vacuum chamber must be increased.

6.3 Discussion of Stationary Brake Design

Stationary brake sizing does not look (entirely) at the influence of what surrounds the vehicle. It is assumed that the applied brake torque can be completely transferred to the street. In the case of simulation of brake systems the following subsystems

- suspension
- tire-road interface

are not considered, and

- control devices (e.g. pressure control valves, ABS, ESP, EHB)
- brake lining
- driver and brake actuation equipment (pedal and vacuum booster)

are only insufficiently considered regarding their real, dynamic behavior.

The question whether the classic stationary approach is still up-to-date, or whether the future of the brake system design completely lies in dynamic simulation, or whether intermediate solutions, as presented in the chapter and already implemented in CoBrake, are sufficient, is discussed in the following paragraphs of this paper. To do so, the boundary conditions, which influence this question, are systematically listed.

6.3.1 Place in the Development Cycle

Brake design takes place in the development cycle of brake systems at a very early stage - usually at the latest, in the early quotation phase, i.e., some years before job one of the planned vehicle. Normally that means only a few secured parameters are available about the vehicle. The stationary brake sizing is mandatory in every planned project, independent of whether the brake system supplier is awarded the business or not. After a project is awarded the sizing calculation is updated at least for each major design change. The available parameters gain in quantity as well as in quality with increasing development time. Additionally however, particular questions, that can only be answered using dynamic approaches, can also appear with increasing development time.

From that boundary condition it can be derived that:

- The simulation parameters must be chosen so that they are available and as robust as possible in a very early phase.
- The acquisition of parameter sets and their change must be easy.
- A powerful simulation takes into account an increase in quality and quantity of parameters over the development time.

6.3.2 Influence of Legislation

A minimum requirement for brake sizing is proof of meeting all legal requirements (ECE R13, R13H, and FMVSS). Parts of these laws are text-based and allow room for interpretation. Explicitly stated requirements postulate stopping distances or decelerations. The conditions under which these values must be achieved are not, or only vaguely, formulated. For example, there are no explicit rules about center of gravity heights, no exact definition of the tire-road interface or of the condition of the brake lining. For the classic brake system sizing it must be additionally noted: the proof of meeting the legal requirements based on single, nominal values of parameters is valid only for this nominal case.

From that boundary condition it can be derived that:

- A more elaborate calculation to check legal requirements is meaningful and necessary only to such an extent as those necessary, more detailed boundary conditions are part or will be part of the legislation. Otherwise the calculated system is under-determined and the desired solution could be pseudo-optimized by changing those uncertain boundary conditions.
- Legal requirements refer to a test with the real vehicle. In reality however there are no nominal values, i.e. the simulation must take the variation of parameters into account.

6.3.3 Requirements for Controlled Brake Systems

The sizing of wheel brakes by means of classic brake sizing is still based on times when pressure-control systems did not exist. Achieving brake stability, i.e. making sure the front axle has the greater brake force for almost all decelerations was obsolete. Achieving certain decelerations with limited pedal forces in case of circuit failures or vacuum booster failures drove the dimensions of both the master cylinder and the booster. Such sizing constraints could be partially eliminated when using passive control systems (ABS), and almost entirely eliminated for active control systems (ESP, EHB). The sizing strategy could be driven to an essentially larger extent by other considerations, such as the goal of introducing standard platforms. The market penetration of ABS in Germany has surely exceeded 90 percent already today; based on of the self-obligation of the automobile manufacturers it will be 100 percent in Western Europe by 2004. Worldwide approximately 60 percent of all new cars are equipped with ABS, with a growing tendency. Vehicle dynamics control systems clearly have a more inferior penetration up to now (Germany approx. 45%,

Europe approx. 15% and worldwide approx. 8%), however, there is also a strong growth potential. Therefore the influence of control systems must be adequately taken into account in brake design.

From that boundary condition it can be derived that:

- system parameters for control systems must be determined
- reality-like characteristics of control systems must be taken into account in brake sizing
- evaluation of the vehicle dynamics control systems necessitates the possibility to be able to simulate different dynamic driving maneuvers

6.3.4 Influence of Suspension, Brake Lining and Tire-Road Interface

The predominant requirements for controlled brake systems, which require essential development capacity, depend on suspension, brake lining, and tire qualities.

From these boundary conditions it can be derived that:

- robust sizing parameters for these subsystems must be determined
- reality-like characteristics of subsystems must be taken into account in brake sizing

6.3.5 Interfaces Between Automobile Manufacturers and Suppliers

Over the long-term, the interface between automobile manufacturers will be a computer-aided one. The exchange of data and models could already take place using computer networks today.

The challenges are:

- know-how protection
- development and product liability

6.4 Dynamic Considerations

In this section, a set of first investigations is presented to help clarify the necessity to consider dynamic processes in brake sizing.

6.4.1 Influence of Suspension Qualities on the Stopping Distance Calculation

Only the total mass of the vehicle, the piston diameters of the wheel brakes, the effective radii of the wheel brakes, and the friction coefficients are used in the stationary calculation of the stopping distance. Additionally only the "Dead"- and "Rise-Time" of the actuation can be introduced. The stationary calculation assumes that the generated brake torque can be transferred fully, i.e., that a sufficiently large coefficient of friction is available between tires and the road. In the stationary calculation of the stopping distance, suspension qualities (for example dyn. axle loads, or changes in the location of the center of gravity) have no influence. To compare the two situations, dynamic simulations were done, assuming a brake pressure and road-condition such that the wheels would not lock up. First, brake maneuvers without control system interventions are looked at (**Table 6.1**).

Table 6.1: Comparison Between Stopping Distances Using Stationary and Dynamic Calculation

Actuation	Press. Grad. [bar/s]	Stopping Distance [m]			Deviation			
		stat.	dyn.	dyn. (& small wheel inertia)	s–d [m]	s- d/d [%]	s-dg [m]	s-dg/dg [%]
slight	100	50.51	49.24	47.97	1.27	2.6	2.54	5.3
full	300	45.08	45.29	44.02	-0.21	0.5	1.06	2.4
panic	600	43.70	44.28	43.01	-0.58	-1.3	0.69	1.6
maximum	1200	43.01	43.77	42.49	-0.76	-1.7	0.52	1.2
step-wise	∞	42.32	—	—	—	—	—	—

(v=100 km/h, final pressure 60 bar, no control sys. interventions, $\mu = 1.1$, C*-values const.)

The greater the pressure gradient becomes the smaller the error. Since the stopping distance calculation is only of interest in case of a hard stop, the error is reduced with the correct tendency. As a conclusion it can be said that: for a stop without control system interventions or locked-up wheels, i.e., assuming a sufficiently large coefficient of friction between tire and road, the influence of the suspension qualities as well as the dynamic pressure rise can be neglected for the stopping distance calculation.

Having said that, no statement about the influence of the following facts are made:

- hard stops usually occur at higher pressures and show control system interventions
- normally changes in friction coefficient of the brake lining occur during a stop
- dynamic viscosity of the brake fluid fluctuates considerably (approx. factor 1000) between –20 and +120°C and therefore influences the times in which certain pressure changes can be achieved.

6.4.2 Influence of the μ-Slip Curve on a Locked-Wheel-Stop

The influence of the μ-slip curve on a locked-wheel-stop was investigated in another simulation experiment. **Table 6.2** shows the stopping distances for various longitudinal force-longitudinal slip curves.

Table 6.2: Dynamic stopping distance calculation using various μ-slip curves

Friction curve (linear-linear)			Friction curve (linear-constant)		
μ @ $\lambda = 0.2$	μ @ $\lambda = 1$	Distance [m]	μ @ $\lambda = 0.2$	μ @ $\lambda = 1$	Distance [m]
0.8	0.8	50.39	0.8	0.8	50.39
0.9	0.8	50.10	0.9	0.9	44.97
1.0	0.8	49.73	1.0	1.0	40.68
1.1	0.8	49.38	1.1	1.1	37.17

(v=100 km/h, 600 bar/s, final pressure120 bar, no control sys. interventions, C*-value const.)

When calculating the stopping distance for locked-wheel-stops (no control system interventions), the spike at the maximum of the μ-slip curve can be neglected (**Table 6.2**, left). Only the coefficient of friction of the locked wheel is of crucial importance (**Table 6.2**, right).

Having said that, no statement about the influence of the following facts are made:

- changes in coefficient of friction of the locked tire during the stop

– straight-line-stops with control system interventions happen around the spike at the maximum of the μ-slip curve

6.4.3 Influences of Brake Lining-Friction Changes during Braking

Usually a nominal value for the lining friction coefficient – determined from dynamometer tests – is used in a stationary calculation of a brake application. In the best case this nominal value will be determined from a row of similar brake applications (e.g. from a pressure-controlled section of a dynamometer test). More frequently however, all average values of an entire test (e.g. AK-Master) are combined to one nominal value, with which all brake maneuvers are expected to be correctly reproduced in a simulation. It is questionable whether results based on such nominal values exhibit sufficient accuracy, independent of whether the vehicle is reflected correctly in the model. The influence of changes in friction during the brake application was investigated in a study. As a result a statement could be made whether one single nominal value is sufficient for the brake lining, or whether and if so, which simplifications can be done to the friction coefficient without negatively affecting the accuracy of the result.

Using dynamometer test measurements and a MATLAB/SIMULINK-model of the dynamometer, the influence of the change in friction coefficient during a brake application was examined in a dynamic simulation. The brake pressure during the dynamometer test was held constant by a pressure controller after an initial pressure gradient (pressure-controlled test). Since the pressure control of a dynamometer is not ideal it was possible to examine the influence of the natural pressure fluctuations during a brake application at the same time, in such a way that the pressure was linearly approximated for some simulation results. **Table 6.3** states the results for an exemplary brake lining.

Table 6.3: Comparison between Dynamometer Test Measurements and Dynamic Simulations of the Braking Time Using Different Inputs for Pressure and Friction

	Simulation Inputs		Avg. Deviation Braking Time [%]
	Friction Coefficient	Pressure	
(a)	Measured values	Measured values	0.394[1]
(b)	Avg. of single stop data set	Measured values	0.852
(c)	Avg. of single stop data set	linearly approximated sections	1.598
(d)	Measured values	linearly approximated sections	1.196
(e)	Average of all stops	linearly approximated sections	11.41
(f)	Maximum of all stops	linearly approximated sections	9.51
(g)	Minimum of all stops	linearly approximated sections	15.79

(Disc Brake, same brake lining, 505 stop data sets)

Assuming that errors and deviations tend to be greater in vehicle tests than in a dynamometer test, the following uncomfortable conclusions must be drawn: nominal values for the lining friction coefficient (**Table 6.3**: cases e, f, g) can lead to deviations in braking time of more than 10 percent and therefore cause considerable errors in a stopping distance calculation. Using a separate friction coefficient curve taken from a single stop (**Table 6.3**: cases a, b, c, d) on the other hand, yields good results. If doing so it is not necessary to use the dynamic friction coefficient (**Table**

[1] Deviation attributed to numerical integration-errors of Simulink-model in comparison to measured braking time

6.3: case d), despite the fact that this run over time is known for each single stop. But it is merely sufficient to employ the average of each single stop (**Table 6.3**: case c). The small difference between the results when using an average for the friction coefficient of one stop and a linearly approximated pressure curve (**Table 6.3**: case c) additionally proves reality-like pressures are not necessary. Instead a linearly approximated pressure curve with "Dead-" and "Rise-Time" is sufficient – as long as an average for the friction coefficient is also available for each stop. The main problem for a predictive simulation however is to obtain such an average.

6.4.4 Influence of Driving Resistances

The legal requirement for vacuum booster failure necessitates a certain deceleration (z = 0.3 g) at a defined pedal force (ECE R13) or alternatively a certain stopping distance (FMVSS). A small simulation maneuver can show that driving resistances should not be neglected any longer, especially for all-terrain vehicles, since they normally feature large projected surfaces and cw-values.

For a typical SUV (Sport Utility Vehicle, e.g. m = 2176 kg, c_w=0.39, A = 3.3 m^2) an additional deceleration of 0.29 m/s^2 is induced at a speed of 100 km/h simply caused by the driving resistances. This deceleration already amounts to approximately 10% of the deceleration required in case of a vacuum booster failure. For the stopping distance (assuming a deceleration created by the brake torque z_{brake} = 0.3 g) this corresponds to a difference of 11.5 m.

6.4.5 Influence of the Brake Actuation

Another area in the brake system where dynamic influences can be found is the brake actuation (pedal, booster, and master cylinder). If small inertial influences and the dead travel[2] of the pedal installation are neglected, it is a very good approximation to say that a force exerted onto the pedal may be weakened by friction and perhaps spring resistances, but nevertheless transferred simultaneously (time-wise). Within the vacuum booster however, a dynamic process is started

[2] The dead travel of the pedal is usually eliminated almost completely by how the measurement for pedalforce/-travel is carried out

where masses move relative to each other and air is exchanged through a valve between chambers. Depending on its design and manufacturing as well as ambient conditions, a vacuum booster can react slower or faster depending on the actuation rate. **Fig. 6.13** shows this effect using a measurement as an example. The measurement was done with a tandem vacuum booster of an upper-level SUV. The curve depicting the relationship between the input force and the master cylinder pressure clearly shows that at the highest actuation rate less boost could be reached within the measured time than with the two slower actuation rates. Moreover, the slowest of the three actuation rates shows quite clearly how the "Crack-Point" affects the output force – and by that, the pressure curve. For the two faster actuation rates the driver overrides the "Crack-Point", which is why the master cylinder pressure starts to climb almost without any time delay.

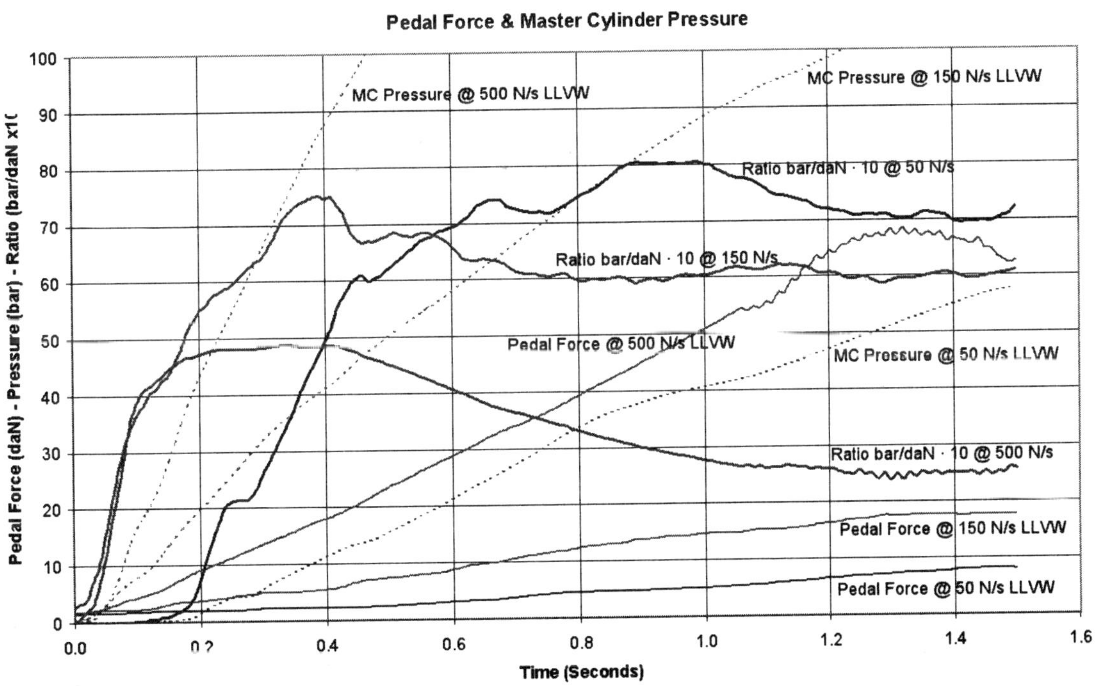

Fig. 6.13: Dynamic Reaction of Vacuum Booster and Master Cylinder (LLVW = Lightly Laden Vehicle Weight)

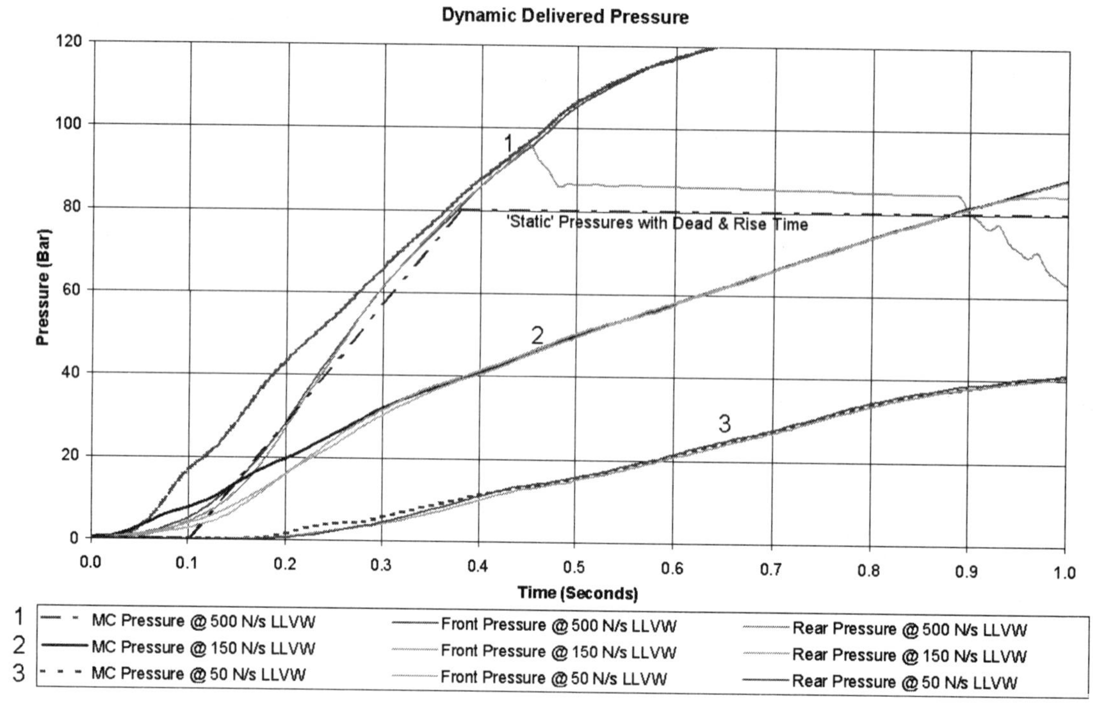

Fig. 6.14: Dynamic System Pressure for Various Actuation Rates

In addition to the negative performance of the vacuum booster, other losses (e.g., at orifices) in the hydraulic components (tubes, pressure control device(s)), which are connected to the master cylinder, cause a delayed pressure rise at the wheel brake. **Fig. 6.14** shows the influence of various pressure gradients on the pressure that is available at the wheel brake. The measurements were done on the same vehicle as in **Fig. 6.13**. The temperature was the same for each measurement (approx. 20 °C ambient temp.). It is easy to see that the ability of the wheel pressures to follow the pressure target of the master cylinder decreases noticeably with an increasing pressure gradient. If the temperature dependence of the brake fluid is also taken into consideration (see chapter 6.4.1) this effect can become even more drastic with decreasing temperature.

A stationary calculation would follow approximately a pressure curve as shown in **Fig. 6.14**, provided a "Dead-" and "Rise-Time" is used. However, a statement about the system behavior of the actuation equipment at different actuation rates and ambient conditions would still be insecure.

The following conclusions can be drawn from the above observations:

- For a detailed look at the actuation equipment (especially for questions about pedal feel), a dynamic calculation is necessary.
- A dynamic model of the actuation equipment must be described well enough (at least for the vacuum booster component and hydraulic system) that a change in ambient conditions and main design characteristics leads to the accordingly correct result.

6.5 Conclusions and Outlook

The following conclusions can be made from the above investigations about the necessity to look at dynamic processes when doing brake sizing:

- Even though a dynamic model of the vehicle shows no considerable improvement over stationary calculation for stops without control system interventions, the circumstance that almost 100% of present-day vehicles are equipped with brake control systems forces a more exact examination of the effects of these systems on brake sizing.
- A similar conclusion can be drawn from the influence of the tire. It is also true here that the use of brake control systems necessitates a separate examination since the tire is used near the maximum of its longitudinal and/or lateral force capabilities.
- The influence of brake lining friction coefficient changes does not necessitate any dynamic considerations. However, obtaining valid average values for single, particular brake maneuvers remains a challenge. Whether such an average value can be determined through measurements or through a component simulation must be examined.
- Driving resistances necessitate a dynamic calculation, especially for low decelerations. It is possible however, to adjust for the effect of driving resistances in a stationary calculation with a good approximation.
- The dynamic behavior of the actuation equipment can be neglected for brake sizing focused purely on deceleration capability. For questions about the expected pedal feel of the future brake system, dynamic models must be used

early in the development process, provided that simplifications imposed by stationary treatment are not tolerable.

Therefore, the future work on CoBrake will have the following main focuses on the technical side:

- improve statements regarding pedal feel
- further examine possible descriptions of friction couples (brake lining, tires) beyond just nominal values
- include behavior of brake control systems into the system design process
- introduce automated methods for analysis and optimization

The current approach of having the user interface(s) intranet-based and globally available will still be pursued further on. In doing so, the tool development has to remain flexible with respect to the evolution of available databases as well as internal and external interfaces.

6.6 References

[6-1] Weller, N.: „CoBra – Architektur und Konzepte“, interner Bericht, Robert Bosch GmbH, 1998.

[6-2] Sailer, U.: „CoBra – ein Tool für den globalen Einsatz in der Bremssystementwicklung“, in: 3. Stuttgarter Symposium, Kraftfahrwesen und Verbrennungsmotoren, (Hrsg.) Bargende, M.; Wiedemann, J., Expert-Verlag: Renningen, 1999. (ISBN 3-8169-1751-8)

[6-3] von Glasner, E.C.: „Beitrag zur Auslegung von Kraftfahrzeugbremsanlagen“, Universität Stuttgart, Dissertation, 1973.

[6-4] Burckhardt, M.: „Bremsdynamik und Pkw-Bremsanlagen“, Vogel-Fachbuch: Würzburg, 1991. (ISBN 3-8023-0184-6)

[6-5] N.N.: „Bremsen-Handbuch“, Autohaus Verlag: Ottobrunn, 1993. (ISBN 3-89059-008-X)

[6-6] Kappelmann, P.; Klenk, T,; Wiedemann, J.: „Objektorientierte Modellierung einer Pkw-Bremsanlage", in: Fortschritte in der Simulationstechnik, (Hrsg.) Hohmann, G., SCS –European Publishing House: Ghent, 1999. (ISBN 1-56555-130-3)

[6-7] Klenk, T.; Kappelmann, P.; Wiedemann, J.; Sailer, U.: „Effiziente Behandlung mehrdeutiger Abbildungen am Beispiel der Auslegungsrechnung für Kfz-Bremsanlagen", in: Fortschritte in der Simulationstechnik, (Hrsg.) Möller, D., SCS–European Publishing House: Ghent, 2000. (ISBN 1-56555-189-3)

[6-8] Brassard, G.; Bratley, P.: „Algorithmik: Theorie und Praxis", Prentice Hall, 1993.

[6-9] Kappelmann, P.; Klenk, T.; Wiedemann, J.: „Simulation eines Unterdruckverstärkers im Verbund mit dem Gesamtbremssystem und dem Fahrzeug", in: VDI-Berichte Nr. 1559, 2000. (ISBN 3-18-091559-5)

7 Controlling the Lateral Dynamics of Vehicles by Braking

Dr. T. van Zanten, Dr. R. Erhardt, Dipl.-Ing. K. Landesfeind, Dipl.-Ing. G. Pfaff, Robert Bosch GmbH

Schwieberdingen, Germany

Abstract

The problem of vehicle handling during limit maneuvers will be explained. This problem is then used to formulate the goal for the vehicle dynamics control system ESP (Electronic Stability Program, alternatively called VDC). The approach chosen to reach this goal will then be shown. Using two examples it will be shown that ESP can significantly improve vehicle handling in extreme maneuvers by automatic control of the brakes and the engine. Both under-steer, over-steer and even vehicle spin can effectively be suppressed if sufficient information on the vehicle motion is available. It will be shown that the vehicle slip angle is a crucial indicator for the maneuverability of the automobile. For the automatic control of the slip angle the required yaw moment can be generated by individual wheel slip control. Since the complete vehicle state is not readily available, estimation algorithms are used to supply the control algorithms with sufficient information. A necessary requirement for the safe operation of ESP are reliable sensor signals. The testing and monitoring of the yaw rate signal is shown as an example.

7.1 Introduction

For a driver with average driving experience, vehicle behaviour in the limit area, i.e. at maximum friction between tire and road, is almost unknown. Caused by the non-linear, even unstable behaviour of the tires when maximum positive engagement is reached, the vehicle may break away, which is incomprehensible for the driver and a situation which cannot be recovered. As a result of the side friction coefficient with wide steering angles, the driver involuntarily increases the steering wheel angle, which serves to worsen the situation. In both cases, the ESP (Electronic Stability System) helps the driver to control the vehicle, so that the vehicle behaviour, even in the limit area, corresponds to the driver's level of experience. [7-1, 7-2, 7-3].

Förster [7-4] believes that the role of the driver and the consideration of human behaviour in relation to the concept of tires, chassis and road, which determine driving behaviour, should be considered the most important issues. This also applies to the concept of ESP, which interferes with driving behaviour. He also claims that the average driver is unable to ascertain the frictional value of the road, has no idea of his positive engagement reserve and is therefore taken by surprise if the driving behaviour changes. He also believes that in panic situations, the driver should not be expected to perform considered actions. Very often in panic situations, such as skidding, drivers react incorrectly. An ESP must therefore be designed in order to avoid panic, e.g. when the driver accelerates quickly or oversteers after a shock. It is to be expected that simple systems will not be successful on the market.

The reason for vehicle control in the limit area being so difficult is explained by Shibahata [7-5] (**Fig. 7.1**). He has developed a simple method, the β-Method, which makes justification easy. When a vehicle is steered, a yaw moment occurs on the vehicle, which leads to a change in movement. It appears that the yaw moment achievable by braking is dependent upon the float angle. With wide float angles, the yaw moment does not notably change. Typically, when the float angle reaches the value at which the slip angle of the tires is at its maximum, this steering capability is practically lost. This characteristic value is dependent upon the frictional value of the road. On dry asphalt, this amounts to approx. $\pm$ 12°, whilst on ice, it is approx. $\pm$ 2°. From this characteristic float angle value, the average driver can no longer control his vehicle, and his chances of avoiding an accident are reduced drastically. In terms of the driver, it must be considered that for the average driver, the float angle does not exceed $\pm$ 2°. In consideration of this fact, during ESP operation, the float angle should not be too large. The task of limiting the float angle must not be left up to a driver who is not familiar with such functions.

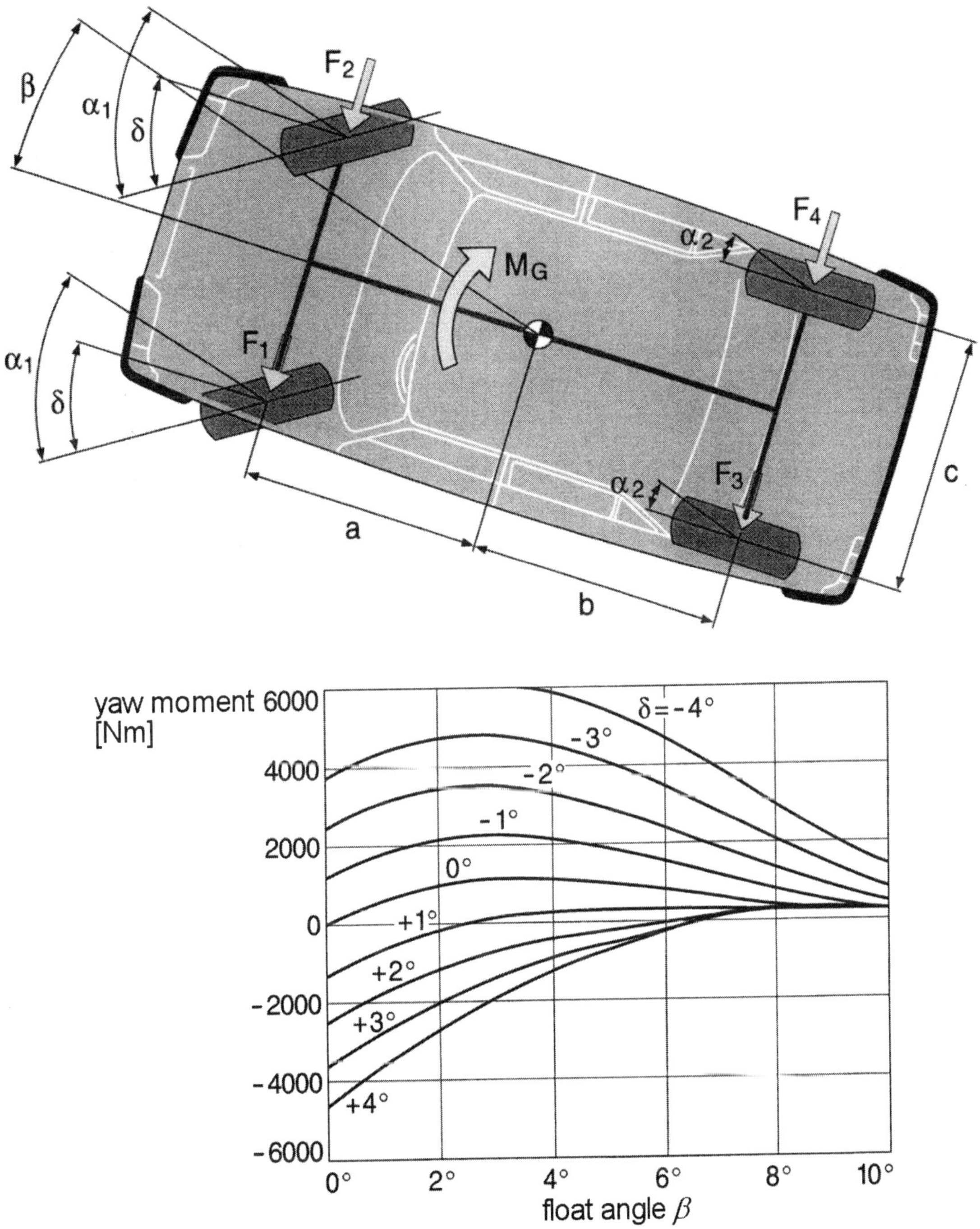

Fig. 7.1: The β-method

The main function of the ESP therefore is to limit the float angle to at least the characteristic value. Until the float angle reaches this value, the driver cannot control his vehicle, and ESP intervention is vital. On reaching or exceeding the characteristic float angle, when the driver realises that he is losing control of the vehicle, the ESP must intervene. Within the range of characteristic angles of floatation, the ESP can produce even larger yaw moments by means of (actively) controlling brake and drive slip on the individual wheels. This can be demonstrated by the effect of brake slip λ_0 on the left front wheel during free right-hand cornering (**Fig. 7.2**).

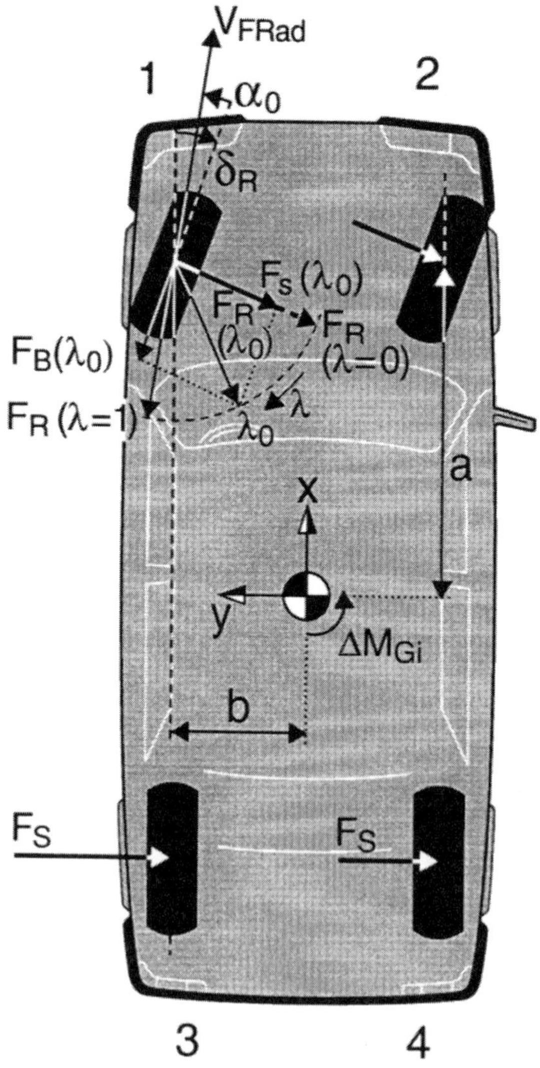

Fig. 7.2: Incrementing the yaw moment by slip control

$F_R(\lambda=0)$ is the lateral force on the tires, when slip is zero. As a result of brake slip λ_0 (with equivalent slip angle α_0 and equivalent contact force F_N) the lateral force is reduced to $F_S(\lambda_0)$. At the same time, the braking force $F_B(\lambda_0)$ occurs. $F_R(\lambda_0)$ is the resultant force on the tires created from the vector total of $F_S(\lambda_0)$ and $F_B(\lambda_0)$. If the tires are saturated, i.e. in the limit area, the values of $F_R(\lambda=0)$ and $F_R(\lambda_0)$ are approximately equal. The effect of a brake slip change is now obvious. Changes in the brake slip cause the resultant force on the tire to turn and this then leads inevitably to changes in the yaw moment, the lateral force and the longitudinal force on the vehicle. As a consequence, the yaw moment can be adjusted through slip adjustment. The control concept determines the degree to which the slip on each tire is changed, in consideration of the required framework conditions such as constant acceleration or speed of the vehicle and the accuracy to which the working point (α_0,λ_0) is known.

Unfortunately, with the sensors currently available, neither the float angle nor the road frictional values can be measured. These values must be derived by estimation. Most importantly, a reliable estimation of the float angle is very difficult to deduce and can lead to serious errors. For this reason, a pre-control is introduced on the basis of measured speed. Thus, positive pre-control creates positive conditions for the subsequent float angle control. However, this creates additional problems, such as the yaw rate at which control now takes place. For this purpose, the intrinsic yaw behaviour of the vehicle is measured and stored in the control programme as a formula (reference model, derived from the linear, single-track model). However, the intrinsic steering behaviour of the vehicle can now change over time, e.g. through tire change or wear, which means that the target value for the yaw rate must be calculated carefully.

7.2 Control concept

In comparison to ABS and ASR, the ESP on cars has additional components, on the one hand in respect of the driver's intentions and vehicle movement, and on the other, in respect of the high-level requirements of the speed at which the required braking pressure can be actively set in the wheel brakes. It is also possible to reduce the torque by means of an interface with the engine control unit.

As measured variables for the ESP, the steering wheel angle, the yaw rate, the lateral acceleration, the rpm of the wheel, the engine torque, the driver default torque and the braking pressure in the main brake cylinder, can be used (**Fig. 7.3**).

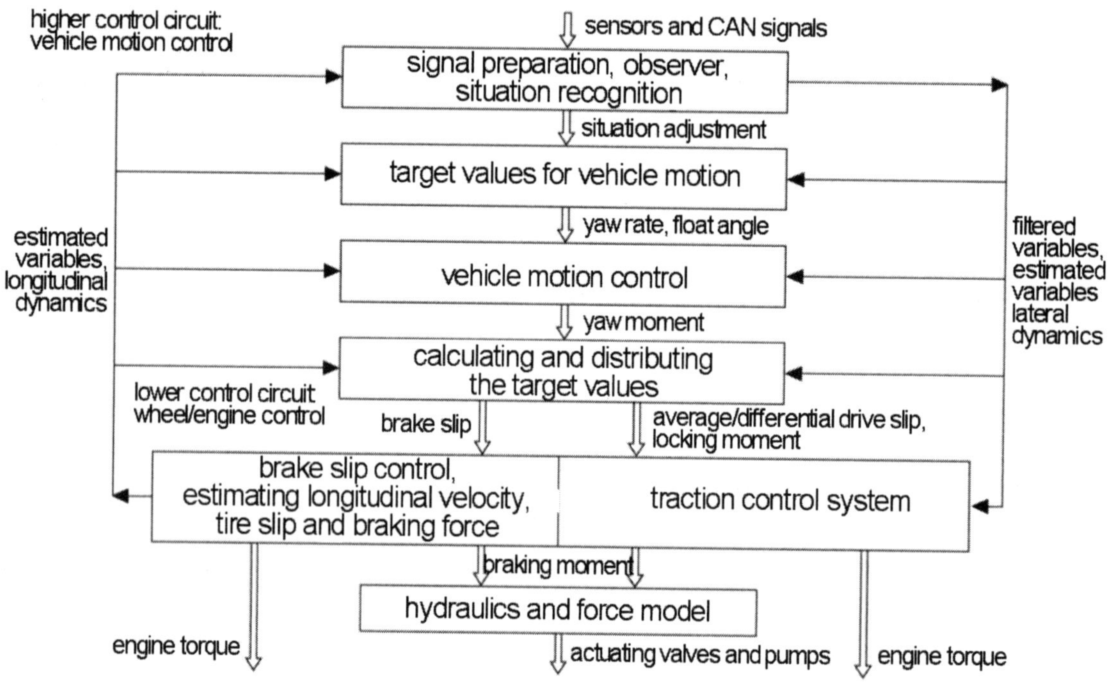

Fig. 7.3: Simplified block diagram of the ESP controller

The brake and drive slip regulator estimates vehicle speed, deceleration, tire slip and tire braking force. Based on a simple two-track model, this information is used to assess the float angle, the tire slip angle, the lateral speed, the lateral tire forces, the tire contact forces and the resultant forces. To estimate the float angle, the movement equations for the two-track model are used for the yaw rate and the lateral speed. After linearisation and discretion, these equations are used as the initial basis of a Kalman filter. To calculate the side forces, the relationship between lateral steering forces and lateral forces is used by means of the HSRI tire model [7-6]. This eliminates the need for the non-linear equations of the model. Thus, for each wheel, the lateral and longitudinal forces of the tire can be calculated, the condition being that the slip is not too small. The observer at full brake control

(ABS) can then supply optimum results and in the case of controlling the free cornering vehicles, less positive results for the float angle assessment. In fact, the float angle estimate at full brake control is very good. In the absence of slip, only the total of the lateral forces for the four tires can be estimated from the lateral acceleration of the vehicle and not the distribution of this total over the four tires.

Depending on driving conditions, the float angle estimate can have greater or smaller margins of error. For this reason, as already mentioned, the yaw rate is also regulated. In the next block, the target values for the float angle β_{So} and the yaw rate $\dot{\psi}_{So}$ are specified. Using the linear single-track model, an initial target value for the yaw rate $\dot{\psi}_{So}$ is derived from the steering wheel angle set by the driver.

$$\dot{\psi}_{So} = \frac{v_x \cdot \delta}{l \cdot i_S \cdot \left[1 + \left(\frac{v_x}{v_{ch}}\right)^2\right]}$$

The wheelbase l, the steering translation i_S and the characteristic speed v_{ch} are vehicle-dependent parameters.

Depending on the estimated maximum friction value between tires and road μ_0, this value is limited to the second target value of the yaw rate $\dot{\psi}_{So}$.

$$\left|\dot{\psi}_{So}\right| \leq \left|\frac{\mu_0 \cdot g}{v_x}\right|$$

An initial target value for the float angle corresponds to the characteristic value described above, which depends on the friction value μ_0. The initial target value for the float angle is reduced to the second target value $\beta_{So,}$ depending on the speed, in order to offer the driver additional assistance at higher vehicle speeds.

If the actual value of the yaw rate and the float angle deviate from their target values by more than their specified thresholds, the required yaw moment change on the vehicle is calculated in the condition control, in order to reduce the differences. In this case too, the behaviour of the driver is taken into consideration. For example, on a smooth road, the vehicle responds sluggishly to the steering. The consequence is that the driver involuntarily over steers and the vehicle responds by continuing to decelerate rather than accelerate. In order to avoid this involuntary reaction, ESP speeds up the vehicle's reaction to the steering and turns the vehicle immediately into the curve ("leaning the vehicle") without exceeding the float angle threshold. This "vehicle leaning" technique is well-known amongst test drivers.

By changing the slip on the tires, every tire can contribute towards the required yaw moment change. Since each tire makes a different contribution and the wheel slip control leads to an involuntary braking action by the driver, the degree of change to the target slip on each tire must be chosen carefully. The choice is made in accordance with the draft regulation, which was created with the aid of simulation studies with precise dual-track models. Thus, preferably on the vehicle, which is oversteering at full braking (ABS), the tire-slip on the front offside wheel is increased and decreased on the rear offside wheel, whilst on the other tires, the braking slip values are not adjusted, and the values at which the adhesion friction values are achieved are maintained.

When driving on a road which is smooth on one side, the traction can be improved to the lower frictional value by braking the driven wheels (differential effect through braking). This leads to the driver creating unintentional yaw moments, which he must compensate by steering. If these yaw moments are too high, according to the vehicle manufacturer, the braking effect on the driven wheels, and at the same time, the engine torque must be reduced (so as to avoid wheel spinning).

The required brake slip values are set by the brake slip regulator. The slip values are only set by the drive slip regulator during motion and only then when the wheels are turning. In order to modulate the brake pressure, the valves on the hydro-unit are actuated, whilst the engine control is actuated to adjust the engine torque.

7.3 Examples of result

If a car is driven at constant speed in a slalom movement with increasing steering amplitude, the entire lateral dynamic driving range is covered in one single test run (**Fig. 7.4**). In order to be able to guarantee a constant vehicle speed, the engine torque must be continuously increased, which leads to an increasing drive slip on the driven rear axle. After each jump in the angle of the steering wheel, a constant yaw rate and a constant float angle is set. If the vehicle's lateral acceleration achieves the maximum frictional value between tires and road, the situation then changes. In contrast, the float angle value is increased through the decreasing lateral force on the front wheels to almost double the stationary value. The vehicle responds slowly to the change in steering angle and the stationary cornering is achieved with a long time delay. By the fourth negative angle jump of the steering wheel, the time delay is so long that the vehicle becomes unstable without ESP and no longer responds to the steering. The float angle has thus achieved the characteristic value of the road. The float angle continues to increase and the vehicle skids.

The ESP intervenes one steering cycle before instability occurs. Since in this cycle, the vehicle movement does not respond quickly enough to the steering movement, due to the large float angle (see section 7.1, the β-methods) and the yaw rate changes too slowly, the drive slip control is triggered, both by accepting the engine torque and also with an active brake intervention on the two driven rear wheels, in order to increase the lateral forces on the two tires and to stop the unwanted rotational and lateral movement of the vehicle. During further movement, the braking pressure in the right driven wheel is kept constant, whilst in the left, it continues to increase. This creates a yaw moment on the vehicle, which supports the steering and which is increased further by brief, pulsed braking intervention on the left front wheel, which means that the vehicle now follows the steering in a more controlled manner and the float angle is quickly reduced. In terms of comfort, the driver hardly notices anything from the supported yaw moment on the rear axle, giving him no cause to adjust his driving behaviour. He clearly senses only the braking intervention on the left front wheel, which tells him that it is time to adapt his driving behaviour to the conditions.

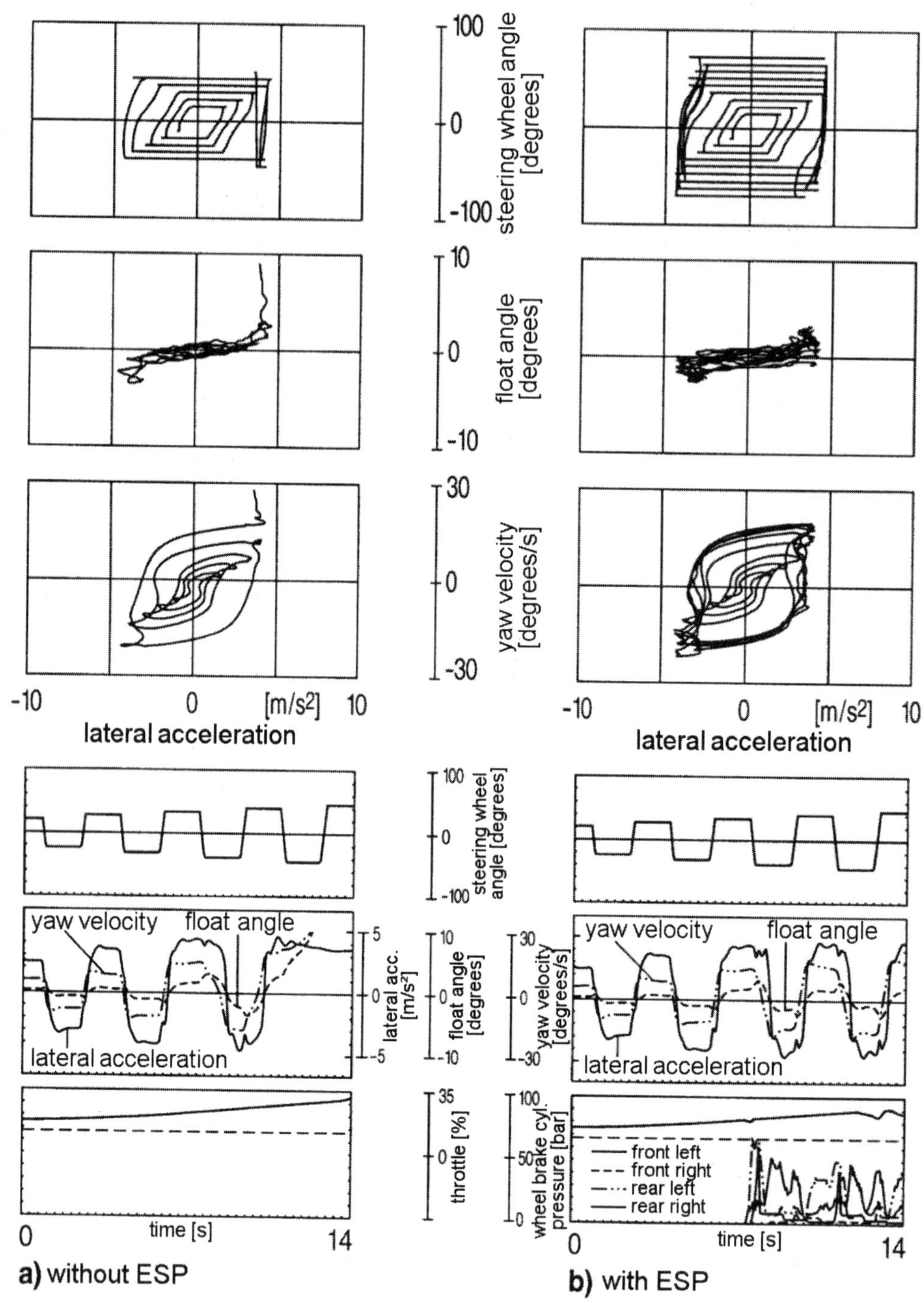

Fig. 7.4: Slalom manoeuvre without ESP (a) and with ESP (b) at a constant velocity of 72 km/h on packed snow ($\mu_0 \approx 0.45$)

Fig. 7.5 shows the results of a driving test during full braking. From an initial speed of approx. 50 km/h, on a smooth road (ice, $\mu_0 \approx 0.15$), a simple lane change occurs during full braking, which is necessary due to vehicles being parked on the right hand lane. **Fig. 7.5**a shows the manoeuvre with an ABS vehicle. Without ABS, this manoeuvre would not be possible and the vehicle could drive straight into the parked vehicles. Even after the first steering action with ABS, the float angle and the yaw rate are so great that the driver must exaggerate the counter-steering. Thus, the early signs change the float angle very quickly, which is increased immediately. The driver is forced to counter-steer again quickly. He still manages to stabilise the vehicle and stop in the left-hand lane. The ESP-controlled vehicle remains stable (**Fig. 7.5**b).

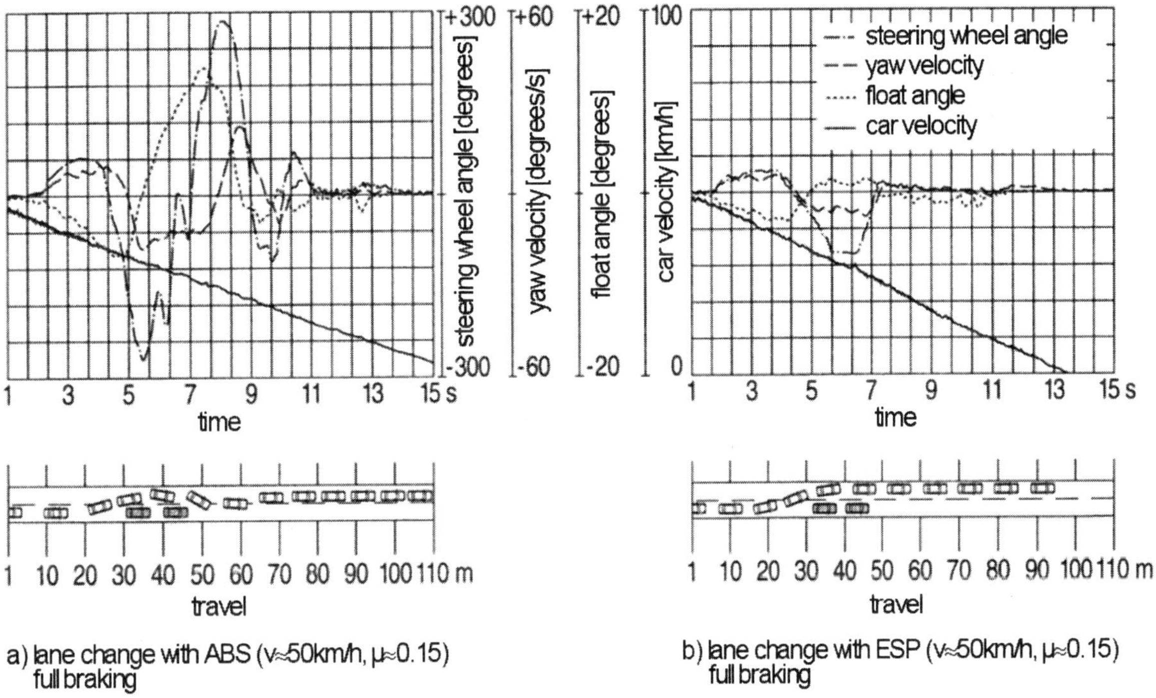

Fig. 7.5: Lane change manoeuvre while full braking with ABS (a) and with ESP (b) from an initial velocity of approx. 50 km/h on polished ice ($\mu_0 \approx 0.15$)

At the start of the manoeuvre, as described for the slalom, the braking expenditure is less. The driver notices that the vehicle responds to the steering movement immediately. Therefore, he is less inclined to exaggerate his steering, i.e. oversteer. Both the degree exceeds the characteristic value of approx. 2° only slightly, and then only at the outset. The result leads to conclusions on the very high degree of accuracy, at which the angleof floatation can be estimated at full braking control.

Under normal driving with this manoeuvre, the float angle remains below 3°, the driving behaviour tends to match the level of experience of the average driver.

7.4 Monitoring the components

ESP is an active, dynamic control system, which monitors the vehicle movement at every speed and every manoeuvre, to match the driver's requirements and makes any necessary adjustments. Incorrect sensor signals can therefore have serious consequences, in the absence of an appropriate monitoring system. Complete sensor redundancy should therefore be avoided for financial reasons. For this reason, in addition to sensor self-tests [7-6, 7-7], the principle of the analysed redundancy already applied to air travel, is used. In this respect, on the basis of the model vehicle, a link should be created between the various sensor signals: model-supported monitoring [7-8].

To demonstrate a sensor self-test, we shall describe the monitoring of the yaw rate sensor. This consists of two sprung, suspended, coupled micro-mechanical mass sensors, which are excited within their resonance frequencies by means of Lorenz forces (**Fig. 7.6**).

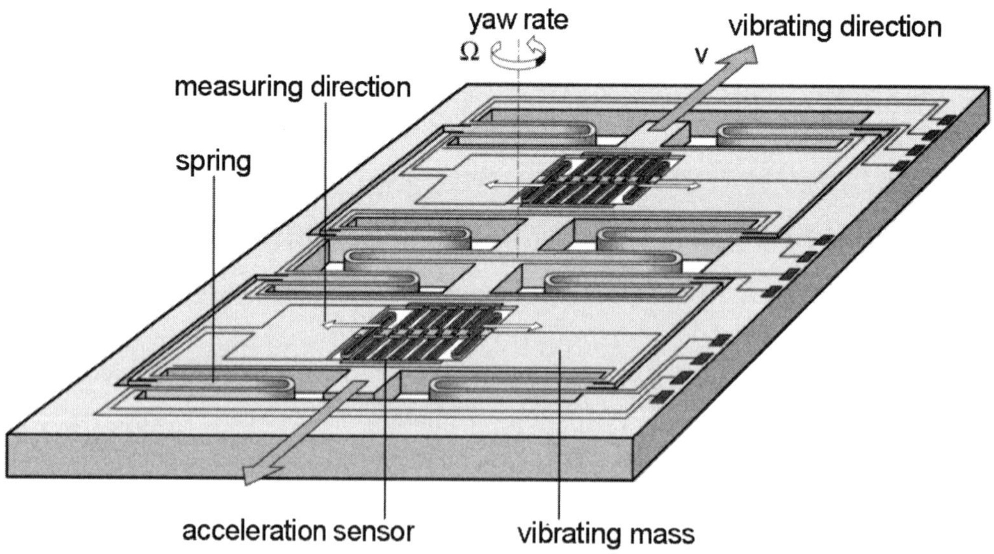

Fig. 7.6: Principle of the micro machined yaw rate sensor

If the sensor is turned around the axis, vertically to the vibration level, each mass accelerates in the vibration level vertical to the vibration direction; the so-called Coriolis acceleration. This Coriolis acceleration is measured by an acceleration sensor fitted to each vibrating mass and is proportional to the speed of rotation. Now, the Coriolis acceleration is increased by an interference force triggered by the control unit. With a precisely defined interference force, the sensor signal can be increased or reduced by a precisely defined amount (**Fig. 7.7**). Checking the degree of change therefore allows the electrical, electronic and mechanical parts of the sensor to be checked. Since the sensor signal is used to pre-control the dynamic control, faulty signals must be revealed immediately. For this reason, the control unit transmits the exciter signal every 40 milli-seconds. So that the sensor can still be checked when the measurement limit is reached, the self-test is performed alternately. Errors in the signal transfer chain from the sensor to the ESP control unit can also be quickly detected using this self-test.

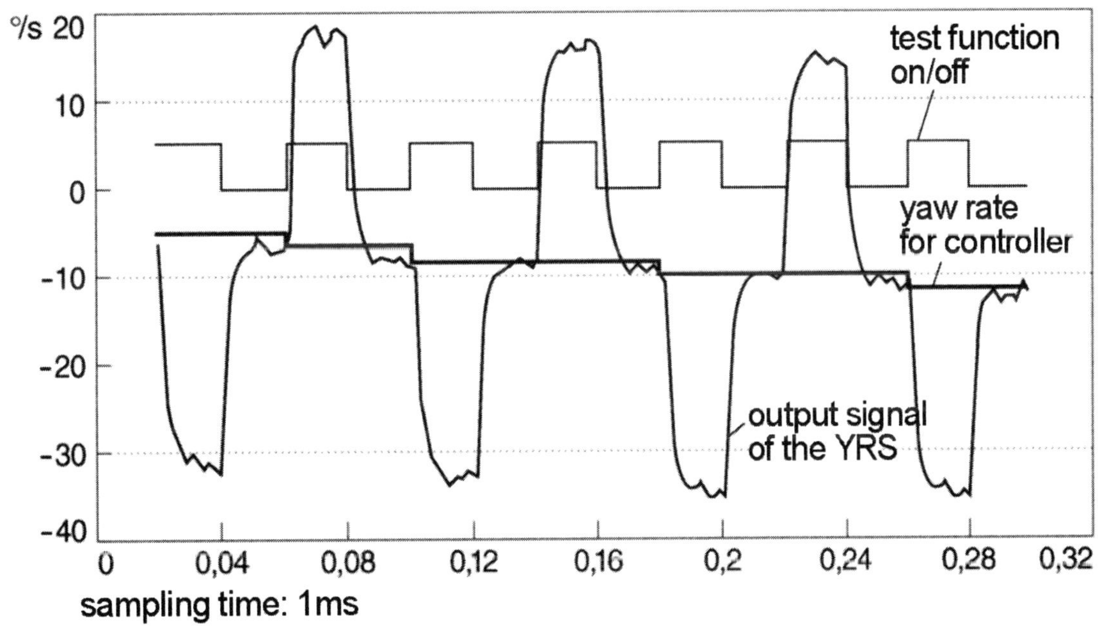

Fig. 7.7: Yaw rate signal

The model-supported monitor is now demonstrated by explaining the yaw rate sensor (**Fig. 7.8**). Apart from the control, when the vehicle moves in accordance with the target behaviour, the relationship between steering wheel angel and yaw rate applies in accordance with the self-steering behaviour of the vehicle (Model 1). In the case of "steady state" cornering, another value for the yaw rate can be derived from the vehicle's lateral acceleration using equation $a_y = \dot{\psi} \cdot v_x$ (Model 2). Ultimately, from the differences in wheel speed signals on the left and right hand sides of the vehicle, the yaw rate during "steady state" cornering can be calculated. Thus, four values are derived for the yaw rate, which are used to detect errors using the selection procedure (e.g. the "3 from 4" control). The pre-condition for this procedure is that the offset comparison for the yaw rate signal, the steering wheel signal and the lateral acceleration signal, the sensitivity comparison for the yaw rate and the tolerance comparison for the wheel radii (RTA) have already taken place. From "estimating the yaw rate (selection algorithm)", a reference yaw rate signal is given. By converting this signal by means of an appropriate inverse model, the steering wheel angle, yaw rate and lateral acceleration can be checked for errors.

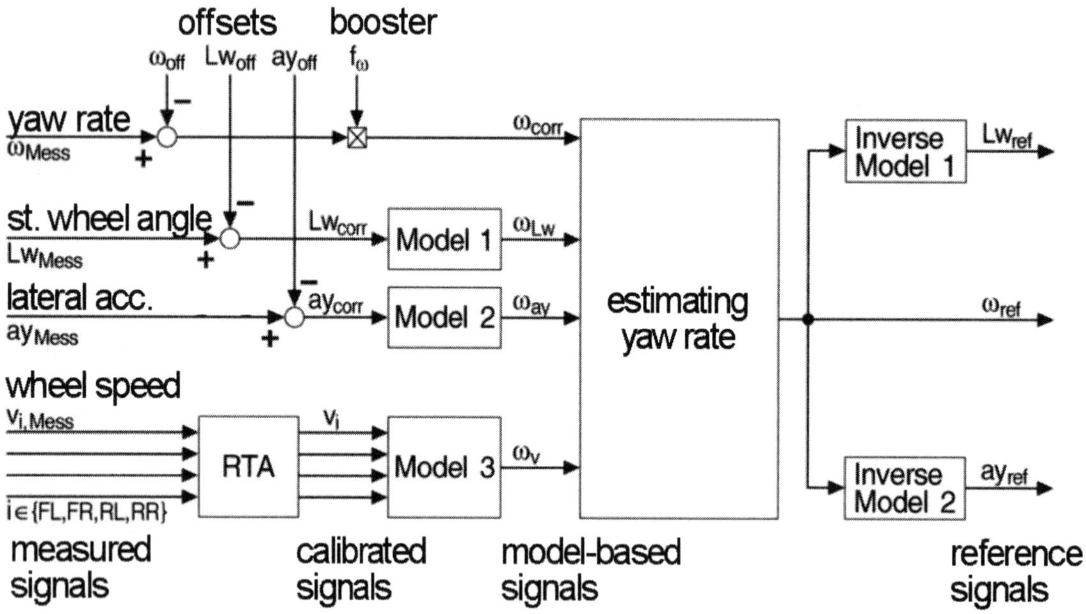

Fig. 7.8: Model-based sensor monitoring of the yaw rate sensor signal

7.5 References

[7-1] Müller, A.; Achenbach, W.; Schindler, E.; Wohland, T.; Mohn, F.-W.: The New Vehicle Safety System, the Mercedes Benz Electronic Stability Program, ATZ Technical Automobile Publication 96 (1994) 11, pp. 656 – 670.

[7-2] van Zanten, A.; Erhardt, R.; Pfaff, G.: FDR – The Bosch Vehicle Dynamic Control System, ATZ Technical Automobile Publication 96 (1994) 11, pp. 674 - 689.

[7-3] Fennel, H.; Gutwein, R.; Kohl, A.; Latarnik, M.; Roll, G.: The Modular Controller and Control Concept with the ITT Automotive ESP, 7th Aachen Examination of Vehicle and Engine Technology, 5th – 7th October, 1998, Aachen, S. 409 – 431

[7-4] Förster, H. -J.: The Driver as the Link between Tires, Chassis and Road, VDI Reports, No. 916, 1991

[7-5] Shibahata, Y.; Shimada, K.; Tomari, T.: Improvement of Vehicle Maneuverability by Direct Yaw Moment Control, In: Vehicle Systems Dynamics, 22 (1993), pp. 465 – 481

[7-6] van Zanten, A.T.; Erhardt, R.; Pfaff, G.; Kost, F.; Hartmann, U.; Ehret, T.: Control Aspects of the Bosch-VDC, AVEC'96, International Symposium on Advanced Vehicle Control, Aachen, June 24 - 28, 1996, pp. 576 – 607

[7-7] van Zanten, A.T.; Erhardt, R.; Landesfeind, K.; Pfaff, G.:VDC Systems Development and Perspective, SAE98, Nr. 980235

[7-8] Stuckenberg, N.: A Contribution to the Detection and Isolation of Sensor Erros in Flight Control Systems using Observers, DFVLR – FB 81 – 26, 1981

8 Development Tools to Assist in the Objective Description of Brake Pedal Feel

Prof. Dr. K. Augsburg, Dipl.-Ing. R. Trutschel, Ilmenau University of Technology

Ilmenau, Germany

Summary

The task involved in investigations of brake pedal feel is not only of current importance but also academically demanding. It is of such complexity that it must be approached with rigorous and at the same time holistic methodology.

This paper describes examples of results in such an investigation. The results are, in the first instance, those of an attempt at analytical description of what happens in conventional braking systems, using a behaviour model. Particularly close attention is given to the behaviour of the brake linings. There is a description of the simple mechanical model of a driver's leg used in the detailed investigation of the mutual influences at work in the interface between the human (the driver's foot) and the machine (the brake pedal).

Employing a test vehicle with active modification of the brake pedal and an electro-hydraulic brake-by-wire system raises the quality of research on pedal feel. This manner of working is presented in the paper. It permits virtually any relevant pedal feeling to be modelled and to be activated within seconds.

Finally, the findings on the appreciable limits relating to defective pedal functioning and to the subjective evaluation of modified pedal characteristics are given.

8.1 Subject matter defined; limits of paper

"Brake pedal feel" describes how three features interact over time: the pedal effort, travel, and the movement of the vehicle. The interaction of these features in numerical terms is extremely complex, with both static and dynamic ingredients which bear, on the whole, a non-linear relationship to each other [8-2], [8-5], [8-6], [8-9]. "Feel" is thus, in the terms of this paper, not a subjective matter, but the word used for a technical value.

The brake pedal feel is of equal significance for both safety and comfort. In view of the fact that customers are tending increasingly to reward with their custom manufacturers of that complex system "the vehicle" where they find the easiest driving, brake pedal feel is growing in importance. The question is of even greater current significance because to answering it is vital in the layout of both conventional braking systems and of "brake-by-wire" system components. (**Fig. 8.1**).

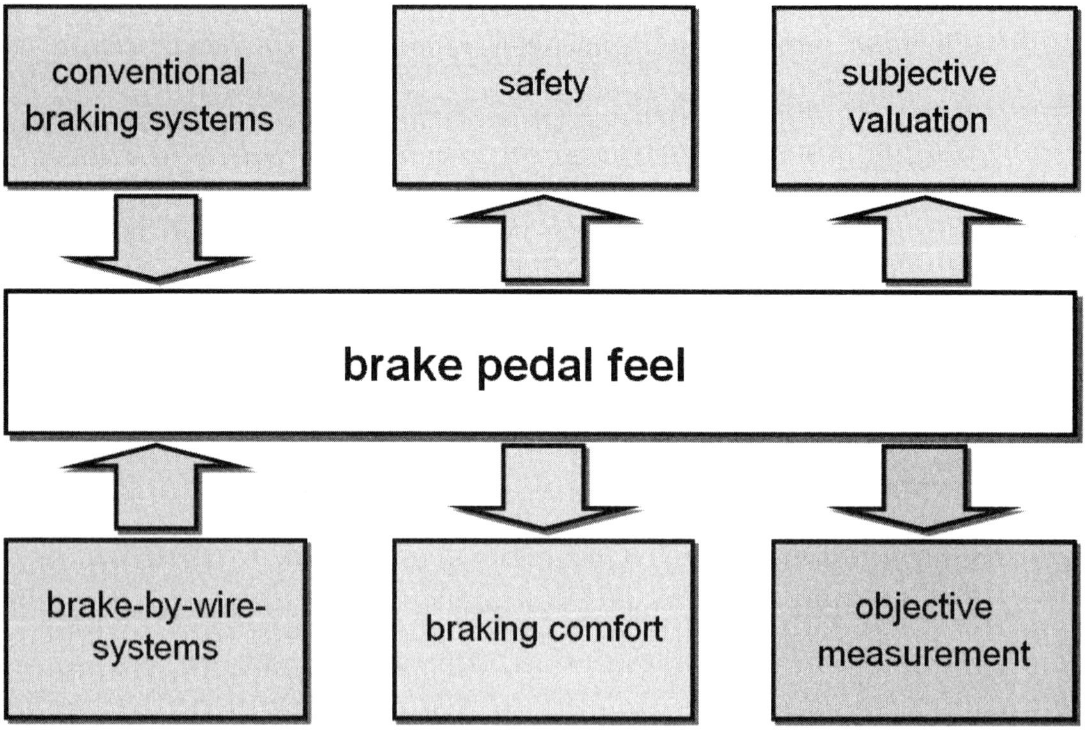

Fig. 8.1: The brake pedal feel defined

Furthermore, if one observes that pedal feel is a phenomenon lying on the interface of the objective measurement (i.e. engineering) and the subjective perception (i.e.

physiology, bio-mechanics, medicine), one will appreciate that classic methods of engineering are not enough to give the full picture.

People tend to say that engineers are hesitant to extend their horizons and embrace the academic disciplines with a high subjective content, such as psychology. Such observations are without foundation. Debate at the intersection of the various scientific disciplines challenges the teacher and spurs research into today's technical developments, helping in the creation of new didactic and methodological approaches for the benefit of the next generation of engineers. This generation will take for granted that they work in multi-disciplinary teams.

The literature has already reported certain approaches to objective description of brake pedal feel and how it relates to vehicle behaviour and to subjective aspects [8-2], [8-6], [8-7], [8-11]. The experts gathered at the "μ Symposium" have already heard about some relevant research [8-3]. Indeed, the question might be asked, "Why have yet another paper on the subject?".

The answer is quite simple but comes in three parts:

1. Almost any engineer concerned with the design of vehicles or braking systems or brake components is bound to be confronted directly or indirectly with the problems attached to the human-to-machine interface, here the brake pedal feel.
2. As brake-by-wire systems offer the designer more free play in the form the interface takes, both the car manufacturers and their suppliers are forced to tackle the questions involved if they are going to make best use of the new potential and be sure to avoid design faults.
3. The speed of developments in mechatronics, sensor science, computers and software has contributed to the academic debate on this subject, supported by staff who are ready and willing to work across the disciplines on new, efficient approaches.

Most publications focus on the technological side, though a holistic view actually demands that such things as bio-mechanical sub-systems be taken into account (**Fig. 8.2**).

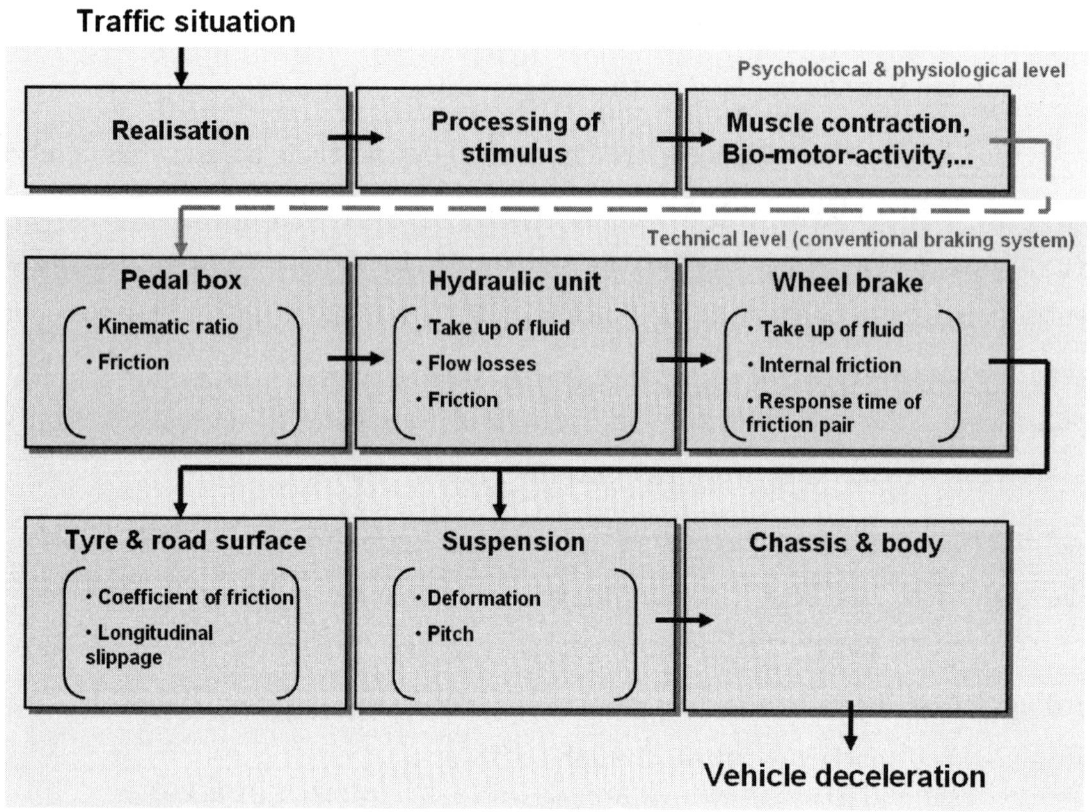

Fig. 8.2: The driver and vehicle as a closed circuit

At the Technische Universität Ilmenau, the Department of Automotive Engineering has opted to take the holistic view from an academic standpoint. In this context, such a view is basically the linking of the psychological and physiological levels to the technical level (**Fig. 8.3**). The main challenge in the investigations is to ensure that the interactions mentioned are illuminated by methodical research which permits actual guidelines for the design of braking system components to be written and experimental design tools to be created. Doing so means that a subject high in both current importance and academic challenge is addressed.

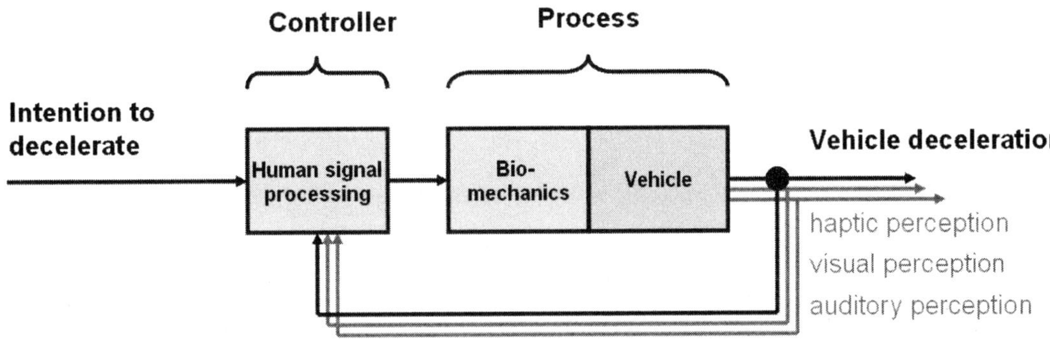

Fig. 8.3: Information and signals flowing in an open loop with reference to brake pedal feel

The Department has, with FORD-Werke AG in Cologne, developed methods of experiment and analysis to permit pedal feel, or the numerical values influencing it, to be described. The outcome of using these tools is described in the rest of this paper.

8.2 Numerical simulation of the dynamic processes at work in conventional brake systems

Firstly, analysis is used to investigate the whole complex of the pedal characteristic curve. The description can be achieved either by using traditional mathematical (differential) equations [8-1], [8-4], [8-5], [8-6], [8-7], or by applying simulation techniques which will support the physical-mathematical modelling and thus the creation - and the solution - of the various equations [8-4], [8-5], [8-8], [8-11].

As the case being studied required computation of mechanical, hydraulic and pneumatic aspects, the program chosen for the analysis was AMESim.

8.2.1 Autonomous overall model using AMESim

The structure of the model involves first of all the actuation side (drivers foot or a hydraulic system for actuation), the pedal box, the hydraulics and the wheel brake. The brake has to be described with reference to all friction and linkage points, in so far as these may be deformed along their axis (**Fig. 8.4**). The geometric parameters used were those of the hydraulic and mechanical components in a middle-range passenger car. The physical parameters required were either taken from the literature [8-1], [8-2], [8-3], [8-4], [8-6] or determined by taking measurements [8-5], [8-8].

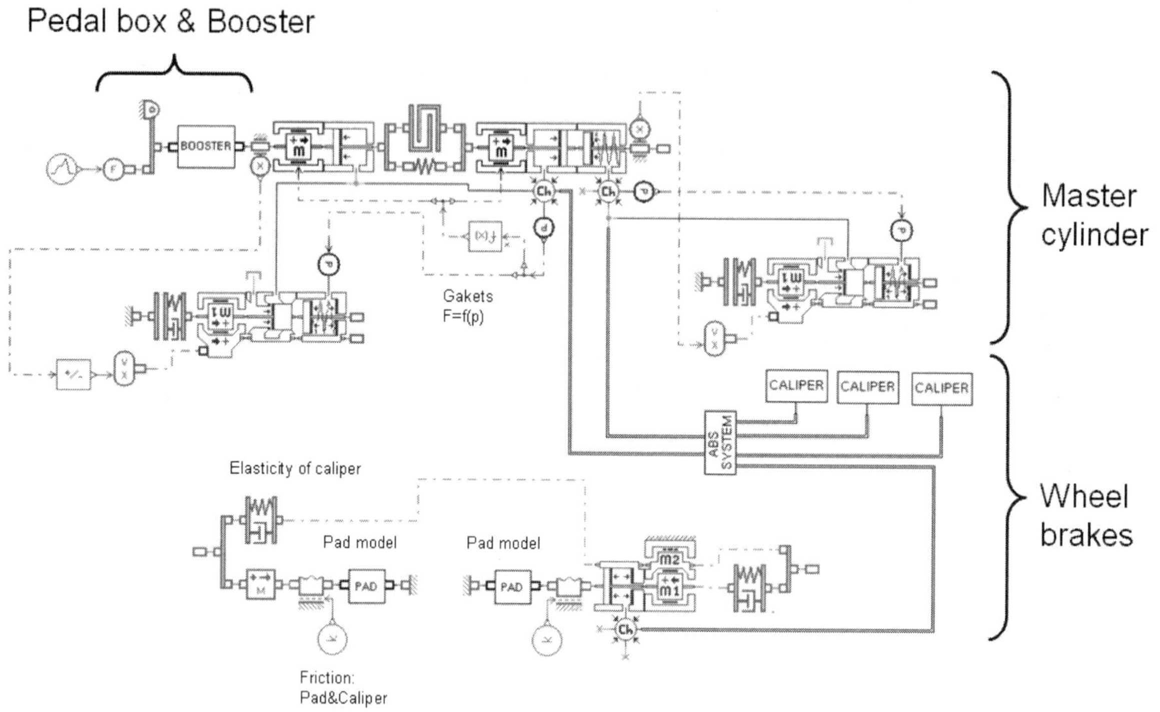

Fig. 8.4: Model of an entire conventional braking system

It is not difficult to depict the relationships, even those which are non-linear. The model of the brake pad is an example of this. The determination of its parameters, was based on experimental investigations of compressibility, using a variety of operating conditions (pressure area, speed of deformation, temperature) and a servo-hydraulic test system. The most significant results on the dynamic compression of the pad in the case of two pad types of different composition are shown in **Fig. 8.5**.

The points of comparison were taken at the pressure area most significant for pedal feel. The graphs shown are those of the 6th to 8th deformation cycle. The speed of deformation was shown to have a negligible influence on the stiffness, but to be significant in relation to the energy absorption, the so called pad hysteresis.

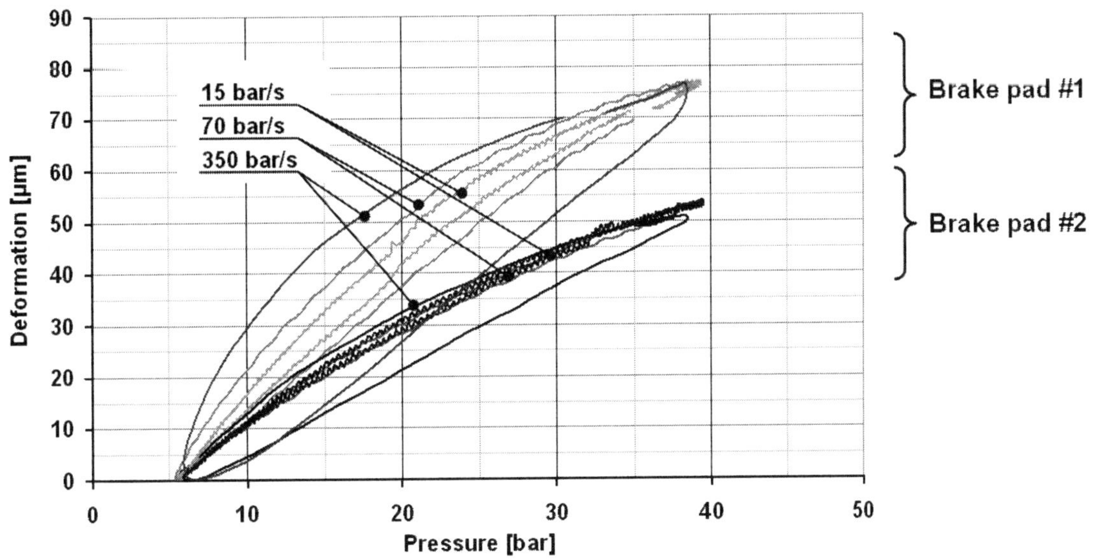

Fig. 8.5: Measurements of the dynamic deformation of the brake pad - cold

Then the pad was modelled using a selection of mechanical substitutes and a specialised program for the identification of parameters.

The model shown in **Fig. 8.6** proved to be particularly suitable for the type of pad and operating conditions under investigation. Intensive experiments are currently being carried out to prove its general usefulness, and a parameter library is being built up.

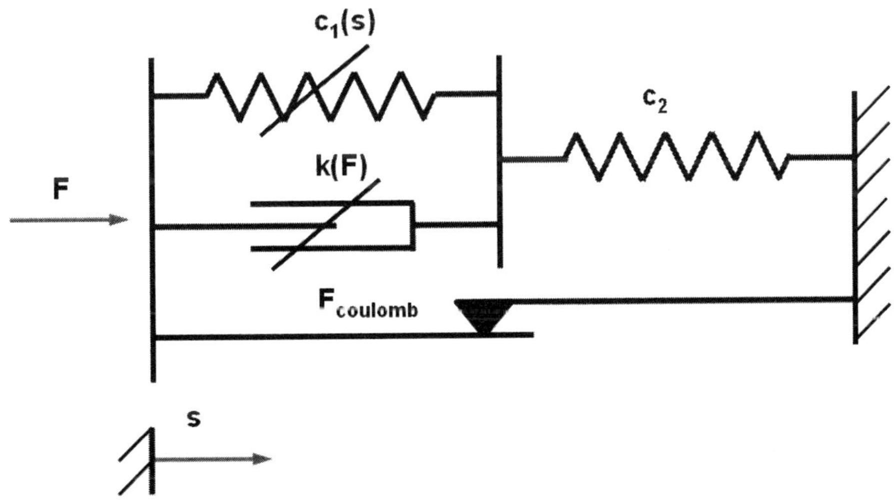

Fig. 8.6: Mechanical model substitute for brake pad

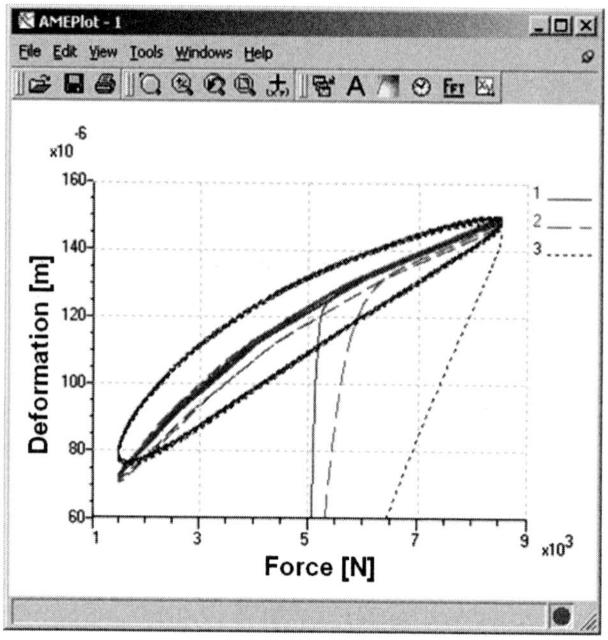

Fig. 8.7: Simulated pad behaviour, pad no. 1

Fig. 8.7 shows the outcome of the simulations regarding the dynamic behaviour of the brake pad no. 1. The consistency achieved is satisfactory.

8.2.2 Experimental validation of the model

The simulation results are validated with the aid of a dedicated test apparatus, built in the laboratory and containing a complete braking system including pedal and seat. Instrumentation is attached to permit detailed statements of the dynamic behaviour of the whole conventional brake system (without the ABS being taken into account).

As already mentioned there is a further research goal within the various projects on pedal feel, and this is the development of a pedal under active computerised control. Because of the energy density thus achievable and the dynamic behaviour the choice fell on a hydraulic actuator system. For later integration into a test vehicle, the various components of the system were selected with the assistance of a purpose-built program and complete pedal unit with hydraulic control as shown in **Fig. 8.8** was put together in a stationary construction and tested.

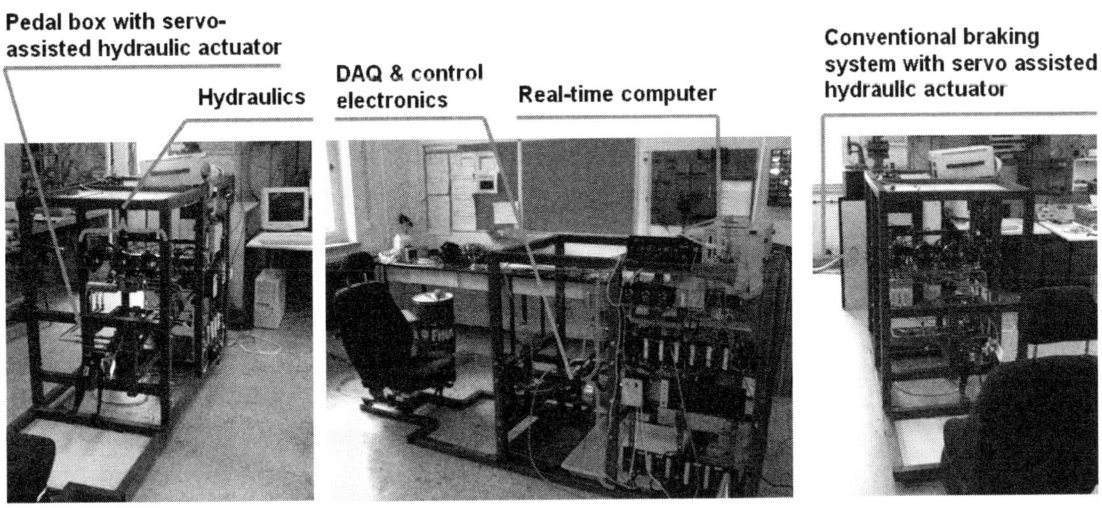

Fig. 8.8: The seat box for the stationary testing of pedal feel

This seat box then permits experimental validation of the various braking system models and facilitates the optimisation of the computerised control for the pedal in action.

This test unit is also conducive to stationary experimentation with other control equipment such as the clutch pedal, the accelerator and the steering wheel following variable characteristic curves.

Fig. 8.9 shows the outcome of the experimental check on the plausibility of the braking system model. The consistency achieved is most satisfactory considering the very early stage of development.

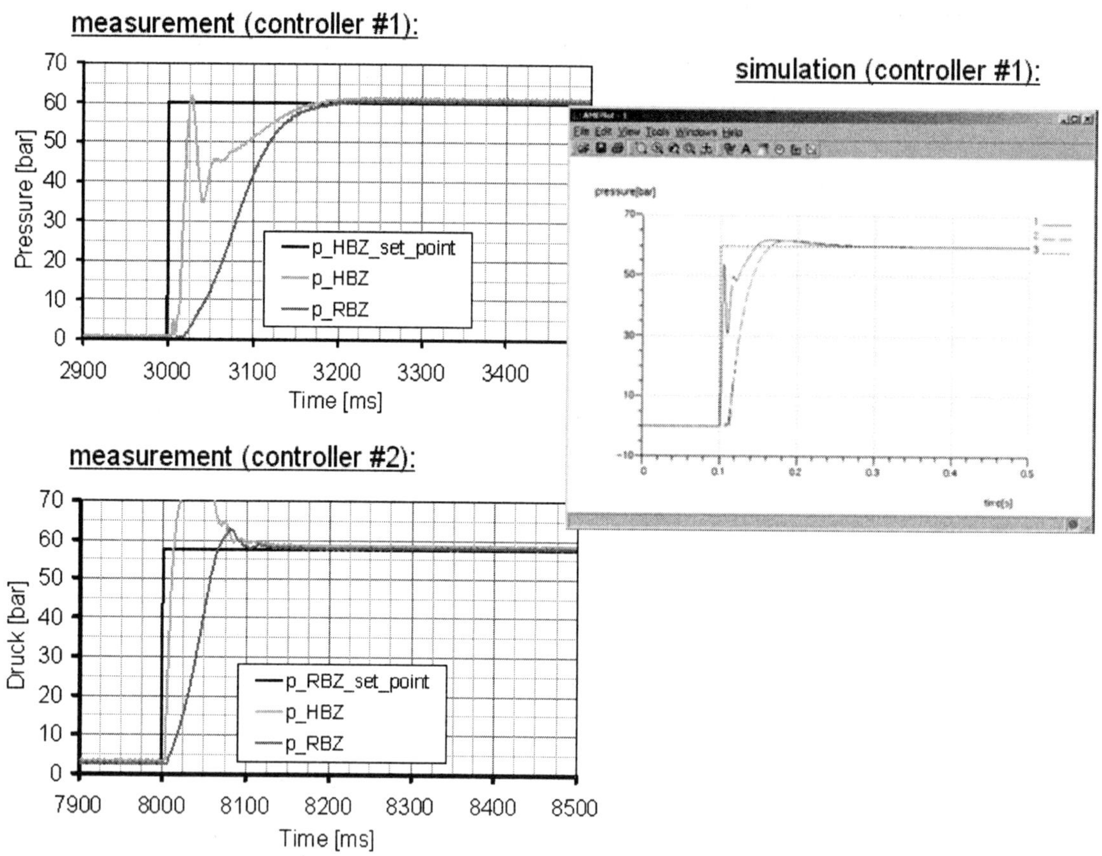

Fig. 8.9: Computation and measurement

8.2.3 The linking of AMESim and ADAMS

To model the whole system, a range of physical domains requires representation, including both stiff and flexible systems with a number of components, and fluid mechanics. While AMESim is a specialised program for hydraulic systems which permits uni-dimensional behaviour modelling, the number of dimensions at the point where the hydraulics link to the mechanics (the pedal and the wheel-brake) is three, which makes it desirable and necessary to use three-dimensional MBS modelling.

There is no tool already existing for comprehensive simulation of multiple domains. For this reason, it was decided to link the two simulation tools AMESim and ADAMS, for which an appropriate interface is available [8-7]. To evaluate how useful this solution would be in practice, a number of examples of the linkage, such as linking the pedal unit (ADAMS) with the hydraulic drive (AMESim) were developed and validated (**Fig. 8.10**). The linkage does basically work satisfactorily, although at the current stage of development of both tools troubleshooting is made more difficult by the inconvenient handling, certain bugs and rather poor documentation. Indeed, it can be impossible to solve the problems for models of complex parts.

In general, this type of real linking of special purpose programs does offer an attractive possibility of overcoming the limitations of autonomous models within one program.

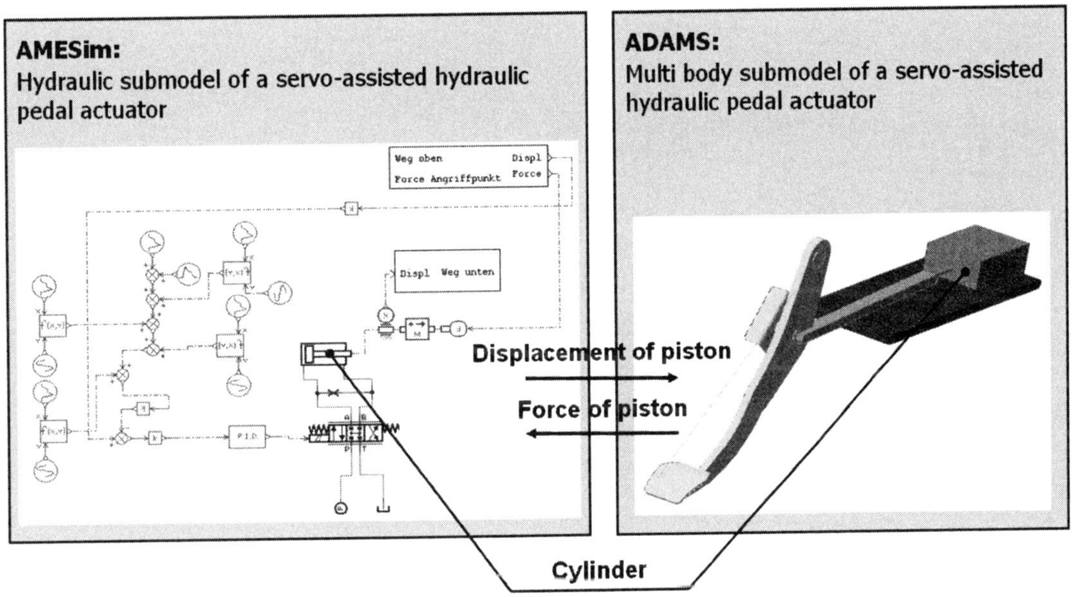

Fig. 8.10: Principle of coupling of different simulation tools

8.2.4 Bio-mechanics

The first step towards the construction of a model of the relevant bio-mechanics was taken for the purpose of developing the pedal under active computer control which is described in paragraph 3 below. If this assembly is to be an authentic reproduction of conventional brake pedal behaviour its dynamics must fulfil sophisticated specifications. The fact cannot be ignored that no pedal applied by the driver and controlled by computer will operate without any feedback effects. These effects may include instances of instability and perceptible rough dynamics. Clearly, the bio-mechanics of the driver's leg must be integrated into any model which is to be reliable.

The AMESim model created in the context of the layout calculations for the actively controlled pedal was therefore extended to include a bio-mechanical sub-model, of which the structure and parameters were decided on the basis of both measurements and plausibility considerations (**Fig. 8.11**).

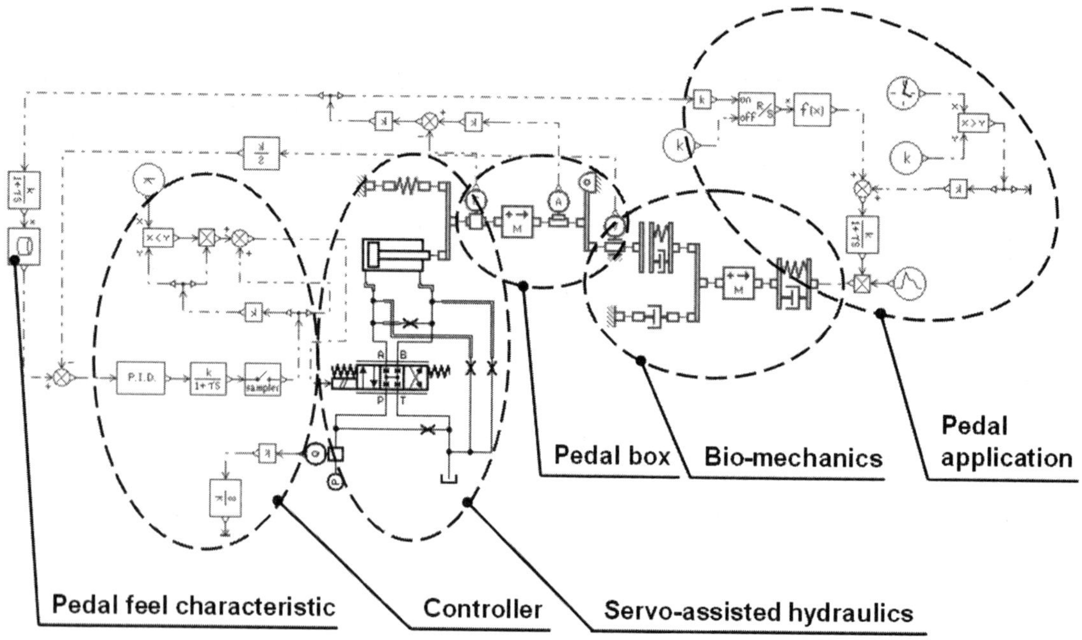

Fig. 8.11: Model representing activation of a hydraulic servo circuit

When in the next stage complete brake apply episodes were simulated, there were early indications of interference in the dead travel region. This was particularly likely during the reproductions of "flat" pedal action. The model was used for quantitative and qualitative adjustment of a variety of preliminary attempts at correcting the hydraulic system and the structure of the regulator. The interference had been completely eliminated by the time the system was realised in the test vehicle, as **Fig. 8.12** impressively confirms.

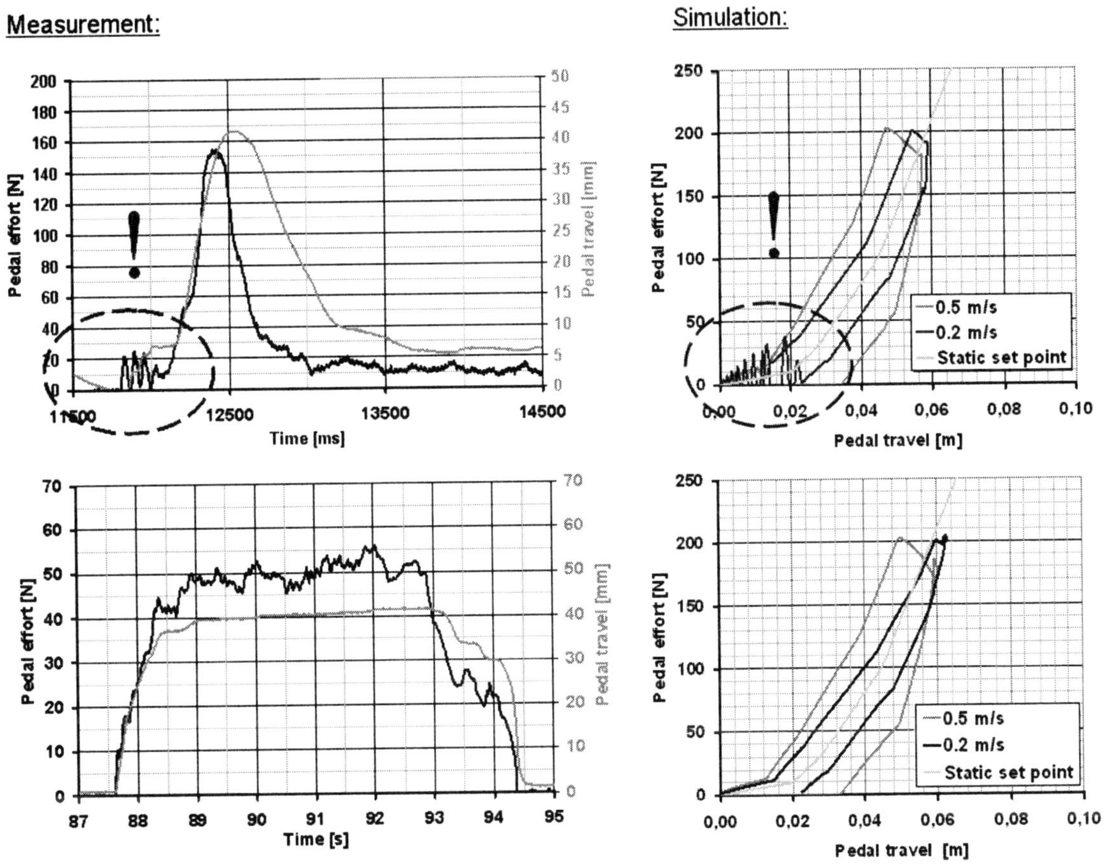

Fig. 8.12: Example of the optimisation potential using a leg model

The results shown have been used as the basis for further intensive investigations [8-11].

8.3 Developing a test vehicle to investigate pedal feel

8.3.1 Design

The project at the heart of the investigations on pedal feel, and also the most ambitious, was the development of a test vehicle and its use for testing of the function. This project was commissioned and closely supported by Ford in Cologne and given the name PEGASYS (from the German for pedal feel analysis system).

The most important specifications include

- Modification of the pedal feel – here the pedal effort – pedal travel – vehicle deceleration – this relationship to be variable smoothly and between wide limits

(**Fig. 8.13** – variability of gradients, of initial values, of static friction [hysteresis] and dynamic friction and of system response times)

- Possibility of changing between one characteristic and the other within seconds
- Will represent authentically the pedal feel of a conventional system
- System will not cause vehicle and vehicle interior to look different

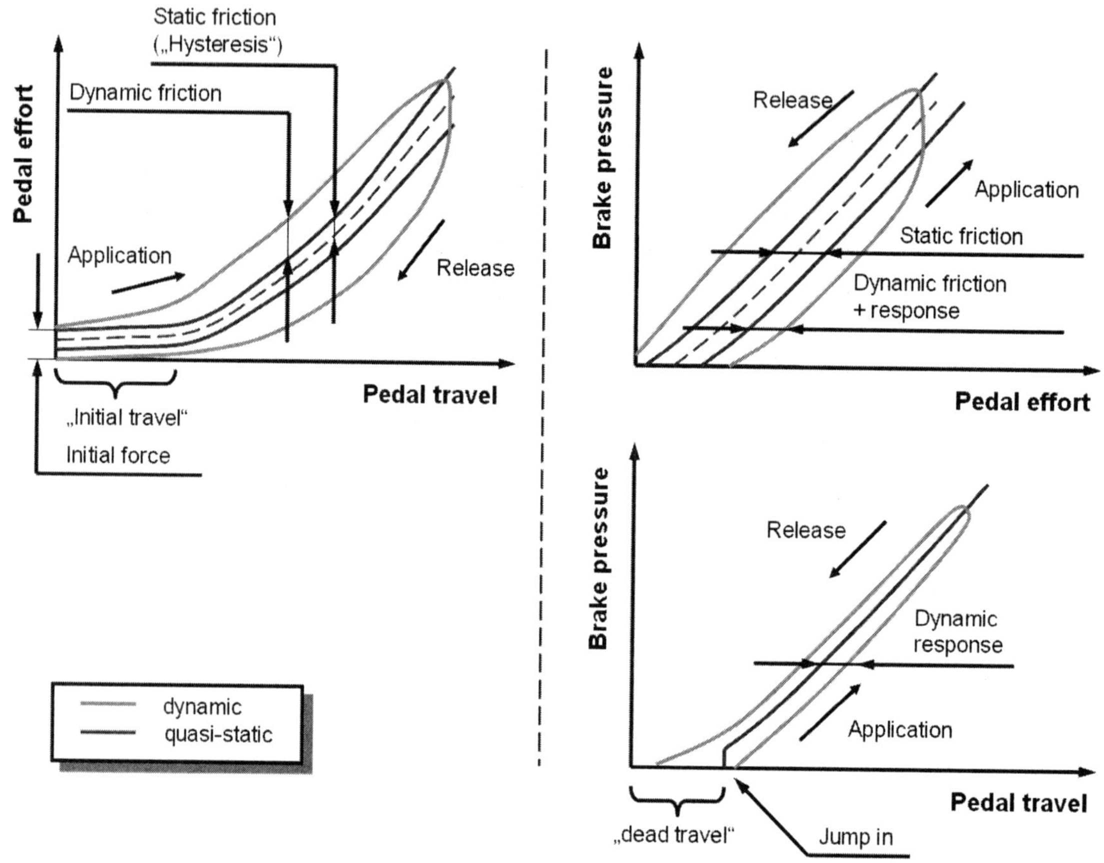

Fig. 8.13: Characteristic curves for brake pedal feel with descriptive terms

A Ford Mondeo was selected to become the test vehicle. It was fitted with the servo-systems which had been tested in advance, with the necessary supply of electric and hydraulic energy and with the computer measurement and control system.

On the wheel brake side, the actuator is a hydraulic servo-cylinder before the master brake cylinder. The entire hydraulic system, including the ABS, remained intact and functioning.

The vehicle deceleration is modulated by regulating the brake pressure. **Fig. 8.14** shows the basic structure of the dedicated electro-hydraulic braking system with the active pedal (EHB-AP), and **Fig. 8.15** gives an overview of the most significant system components in the vehicle.

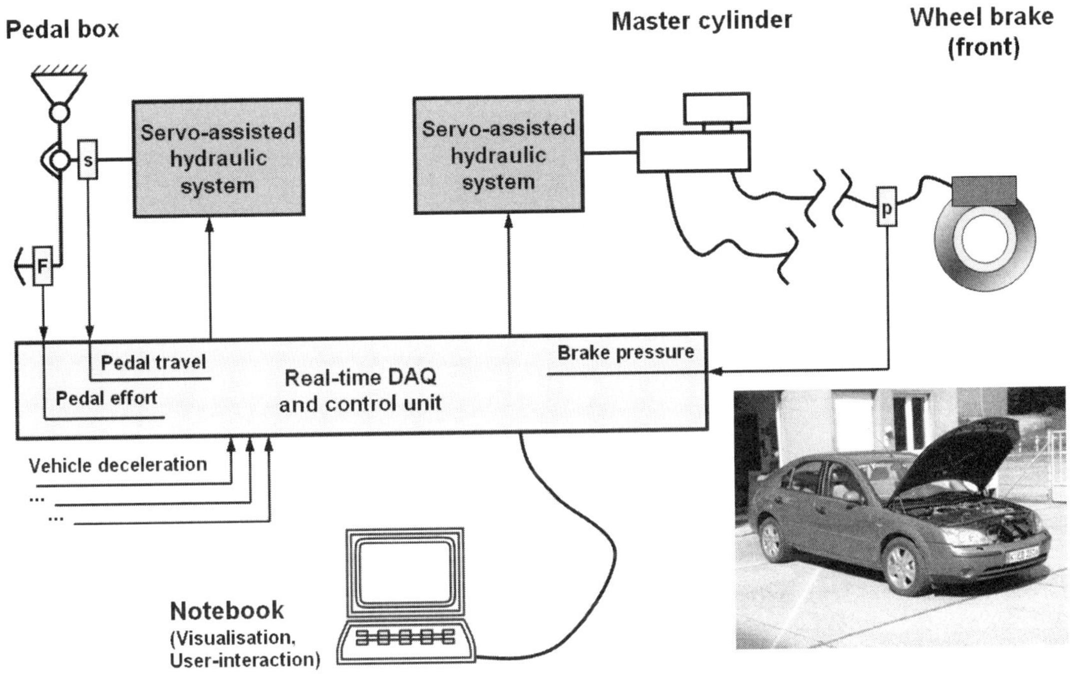

Fig. 8.14: Functions principle of the dedicated EHB with active pedal

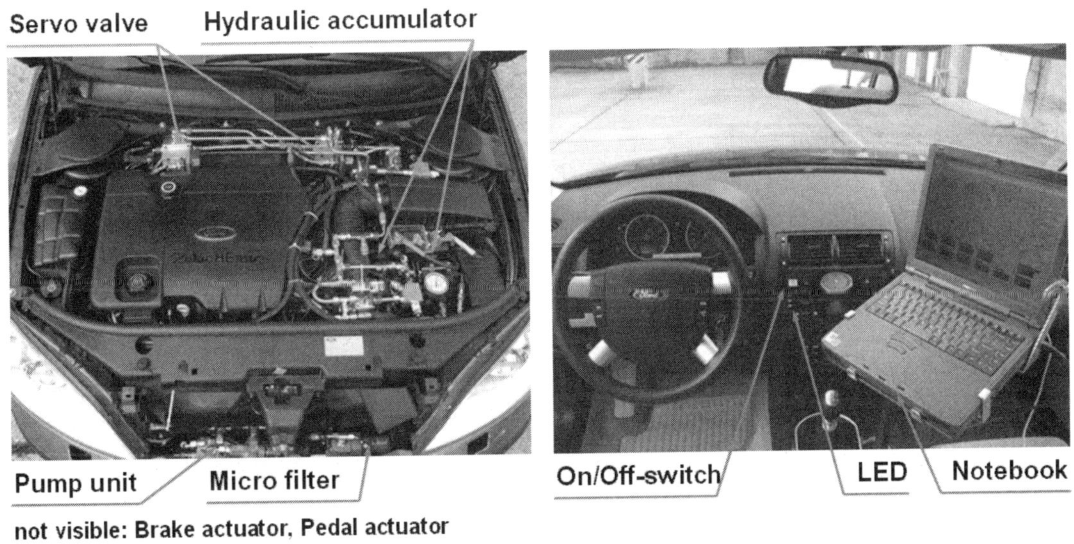

Fig. 8.15: Views of the PEGASYS test vehicle

The theoretical description of pedal feel is based on a physical-mathematical behaviour model, taking as initial value the pedal effort measured. The structure of the model is derived from the simulation approaches described in paragraph 2 above, which are then reduced to essentials. The parameters possible to set for the model are free within a wide range (**Fig. 8.16** + **Fig. 8.17**). This virtual pedal feel is resolved in real time and gives the set points to the servo-hydraulic control circuit, following the so-called master-slave approach [8-10].

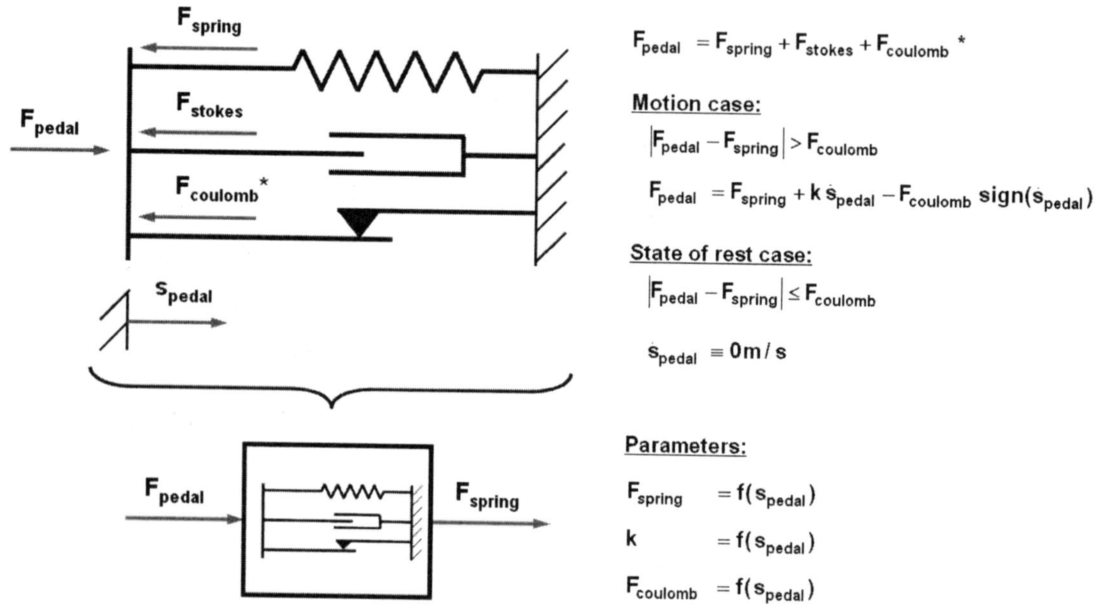

Fig. 8.16: Physical model to assist with simulation of the effort:travel pedal characteristic

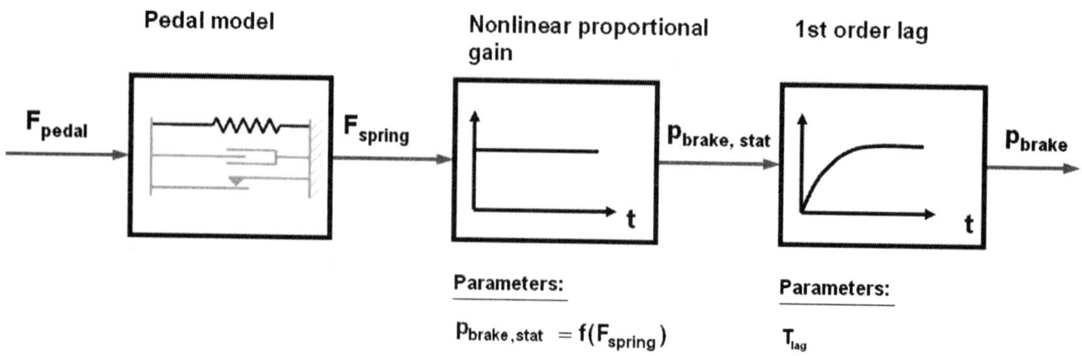

Fig. 8.17: Physical-mathematical model to describe the characteristics of the pedal effort:brake pressure relationship

Modulation of the braking power and braking pressure values is thus absolutely authentic for a wide range of pedal application speeds (**Fig. 8.18**).

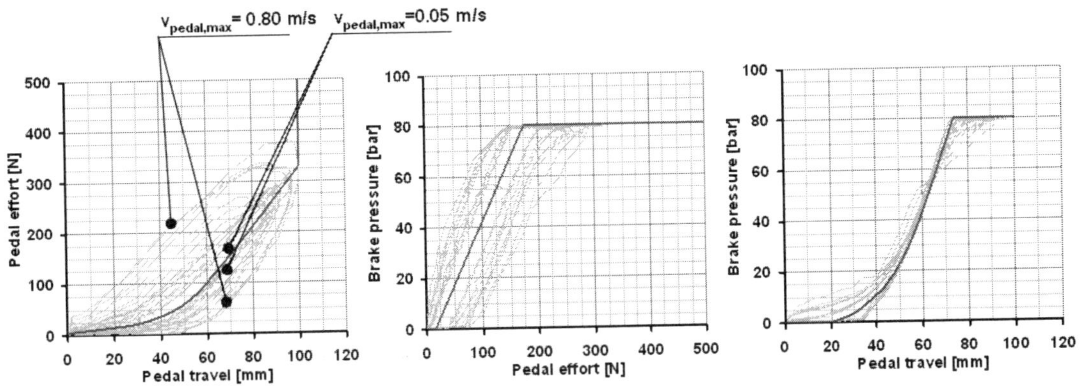

Fig. 8.18: Pedal feel simulation with static and dynamic friction

8.3.2 Perceptibility investigations

In the creation of the PEGASYS test vehicle, the development of the actively controlled pedal produced results in respect how easily the effort-travel relationship of the pedal is perceived.

It was necessary to make the specifications for the quality of the control of the pedal assembly ever more precise, until something that was perceived as an absolutely authentic reproduction of conventional braking apparatus was achieved. As was expected, the perceptibility of the effects of EMI at the pedal is evaluated differently in the significant characteristic curve regions (the region of dead travel and the region of modulation). Additionally, the perceptibility when the pedal is actually being moved is very different from that when the pedal is being held still. In the latter case, as shown in **Fig. 8.19**, a state in which variations of +/-5% to the power (in this case 160 N +/-8 N) when there is virtually no (i.e. less than +/- 0.25 mm) movement of the pedal is already classified as unacceptable.

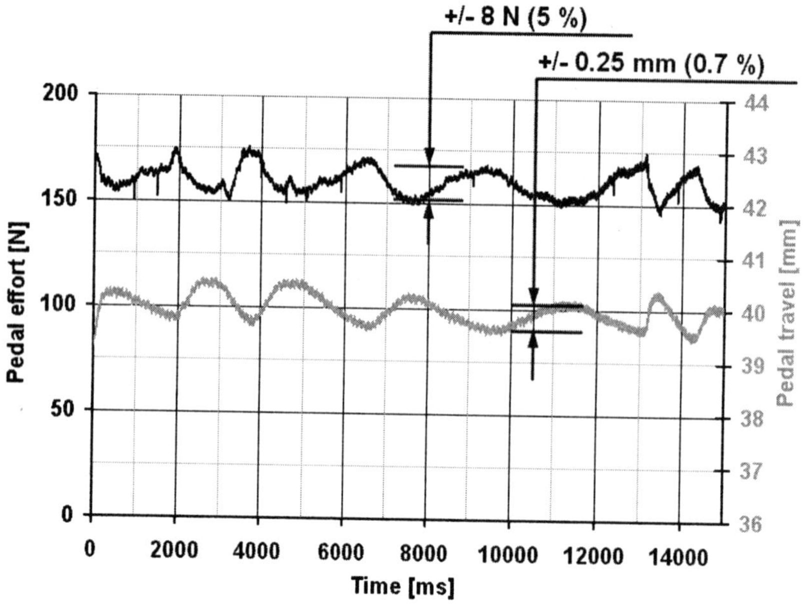

Fig. 8.19: Low-frequency pedal vibration seen as negative

The problems shown were eliminated in the same way as those represented in **Fig. 8.20**. The latter were caused by EMC difficulties, i.e. pick-up from relays or other switches, in this case those associated with the setting of the electric pump assembly for the high pressure supply. Off course these high-frequency problems are not noticeable by drivers. However, the tiny peaks and dips in the force and the travel (lasting less than 10 ms) are felt to be definitely unpleasant. However, the tiny peaks and dips in the force and the travel (lasting less than 10 ms) are felt to be definitely unpleasant.

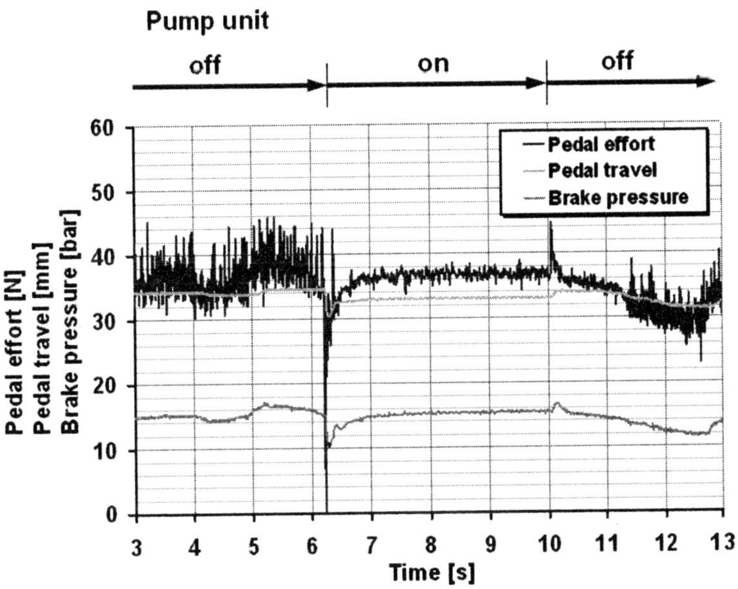

Fig. 8.20: Problematic nominal pedal values shown over time

The behaviour shown in **Fig. 8.12** also attracts negative ratings; short term variations in power of about 20 N at a frequency of 15 Hz at the start of pedal application are seen as neither authentic nor pleasant.

The tests furthermore revealed that it is basically impossible to evaluate pedal feeling unless the vehicle is moving. In the stationary vehicle, the drivers almost always use pressure that would cause an emergency stop in the moving vehicle and is thus unrealistically high [8-10].

8.3.3 Measurement for pedal graphs

It is necessary to have clear definition of the nominal values for the pedal activation in order that pedal feel be evaluated. The reason for this is the fact that the ergonomics and kinetics of pedal use are only one component of the activating force involved in the effective pedal torque. The measurement methods applied must take account of the fact that there is only a clear relationship between pedal torque and activating force when the effective activating radius remains constant. In **Fig. 8.21**, an attempt is made to show this phenomenon by means of a simple kinematic comparison.

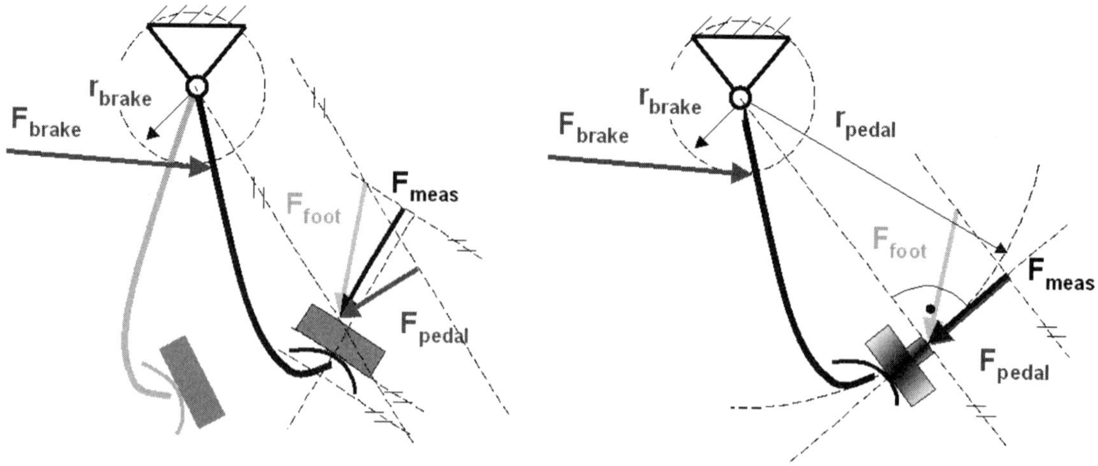

Fig. 8.21: Kinematics of applying the brake pedal

The above fact is of relevance to all methods of evaluation which use "internal" measurement of the pedal characteristics. The usual means of measuring pedal effort must thus inevitably lead to results which must be differentiated in respect of the activation torque for the braking system as against the kinematically balanced method. In addition, it is to be expected that the reproducibility of the activation process will be limited on account of certain changes in the direction of the force. Measured results (see **Fig. 8.22**) fulfil this expectation. As can be seen, the forces measurable using the usual method are 15% weaker in the example.

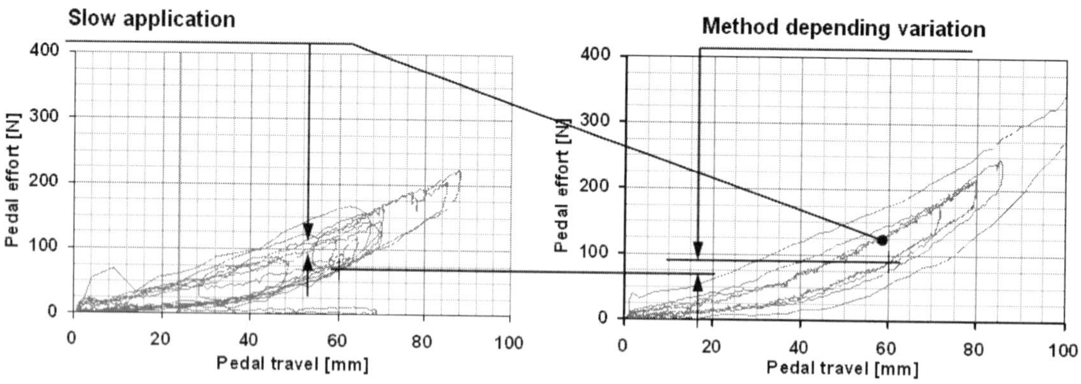

Fig. 8.22: Comparison of pedal graphs for a vehicle with differentiated apply direction

Regarding this results Technische Universität Ilmenau developed a measurement system usual for both methods and with short application time.

8.3.4 Subjective and objective investigations of the pedal feel

A route approximately 1500 m long was marked out on a test site. Each test driver had to drive the route seven times, braking with different pedal characteristics. How this proceeded can be seen in **Fig. 8.23**. In contrast to real life, there was no attempt at simulating traffic ahead and behind, or spontaneously arising situations which cause a driver to brake. It was each driver's task to test 7 combinations, one after another, and at the end of each stretch he or she had to complete a questionnaire. The questions covered the general subjective evaluation of comfort, safety and driving pleasure on the one hand, and on the other, an evaluation of selected objective parameters [8-9]. To avoid the verdict on one combination having a significant influence on a succeeding evaluation, a reference combination was used for the return to the starting point. The first test runs, which are here shown, were conducted by 22 students and staff of the Technische Universität Ilmenau between the ages of 22 and 26.

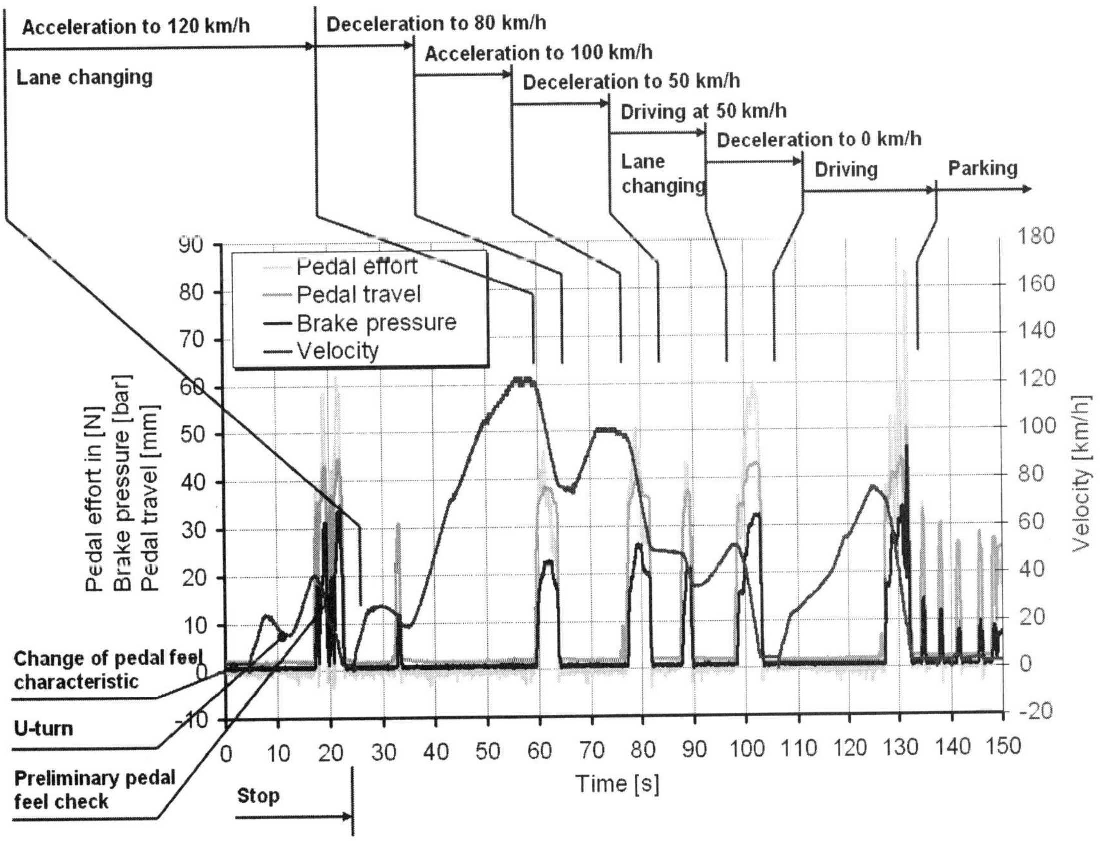

Fig. 8.23: Measured values shown over time for a complete test series

Each characteristic was achieved by targeted modification of one parameter of the nominal reference area selected. Friction, brake performance, and pedal dead travel were all varied to achieve quantitative and qualitative ratings for the perceptibility threshold could be described quantitatively and qualitatively and their acceptability to customers satisfaction.

The drivers never knew which of the parameters had been changed. Neither did they know that the 1st and the 7th combinations were identical with the reference combinations. These duplications were selected to see if evaluation behaviour was modified in the course of the tests. The variables are tabulated in **Fig. 8.24**.

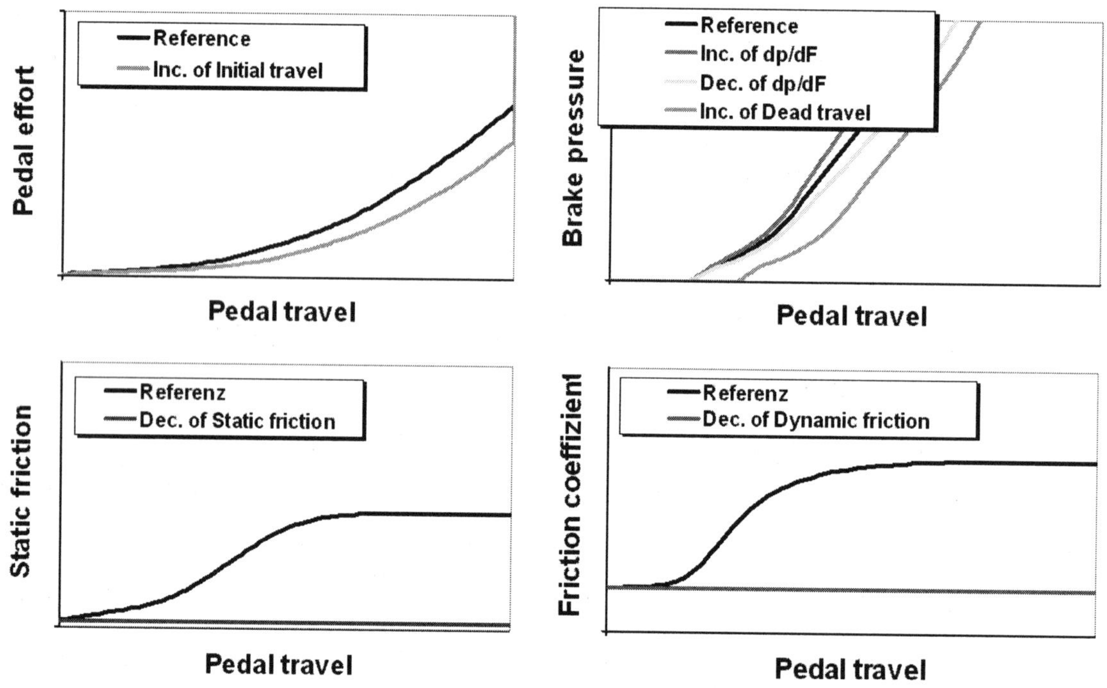

Fig. 8.24: List of pedal characteristics variants for braking experiments

Fig. 8.25 gives as an example of the experimental results the statistical distribution of the evaluations for combinations 2 (reduced static friction) and 7 (reference). The answers to the three questions on general subjective perception are given.

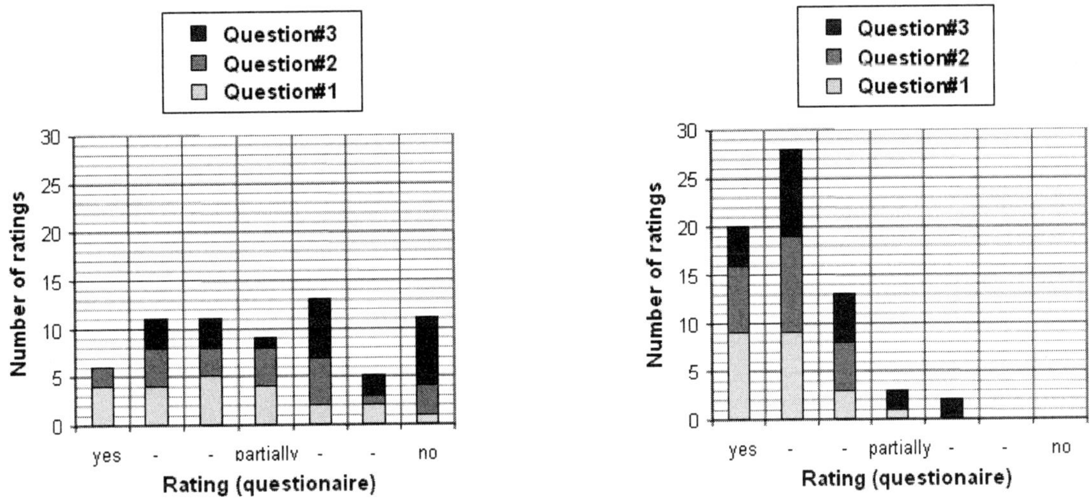

Question No. 1:	Comfort (braking while driving normally) With this combination, were you able to stop or position the car in all situations without difficulty or special effort?
Question No. 2:	Safety (emergency stop) Does this combination give you the feeling you could stop the vehicle safely in any imaginable situation?
Question No. 3:	Driving pleasure (customer acceptability) Does this combination match the type of car and your expectations?

Fig. 8.25: Outcomes from investigations of subjective pedal feeling

The contrast between the ratings given to the reference combination found highly acceptable and those given to the "nervy" combination with reduced static friction is schematically presented in **Fig. 8.26**. This parameter also have main importance regarding the evaluation of an authentic pedal feel.

The description is clearly attributable to the dynamic application region of the brake action rather than to the steady-state region.

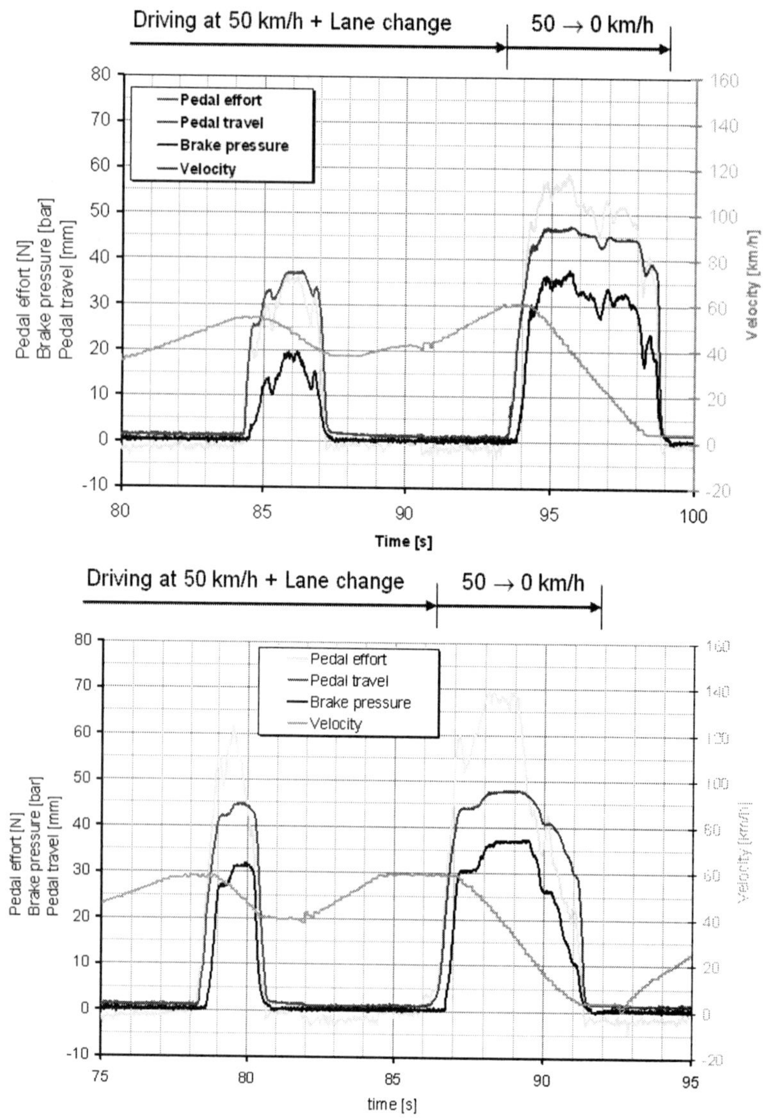

Fig. 8.26: Ratings for "nervy" (top) and "optimal" (bottom) pedal feel

The following conclusions can be drawn from the tests:

<u>Questions relating to general subjective perception:</u>

- There was consistent perception of the reference run, the no. 1 combination and the no. 7 combination as "very good". This implies that there was no observable change in evaluation behaviour over the test period.
- reducing the static friction leads to much more critical evaluation, commonly summed up as "nervy".

- Reducing the dynamic friction tends to improve the subjective evaluation, though the technical reasons for the changed perception are rarely recognised.
- There are ratings at all the offered levels for almost all the combinations. There is thus a broad scatter of perceptions, though definite trends emerge
- The measured data also permit other conclusions on the relationship between the subjective and the objective; these must, however, be omitted on this occasion for want of space.

Questions relating to individual technical parameters:

- The evaluations do not deviate greatly from "exactly right", even though the general subjective evaluation may be much more diffentiated.
- Saying what the objective reason might be for a subjective perception is obviously difficult for the subjects. They sometimes guessed wrong which parameter was relevant, for example guessing the friction had been increased rather than the pedal dead and initial travel.

The subjects are clearly unable to separate out for themselves the various elements contributing to pedal feel. This obviously requires very considerable experience in addition to the technical training that certainly already existed. The conclusion can be drawn that field tests with average drivers will not produce very detailed statements.

A further conclusion is that relevant experiments should relate to genuine traffic conditions.

8.4 Summary

The paper presents a variety of experimental and analytic methods developed at the Technische Universität Ilmenau in joint work with FORD of Cologne for describing and evaluating brake pedal feel. The statements here made should be seen as a status report concerning researches to be pursued further.

A uni-dimensional model is appropriate to the making of basic statements about the dynamics of the pedal characteristics in varying brake systems. It will also serve for the initial description of the bio-mechanics of the driver in the act of braking.

The approach used in the model can be extended by linking it with other tools, especially those such as ADAMS for three-dimensional simulation.

A test vehicle is presented which puts the system of active and variable computer control of pedal and brakes into practice. The vehicle has been developed to the stage where an absolutely authentic pedal feel is conveyed, which can be modified almost instantly.

The investigations at both objective and subjective level can thus now be continued using a broad spectrum of test-drivers.

Currently, three sets of investigations are taking place: into automatic modification of the pedal characteristic curve in relation to the driving situation, into the evaluation of high-frequency interference at the pedal; and into the possibility of influencing driver behaviour by haptic stimulation from the pedal. In addition, a second experimental vehicle from a different range is at present being designed with a hydraulically activated brake-by-wire system and an active pedal.

8.5 References

[8-1] Augsburg, K.: Ein Beitrag zur experimentellen und analytischen Ermittlung des Zeitverhaltens von NKW-Bremsanlagen. Dissertation TU Dresden, 1985

[8-2] Becker, K. (Hsg.): Subjektive Fahreindrücke sichtbar machen II. expert verlag Renningen, 2002

[8-3] Bill, K.: Forschungsansätze und Werkzeuge zur experimentellen Untersuchung des "Bremspedalgefühl" mit Blick auf kommende Brake-by-Wire-Systeme. XVIII Internationales μ-Symposium, Bad Neuenahr 1998. In: VDI Fortschrittsberichte, Reihe 12/373

[8-4] Grochowicz, J.: Experimentelle und theoretische Untersuchungen zu Bremsdruck- und Bremsmomentschwankungen an PKW-Scheibenbremsen Dissertation TU Dresden 1996

[8-5] Heyer, J.: Dynamische Simulation des Bremsanlagenverhaltens Projektarbeit TU Ilmenau, 2002

[8-6] Jahn, M.: Größen und Wirkungen an der hydraulischen Bremsbetätigungsanlage. BVK Sonderdruck TU Dresden, 1982

[8-7] Leber, M.: Radbremse mit mechatronischer Kennwertregelung – Untersuchung von Betriebsverhalten und Fahreranbindungsproblematik hinsichtlich brake-by-wire-Systemen. Dissertation TU Darmstadt, 1998

[8-8] Schalpeter, R.: Multidomänsimulation von AMESim und ADAMS. Projektarbeit TU Ilmenau, 2001

[8-9] Schmidt, O.: Ansprechverhalten von Reibpaarungen. Diplomarbeit TU Ilmenau, 2002

[8-10] Sendler, J.: Durchführung und Auswertung von Fahrversuchen zum Bremspedalgefühl. Projektarbeit TU Ilmenau, 2002

[8-11] Trutschel, R.: Auslegungskriterien für Komponenten von Brake-by-Wire-Systemen. Internationales Wissenschaftliches Kolloquium Ilmenau 2002, Workhop „X-by-wire"

Part II

Brakes and Brake Systems

9 ABS for scooters

Dipl.-Ing. B. Fischenich, FTE automotive GmbH

Ebern, Germany

Abstract

The different performance requirements between passenger car, motorcycle and scooter anti-lock braking systems are described here. The function and design principle of the newly developed power assisted electro hydraulic brake system for scooters is described in detail. This development is characterised by the high level of functionality achieved despite the fact that due to cost and weight constraints, only one channel is anti-lock controlled. It is the first time a combined brake has been realised for this vehicle category.

9.1 Company profile, FTE automotive GmbH

Since 1998, FTE automotive has been part of the Dana Corporation, its registered address being in Toledo/Ohio. Dana has 86,000 employees world-wide, which over 320 locations achieve a US$ 12.5 billion turnover. The first series production deliveries of FTE automotive took place in 1955. These were braking systems for a company called Glas.

Since then, the product range has been expanded and a healthy client structure has continued to grow and be successfully operated. FTE's objective is not to have any one of its clients contributing more than 25 % of its total turnover. FTE's product range includes hydraulic clutch actuation systems, braking systems for vehicles and anti-lock braking systems for motorcycles. FTE was the world's first and for a long time, the only company to accept the challenge of integrating ABS into the motorcycle and producing in series. The first series delivery took place more than 10 years ago.

FTE's five most important customers, in alphabetical order, are BMW, DaimlerChrysler, Ford, General Motors and Volkswagen, each of which make up between 10 and 20 % of FTE's initial product turnover. Throughout the world, FTE has an outstanding position, particularly in the clutch hydraulic and motorcycle ABS markets, with market shares of between 70 and over 90% respectively.

One example of our product application is the concentric slave cylinder for clutch actuation. Depending on quantities, this is designed with either an aluminium or a plastic casing. The master cylinders are now almost exclusively produced from plastic. The link between master cylinder and slave cylinder is a special solution usually made from brake pipe and a special clutch pipe, which is suitable for every application and which frequently often has other "features", such as pedal vibration suppression.

FTE has been manufacturing brake power boosters, which are now constructed from aluminium, since 1972 and the level of production is growing, with permanent one-digit ppm values. Wheel cylinders and brake pipes complete the existing product range.

In retrospect, FTE has developed from a component supplier into an innovative partner to automobile manufacturers with system competence and a global presence.

FTE develops and manufactures the brake and clutch pipe product itself. Rubber components, particularly the sealing element for PTE products, are also part of the FTE product range.

FTE's annual growth in turnover has fluctuated over recent years between 8 and 12%, in a stagnant or 4% growth market respectively. This success is guaranteed through the high reliability and competitive prices of the products and also through the constant self-motivating product improvement. Customer care is provided from 12 locations and customer centres world-wide. Depending on the distance from Ebern, these locations also have their own development teams.

FTE has three production sites within Germany and also in India, Mexico and Brazil and a technical co-operation in Iran. The Brazil site has only been operating for a few months. Mexico has been manufacturing since 1997 and India since 1996.

9.2 Anti-lock for motorcycles

9.2.1 Evolution of the motorcycle ABS

The first anti-lock braking system for motorcycles went into series production in 1988. This significant step towards increasing motorcycle safety not only received particular attention, it became a commercial success for all concerned.

The new-found recognition is partly due to this system being awarded the safety prize by doctors who travel by motorcycle in 1989 and partly because it is constantly assessed as being at the top in many specialist publications. One motorcycle magazine in the USA used it as its title page in 1990; the headline read: “World's safest bike”, this was a BMW K Series with an FTE ABS.

At this point, I wish to point out the quantities sold so far. Approx. 40,000 systems are currently being used in the field and are functioning perfectly. The greatest challenge has for a long time been to satisfy demand, since its market success far exceeded expectations. The very cost-intensive assembly and test equipment could not be expanded without considerable investment. Since 1993, the second generation motorcycle ABS has been produced in series.

The fact that ABS for motorcycles represents a particular development challenge is based purely on the fact that a motorcycle has only two wheels, which can normally be braked separately by the rider. Since a motorcycle has two completely independent brakes, it also has two independent anti-lock braking systems. In short, a motorcycle has half as many wheels as a car, but double the number of anti-lock braking systems. Information on the driving condition is given only once per braking system and not four times as with the car.

On some models, the lifting tendency of the rear wheel is another challenge which makes the life of a designer particularly interesting.

A motorcycle ABS must satisfy more stringent requirements than a car ABS. At the same time, the consequences of malfunction are usually more serious, since when a front wheel locks, the rider's landing is not soft.

9.2.2 Requirements of the safety concept

The differences in respect of both the system and the requirements of the systems mean that a car ABS cannot simply be converted to a motorcycle ABS. In particular, the motorcycle ABS requires specific safety concepts, the most important elements of which, in the case of the FTE ABS, are signal quality and signal reliability. With our systems, the number of pulses is between 40 – 50 per metre of wheel circumference. A typical car system produces approx. 20 pulses per metre of wheel circumference.

For signal quality, signal safety must also be taken into account. Incorrect signals or signals which are interpreted incorrectly must be safely ruled out. The experiences that FTE have collected in this field with the two motorcycle ABS generations being produced in series, are of course being used to develop the latest brake. Active sensors, an electronic control unit with three processes per channel and redundant warning lights help to achieve the most challenging objective of this development.

The three processors per channel are necessary, only one item of speed information is available per wheel. The computer always requires two items of matching speed information. If a processor deviates by more than the tolerance, a corresponding error message occurs.

Another "trademark" of the FTE motorcycle is the unit's automatic self-test at each start-up. The standard electronic test is only one part. FTE systems also check the hydraulic function each time the vehicle is started, through an appropriate load and relief cycle, without the rider having to actuate the brake.

The target values for the pressure increase and decrease are held in the computer and the actual values achieved in the test cycle are compared during each self-test. The warning lights are only extinguished when these values match the target values.

For reasons of redundancy, two warning lights are used and due to the improved signal effect, these flash alternately, unlike in the car, where only one light is used that lights up permanently in the event of a fault. This takes into consideration the special motorcycle requirements, since, due to the design and seat geometry, the warning lights are not in the rider's line of view as mush as they are in the car.

9.2.3 Special features of the scooter

The growing popularity of the motorised two-wheeler is not limited to highly motorised motorcycles, it also applies to the scooter.

Scooters appear mainly in European cities as the popular method of transport for short distances, which satisfies the personal requirements of the rider, arising as a result of high traffic density and lack of parking spaces. Scooters require little more parking space than bicycles, in terms of status, they still precede cars and in some cities, motorised two-wheelers may even use bus and taxi lanes.

Scooters are manoeuvrable and easy to operate. They have electric starters and most are operated by an automatic switch. However, it is often forgotten that the requirement of riding know-how for scooters hardly differs from that of motorcycles. The number of less experienced scooter riders however is significantly higher than that of motorcycles. This can be explained for example through the age structure, certain scooters can be ridden from the age of 16, and also by the shorter distances to be covered.

The load distribution between front and rear wheels differs significantly between the two vehicle categories. Motorcycles have 60 to 65% of their weight on the front wheels, whilst scooters have 35 to 40 %.

When a scooter is ridden by two people, the force required for emergency braking for example, is so great, that it cannot always be applied with the left-hand, which would normally actuate the rear-wheel brake. The consequence is an uneven braking path. The relatively small load on the front wheels as compared with the motorcycle increases the tendency to lock.

Tire dimensions, the available suspension travel and attenuation characteristics are further principle disadvantages which prevent scooters from having optimum braking behaviour. FTE willingly accepted this challenge and developed an external composite brake with ABS for scooters. This development began in 1996 and series production is scheduled for the year 2000.

Due to the limited fitting area and predominantly for cost reasons, only the front wheel is ABS controlled.

9.3 Active combined braking system for scooters

9.3.1 Stopping path

In order to come to a standstill on a dry road from 60 km/h, a scooter requires approx. 26 metres, if only the rear wheel brake is actuated. The required stopping distance is significantly shorter, i.e. 13 metres, if both brakes are actuated at optimum braking power distribution.

This is not surprising, since two brakes are being used instead of one and the power distribution differential between front and rear described above is compensated by the dynamic axle load distribution.

An experienced rider will easily be able to achieve this stopping distance at a constant gripping value between road and tire. Friction value jumps however make it impossible to decelerate without ABS.

9.3.2 Benefits of ABS

The best-known benefit of ABS on two-wheeled vehicles is due to the fact that the front wheel in particular does not lock, which prevents the so-called sole-accident. **Fig. 9.1** shows however that braking paths can be minimised, if even an inexperienced rider can optimise the deceleration potential long before a wheel locks.

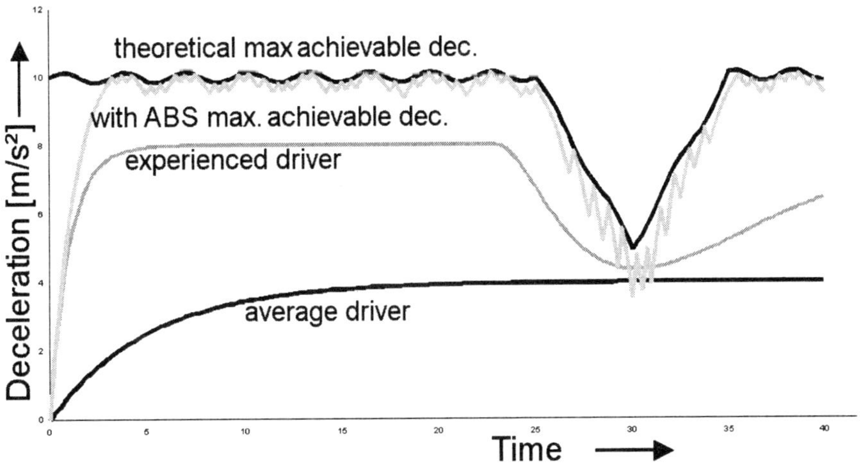

Fig. 9.1: Deceleration with and without ABS

9.3.3 Use of composite brake

Even with ABS, valuable stopping distance is wasted if the rider fails to use the second brake at all, or waits after the first brake is actuated. The so-called composite brake therefore has the advantage, from the onset of braking, to use both wheels simultaneously, without a time delay, for deceleration.

9.3.4 System description

The solution presented by FTE, for creating an optimum braking system for scooters, comprises composite brake, external support and an anti-locking system for the front wheel. The rear wheel lever always actuates both brakes. The front wheel lever still activates only the front wheel, which is now prevented from locking. If a rider inadvertently actuates only the front wheel brake, he will always be able to use the gripping value without risk of locking.

It is however to be expected that the faster the rider uses all brakes, the more confident he will become with the system.

9.3.5 General principle

The pressure of the FTE electro-hydraulic brake is produced through an electrically driven vane pump. The main braking cylinder simply actuates a control valve, which increases the braking pressure required by the rider in the braking system.

The ABS function is performed by an electro-magnet, which works mechanically on the same valve.

The relationships described in **Fig. 9.2** apply to a single-channel external braking system with ABS. Expansion to an appropriate two-channel system is possible in principle through doubling. The task in respect of scooters however was to develop a brake, which represents an optimum in relation to size, cost and function. Taking into account the peculiarities of the application, the idea was to control the front wheel brake using ABS, in fact both with front and rear wheel brake actuation. It is therefore possible to lock the rear wheel, which is more tolerable than on the front wheel, as described above.

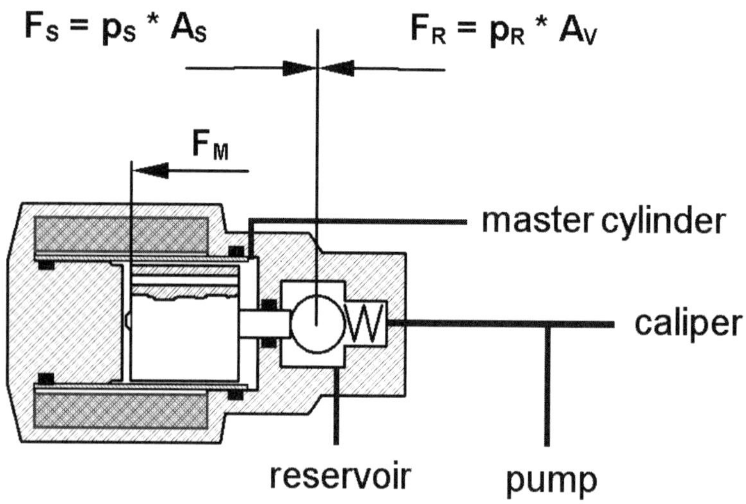

$$F_R = F_S - F_M \qquad p_R = p_S * \frac{A_S}{A_V} - \frac{F_M}{A_V}$$

Fig. 9.2: Effect of the FTE electro-hydraulic brake with ABS

Technically, the compound braking is achieved by hydraulically actuating the front-wheel brake when the composite brake pedal is depressed. The pump to the front wheel brake circuit also runs when the composite brake is actuated. The control valve on the front wheel brake, in the event of compound braking, is impacted by a slave cylinder actuated by the pressure on the rear wheel brake (**Fig. 9.3**).

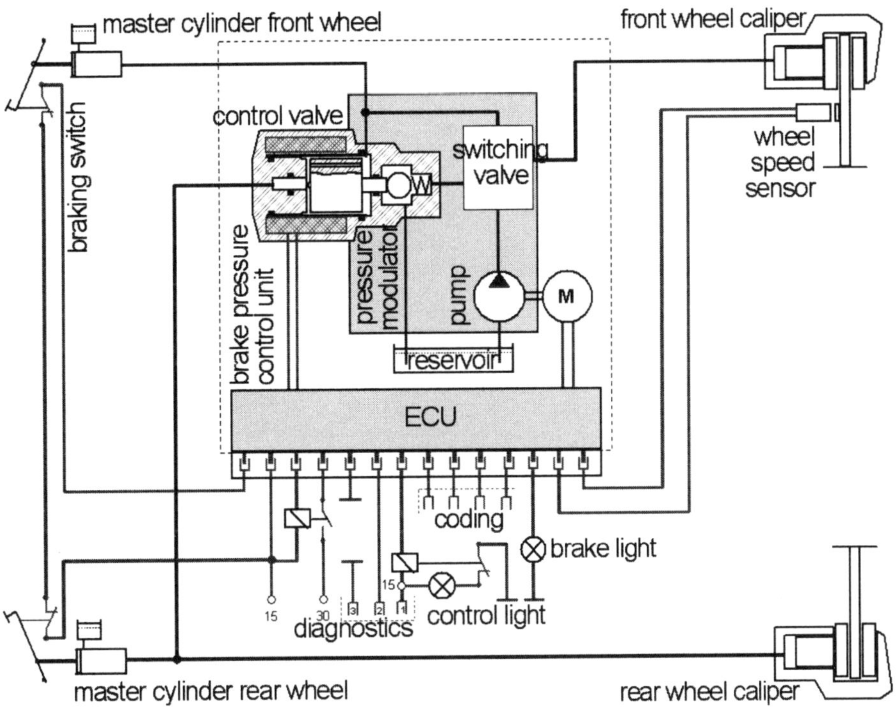

Fig. 9.3: Electro-hydraulic brake system for scooters

9.3.6 Safety concept

It is possible to brake the vehicle if either the ABS or the entire power supply fails. ABS failure simply causes a loss in the relevant counter-force of the electro-magnet , but allows the uncontrolled braking to continue. If the entire power supply fails, an appropriate valve arrangement ensures that the correct main cylinder pressure can actuate the corresponding wheel brake. An increased actuation force is therefore required to continue braking, despite the power failure. In the event of a total power failure, the residual braking force remains on the rear wheel. Hydraulic co-actuation of the front wheel brake is not possible.

9.3.7 Performance data

- Pressure regulation: analogue
- Hand lever response with ABS control: none
- Size: approx. 50 x 120 x 140 mm
- Weight: 2,000 g
- Pressure build-up time: 60 ms
- Power supply: 3 - 12 A

10 BMW Motorrad's New Integral ABS in Practical Testing

Dr. M. Braunsperger, Dipl.-Ing. S. Beulich, Dipl.-Ing. H.-A. Wagner, Dipl.-Ing. G. Heyl, BMW AG

Munich, Germany

Summary

In cooperation with FTE automotive GmbH, BMW Motorrad has now introduced Integral ABS as the third generation of anti-lock systems on the motorcycle. This is the first-ever motorcycle brake system with brake power assistance and adaptive brake force distribution.

After more than one million test kilometers with BMW and the first reactions from customers and the press following introduction of the system on the K1200LT in spring 2001, we are now able to draw the first positive conclusions.

10.1 Introduction

The objectives set right from the start in the concept phase of BMW's third-generation motorcycle ABS were most demanding.

BMW Integral ABS was to offer every customer, even the rider without any experience, a brake system allowing him to enjoy all the benefits and performance of the brakes in full. The new system was therefore to provide a standard of performance even higher than the standard already offered by ABS II in the market:

- Shorter stopping distances when applying the ABS brakes

- Introduction of an intelligent integral brake system capitalising on the frictional coefficient
- Consideration of load conditions
- Reduction of brake operating forces
- Reduction of weight
- Reduction of electrical power uptake while applying the ABS brakes thanks to the reduction of brake activation energy
- No tendency of the pressure point to "wander"
- Complete self-diagnosis ensuring a high level of system safety and reliability

The particular challenge in developing the new system was that it had no predecessor based on similar principles and involves a number of functions networked electronically with one another never seen before on a motorcycle.

10.2 Physical Parameters in Riding a Two-Wheel Vehicle

The motorcycle represents a very complex system in terms of the conditions and parameters interacting with one another when applying the brakes. The specific physical parameters in riding a motorcycle are determined by the relatively high centre of gravity versus the vehicle's wheelbase. A further point is that the stability criteria of a motorcycle differ significantly from those of a car, with effects more dramatic for the motorcycle rider. The motorcycle may for example tend to roll over or the rider may fall due to the front wheel locking. These parameters and criteria are now to be considered in greater detail in the following report.

10.2.1 Tire characteristics

On most of the usual, firm road surfaces motorcycle tires have the same characteristics as car tires. With increasing brake slip we see an increase in the brake power transmitted to the road on a firm surface, until reaching the maximum level at about 10 - 15 per cent brake slip. Then brake power decreases continuously as brake slip increases, until the wheel locks when reaching 100 per cent brake slip. The lateral forces the wheels are able to convey to the road initially drop slowly with increasing brake slip, but then decrease very quickly once maximum brake force has been exceeded, with less than 10 per cent of the maximum lateral forces the wheels are able to convey remaining left over at 100 per cent brake slip.

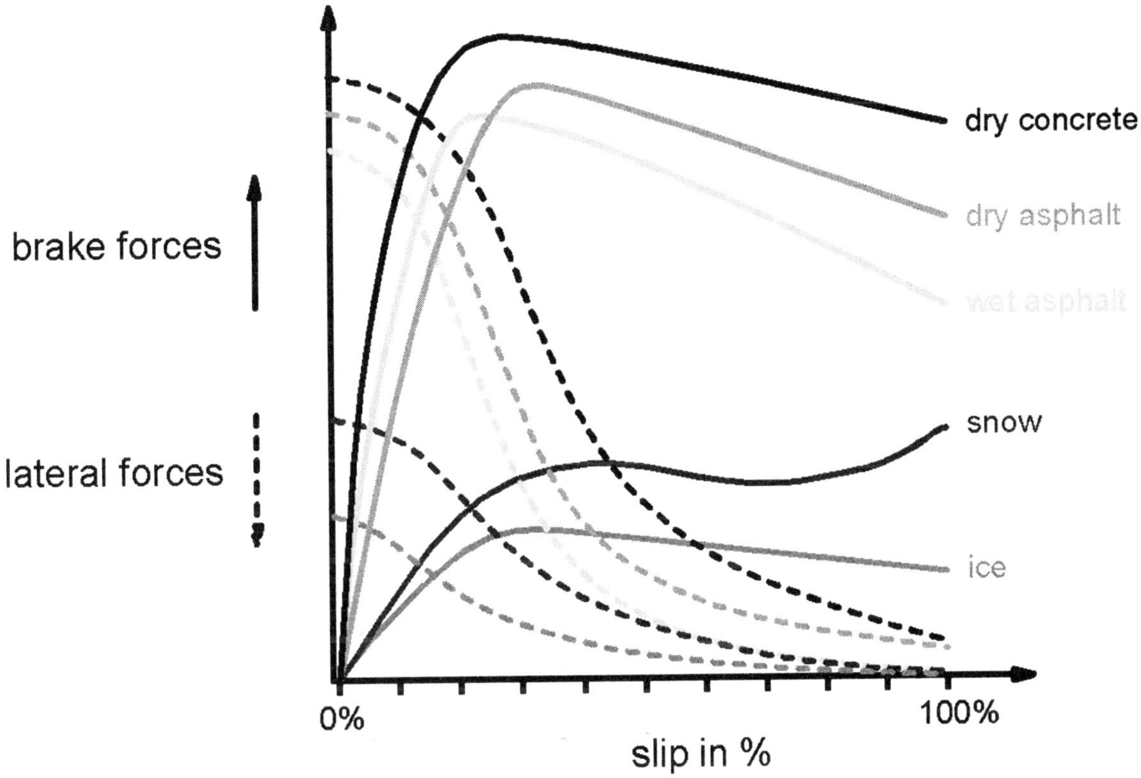

Fig. 10.1: Frictional coefficient of a motorcycle tire

Apart from the characteristics described above, a motorcycle may of course also run on loose ground such as deep sand, gravel or loose snow, where the brake power conveyed by the wheels will increase until brake slip reaches approximately 20 - 30 per cent, then extending along a plateau before reaching its peak at 100 per cent. Side forces, in turn, show the same characteristics and behaviour as on a firm road surface, the increase in brake forces conveyed to the surface with the wheel locked being generated by the wedge effect of the surface material pushed along in front of the locked wheel (**Fig. 10.1**).

10.2.2 Stability

10.2.2.1 Locking wheels

Whenever the front wheels lock on a double-track vehicle, they will merely prevent the driver from steering the vehicle while the stability of the vehicle as such is maintained in full. Should the rear wheels lock, in turn, with no more lateral stability behind the vehicle's centre of gravity, the vehicle will spin by 180° until the rear end points in the direction of travel.

A two-wheel vehicle, on the other hand, builds up its riding stability through the rotational impulse of the wheels turning, the front wheel making the major contribution. Unlike the driver of a double-track vehicle, the rider of a two- wheeler, provided he has sufficient experience and is skillful enough, may keep the vehicle stable even with the rear wheel locking by swivelling and steering his vehicle properly. Should the front wheel of a two-wheeler lock, however, the vehicle will no longer have the stabilising rotational impulse of the front wheel. Together with the very low lateral forces to be seen in the chart presenting the characteristics of the tire (**Fig. 10.1**), this generally means that the rider will fall after 200 - 500 ms when riding straight ahead and with the front wheel also pointing in a straight direction.

10.2.2.2 Rear wheel lift-off

Deceleration of a vehicle causes a vehicle-specific change in wheel load as a function of the deceleration forces exerted (**Fig. 10.2**).

In such a situation the front-wheel force resting on the road increases by the wheel load change factor ΔF

$$\Delta F = m_{ges} \cdot a \cdot h_s / l,$$

the force of the rear wheel resting on the road decreasing by the same factor. This change in wheel load is subject to a specific delay depending on the mass inertia of the vehicle around its lateral axis. Leaving this delay factor and the change in the vehicle's centre of gravity (which results from the kinematics of the running gear subject to the change in wheel load as well as air resistance and lift forces) unconsidered, we receive the first formula approximating the lift-off delay ($a_{lift\text{-}off}$) of the rear wheel:

$$a_{lift\text{-}off} = m_h \cdot g \cdot l \; / \; m_{ges} \cdot h_s = g \cdot l_v \; / \; h_s.$$

At this lift-off point brake forces are generated only by the front-wheel brake, the rear wheel losing its brake effect altogether.

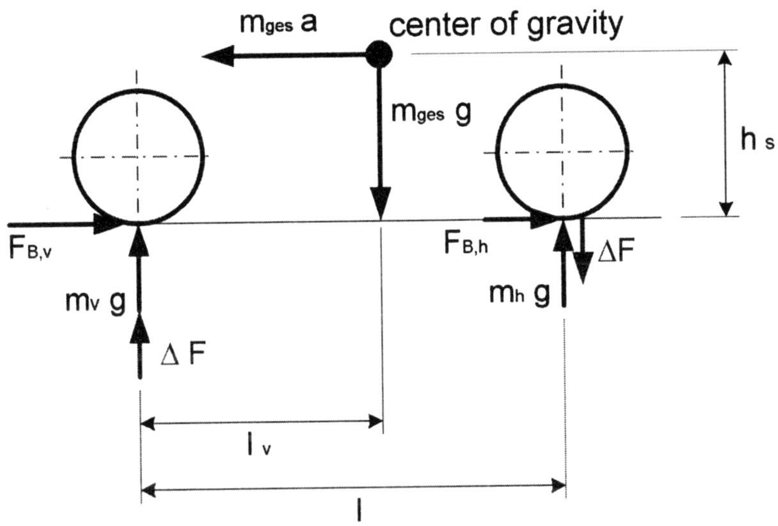

$F_{B,v/h}$ = Brake force front/rear

m_{ges} = Vehicle mass

a = Deceleration

h_s = Centre of gravity (height)

l = Wheelbase (related to the wheel contact point)

m_h = Static share of vehicle mass acting on the rear wheel

g = Earth acceleration

l_v = Longitudinal distance between the centre of gravity and the front wheel contact point

Fig. 10.2: Schematic presentation of wheel load change

10.2.3 Ideal Brake Force Distribution

Upon the start of deceleration (low deceleration in conjunction with dynamic load force still remaining low or not yet effective), optimum brake force distribution is determined by the static distribution of weight on the front and rear wheel.

Due to the change in wheel load already mentioned, the share of rear-wheel brake force in the overall brake force generated on the machine must be reduced consistently with increasing deceleration in order to capitalise on the frictional coefficient between the road surface and the tire. This interconnection is presented in many books on riding and driving behaviour, the diagram in **Fig. 10.3** also applying to a motorcycle in a bend.

Vehicle-specific, dynamic and adaptive brake force distribution on the motorcycle is an entirely new concept ensuring optimum feed of brake pressure to the two wheels in accordance with the *brake forces conveyable at any given point in time.* The ideal brake force distribution for each respective surface and frictional coefficient is determined in a vehicle by its wheelbase and centre of gravity (load condition). Under increasing deceleration, this ideal brake force distribution follows a paraboloid curve and is memorised in the electronic "brain" of each vehicle. The brake force distribution map is adjusted after each change in load as soon as the rider applies the brakes with ABS operating the first time.

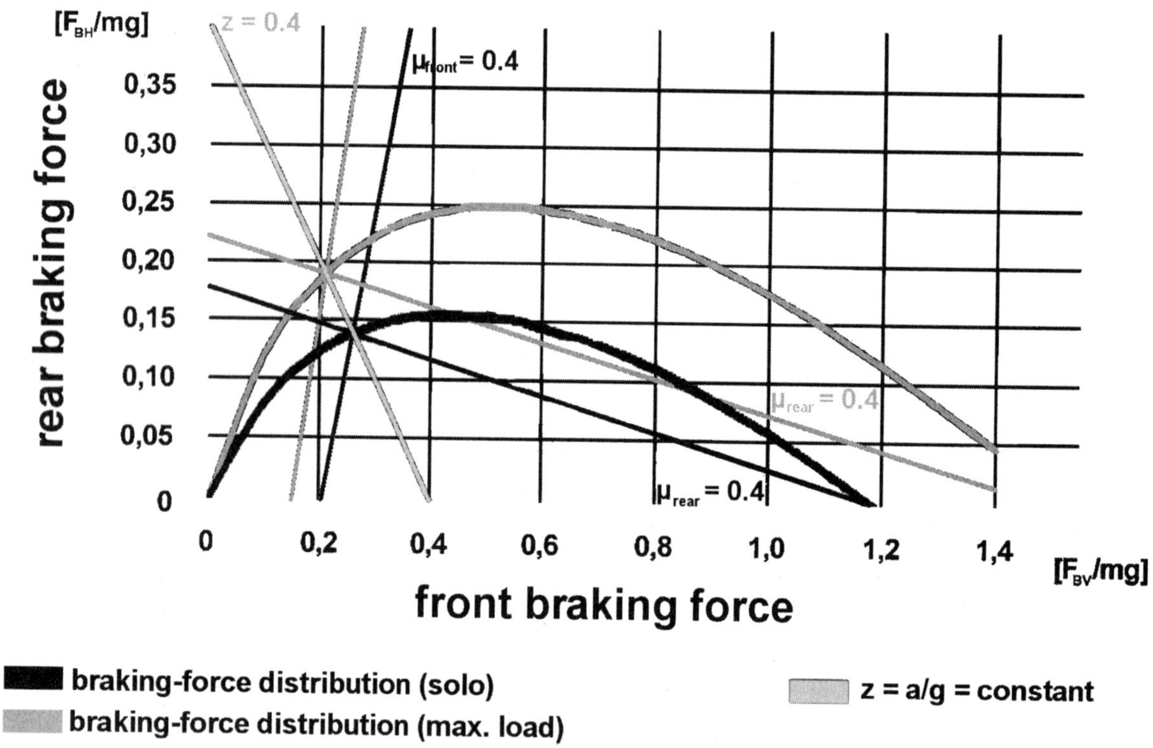

Fig. 10.3: Ideal braking-force distribution on the motorcycle

The curve obtained in the above diagram for a BMW K1200 LT shall be considered in the following by way of three practical examples.

Example 1:

The rider is riding straight-ahead on loose gravel and has to apply to brakes. The frictional coefficient μ between the tires and the road surface is approximately 0.4. In this case the entire frictional coefficient provided by the surface can be used for applying brake force, since the wheels do not have to convey any lateral forces as they would have to in a bend (**Fig. 10.4**).

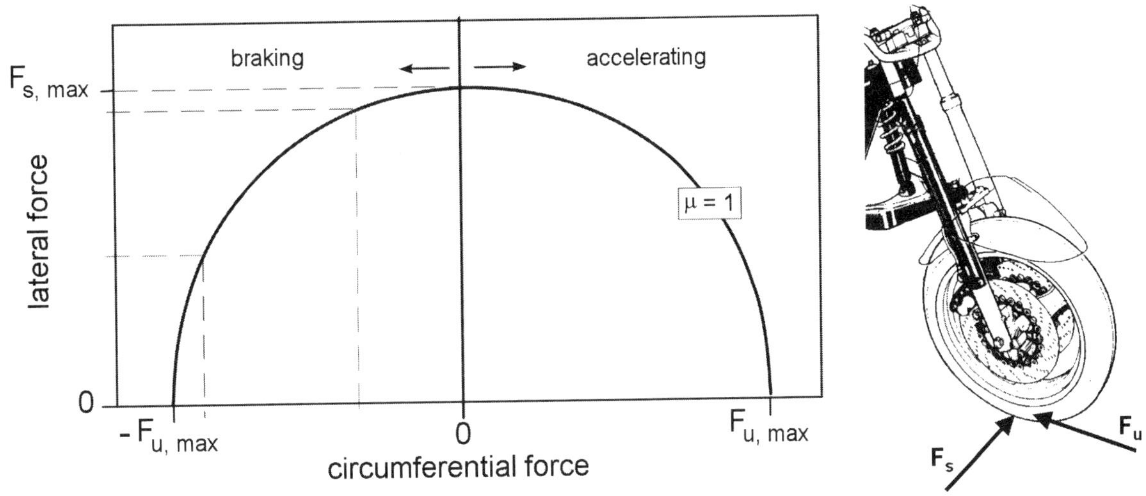

Fig. 10.4: Kamm's circle

If the rider wishes to apply the brakes and keep the motorcycle stable at the same time (without any of the wheels locking or ABS cutting in), brake forces must be spread out on the front and rear wheels in accordance with the intersection point on the Z = 0.4 line and the ideal brake force distribution line for the unladen vehicle. Any deviation from this point towards higher pressure or brake forces would make a wheel lock or would activate the ABS system. Any deviation towards lower brake forces, in turn, would mean a loss of - precious - stopping distance.

Example 2:

The rider is leaning sideways at an angle of approximately 40° on dry asphalt and again has to apply the brakes. Due to the share of brake forces required for counteracting the lateral forces in the bend, a factor of only about $\mu = 0.4$ remains left over for the actual brake forces in a longitudinal direction. Load distribution in this case is therefore exactly the same as in example 1, as is the deceleration built up in the process. It is important to note in this case that the effects of a wheel locking are much more drastic and the effects of ABS cutting in much more unpleasant than on loose gravel when riding straight ahead. The need to spread out brake forces ideally under such conditions is therefore much greater. But what exactly does "ideally" mean in this case? **Ideally**, brake forces are spread out on the front and rear wheels in such a way that both wheels have their maximum lateral stability.

In **reality** the rider will not be able to achieve this ideal distribution, but will apply the brakes, say, only at the front. This means that with a deceleration of x the rider has far less lateral stability on the front wheel crucial to the risk of taking a fall than would be the case if he were able to achieve ideal deceleration on both wheels with an ideal distribution of brake forces. Hence, Integral ABS offers far greater safety when applying the brakes in a bend, preventing the rider from taking a fall.

Example 3:

A rider is riding straight-ahead on loose gravel once again and has to apply the brakes. This time he has a passenger behind him and is carrying luggage on his motorcycle. The frictional coefficient μ between the tires and the road surface is 0.4. To achieve stable maximum deceleration, the driver is required under these circumstances to spread out brake forces front and rear in accordance with the point where the ideal brake force distribution line for the laden vehicle intersects with the $z = 0.4$ line.

Whenever the rider applies brake forces according to this ideal brake force distribution, without making full use of the frictional coefficient μ on the road, each wheel will maintain its maximum lateral stability. And this, in turn, gives the motorcycle optimum braking stability.

10.3 Influence Exerted by the Rider

The conditions and requirements described in the foregoing have to be ensured and implemented by the rider every time when applying the brakes all-out on a motorcycle without ABS. Should the motorcycle have two brake levers operating on one wheel each independently of one another, the rider has to spread out brake forces himself through his own skill. On road surfaces with good grip the rider may do without the rear-wheel brake due to its small share in overall brake forces or may obtain a relatively good result by applying the general rule of "dosing brake forces at the front and locking the brake at the rear". Being afraid of taking a fall with the front wheel locking, however, most riders give up quite a lot of the stopping distance actually at their disposal. A further point is that riders are rightly afraid of rolling over to the front and falling over the handlebars - a phenomenon which may, contrary to the situation experienced when driving a car, occur on a motorcycle under certain conditions (running gear geometry and centre of gravity) (**Fig. 10.5**).

Fig. 10.5: Rear-wheel lift-off without Integral ABS

The human aspect has been described in a number of presentations published, inter alia, by two psychologists, Professor Spiegel and Professor Dr. Eberspächer, both of them motorcycle riders themselves:

"The ideal motorcyclist should actually not have any common sense. For if he did have any common sense and started to think twice, he would suffer the same fate as the centipede who, according to the old fable, was once asked by a biologist how he was able to coordinate his countless legs. Starting in response to watch his own actions, the centipede kept on stumbling more and more. Ultimately, he lost his ability to walk and died a terrible death." [10-1]

The message is clear: The mere "knowledge" of physical conditions when applying the brakes does not make a rider a good motorcyclist. The only way to more or less grasp and command the complex brake control processes when applying the brakes all-out is through intuition and a long period of training and experience.

Clearly, the "average" rider is far from perfect in the way he applies and uses the brakes:

"A short time ago specialists at Vienna Technical University measured the use of the brakes by 110 riders when applying the brakes all-out in a training session on a Honda CB 500. The staggering - negative - result is that on average the riders used only 58 per cent of the maximum brake forces available on both wheels. Again on average, the brake force applied on the front wheel was less than 50 per cent, while the riders often locked the rear wheel. The more miles a rider covered each year on his motorcycle, the better the results became. ... Well-controlled interaction of both brakes is an important factor, many riders finding it difficult to coordinate their use of the brakes. ... The bottom line, therefore, is that practice makes perfect - and ABS helps you in the process ..." [10-2]

A recent analysis of motorcycle accidents shows that some riders had overbraked the front wheel in a panic reaction, then either falling off their machine, failing to avoid the obstacle ahead, or not stopping in time.

Marking the beginning of the "Better Braking" campaign, Professor Eberspächer made the following point recently in a presentation given in Düsseldorf: *"Suddenly confronted in a situation full of stress with an obstacle appearing up front and required to bring his motorcycle to a standstill with two brake levers acting independently of one another on the two wheels, while at the same time running on a surface quite unknown to him (a road surface with frictional coefficients varying dramatically), the human being is simply not able to meet the challenge."* [10-3]

Even experienced racing riders sometimes take a fall when applying the brakes inappropriately, although they are fully aware of the road surface and know the correct brake points well in advance.

The "Best of" test series conducted by motorcycle magazines (and often showing very good brake deceleration results on motorcycles without ABS) always involve highly experienced test riders perfectly prepared in particular for applying the brakes all-out.

Results of this kind merely confirm the potential of the brake system, but do not indicate in any way whether and to what extent the "normal" rider is able to use this potential in practice, particularly when caught by surprise in a dangerous situation or when running on surfaces with a varying frictional coefficient.

10.4 Description of the System

10.4.1 Composition and configuration [10-4]

When the rider operates the main brake cylinder on the handbrake or footbrake lever, the control plunger (**Fig. 10.6**) is moved by hydraulic pressure and passes on the force generated in this way via the control pin to the ball valve. The hydraulic pumps in the two wheel circuits start up at the same time, generating a level of brake pressure in the wheels equal to the ram pressure acting on the ball valve. This, in turn, establishes a balance of forces on the valve, the ratio between the control plunger surface and the ball seat surface determining the actual reinforcement (reinforcement factor) of the active brake function. The forces required for operating the brakes are reduced significantly through interaction with the new EVO front-wheel brake by up to 50 per cent versus ABS II.

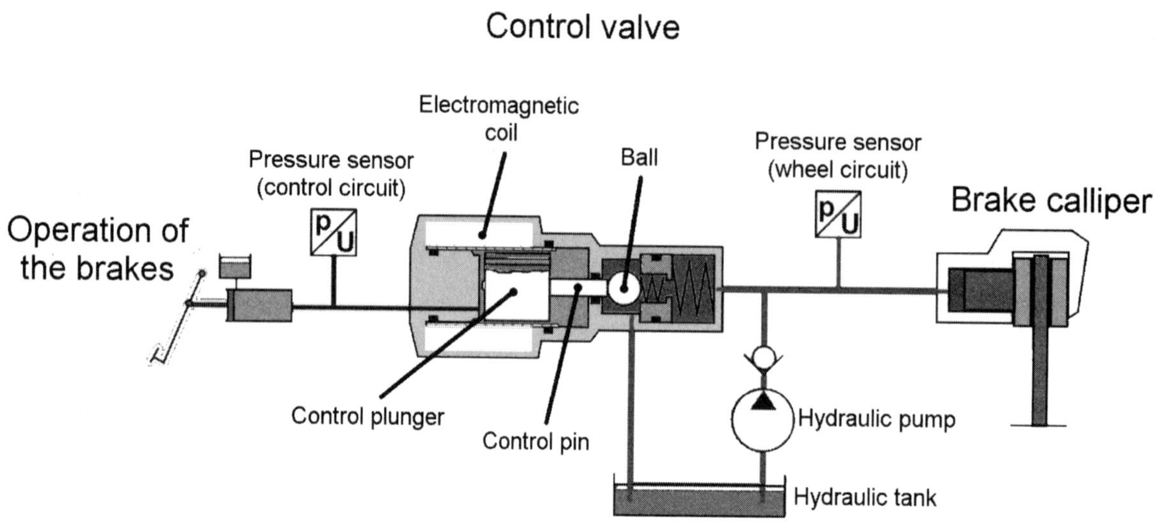

Fig. 10.6: Structure and configuration of the Integral ABS control valve

Since the hydraulic pump also provides the brake volume required in the brake system, not only the force required for operating the brake lever, but also its operation travel is reduced (ie, it becomes shorter) versus a conventional brake system. The pump is also in a position to make up for the volume of brake fluid overheating (formation of bubbles) whenever Integral ABS operates actively.

The integration of ABS into the control valves is much easier than with ABS II. Instead of the components required in the past - the plunger, friction coupling, pull-back spring and electric motor - the new system only requires an electromagnetic coil for modulating brake pressure. This electromagnet acts on the control plunger held back against the pressure applied by the rider, thus reducing brake pressure through the balance of forces.

This illustrates the substantial advantage over second-generation ABS: Incorporating fewer and much lighter components and, as a result, requiring less activation energy, the system can respond even faster to an unstable wheel by building up and subsequently reducing brake pressure.

10.4.2 Integral Brake Function

The integral function enables the driver to apply the brakes on both wheels when pulling the handbrake and/or footbrake lever. A bypass in the control circuit conveys the control pressure in the respective brake circuit through an integral plunger acting on the control pin of the other brake circuit. Through the balance of forces on the valve ball, this force builds up the brake pressure required on the respective wheel. The size and dimensions of the integral plunger depend on the brake specifications. Feeding electric power to the coil on the rear-wheel circuit, the system is able to control brake pressure on the rear wheel as a function of brake pressure on the front wheel, in accordance with the ideal distribution of brake forces in each specific case (ie, for each specific model).

BMW offers fully integral ABS on large motorcycles built mainly for touring such as the K1200 LT and R 1150 RT, both the handbrake and footbrake acting on both wheels. In the case of partly integral ABS, only the handbrake acts on both wheels, making this system more suitable for more sporting, dynamic machines.

10.4.3 Ideal Brake Force Distribution

Electronic distribution of brake forces takes the dynamic distribution of wheel loads when applying the brakes into account, monitors and "learns" the motorcycle's current load condition, and controls rear-wheel pressure in accordance with the ideal distribution of brake forces. Basically, this ideal distribution depends on the respective wheelbase and centre of gravity (current load condition).

Spreading out brake forces ideally on the front and rear wheel gives the motorcycle greater stability when applying the brakes, since both tires retain their maximum possible lateral stability. Wear on the brake pads and tires is more consistent and when applying the brakes in a bend the rider benefits from greater riding stability thanks to the more consistent, balanced use of frictional forces.

Distribution of brake forces is adaptive, meaning that the motorcycle's current load conditions are re-determined every time the rider applies the brakes in the ABS mode, the ideal brake force distribution curve being adjusted accordingly. Load conditions are determined by comparing the speed of each wheel when applying the brakes with ABS. Should one of the wheels show a tendency to lock while the other is still stable under brake pressure, the pressure on the front and rear brakes is adjusted ("adapted") until both wheels start to lock at the same time. With the motorcycle fully laden, for example, rear-wheel brake pressure is increased accordingly relative to front-wheel brake pressure.

10.4.4 Lift-Off Control

The rear wheel may lift off under a high level of frictional coefficient with an appropriate road surface/tyre combination and under high deceleration forces. Recognising certain wheel speed patterns, BMW Integral ABS is able to conclude indirectly that a rear wheel has lifted off and prevents the motorcycle from possibly rolling over by reducing pressure on the front wheel brake. Deviations from the appropriate wheel speed pattern caused by adverse factors (clutch not pulled, very bumpy road surfaces, etc) cannot be ruled out in the case of this indirect detection process. But even in such a critical, motorcycle-specific situation the system helps the rider significantly.

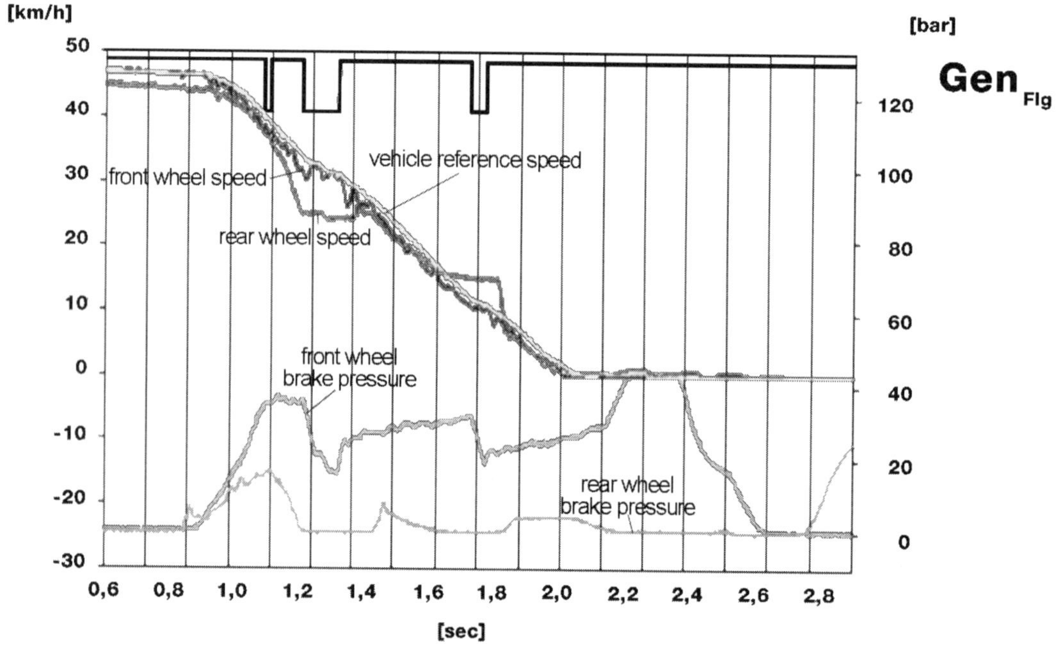

Fig. 10.7: Wheel speed curves with the rear wheel lifting off

Fig. 10.7 shows the rear wheel lifting off twice. Shortly after the deceleration process starts at point t = 1.1 sec, the system recognises the decrease in rear-wheel speed and reduces rear-wheel brake pressure accordingly. With the rear wheel not picking up speed again when pressure has been reduced to zero, the system assumes that the rear wheel has lifted off and brake pressure on the wheel is reduced until the rear wheel picks up speed again when accelerating on contact with the ground beneath. Brake pressure on the front wheel then increases in the process and brake pressure on the rear wheel is successively reduced in accordance with the ideal distribution of brake forces, now at a higher rate of deceleration. Shortly before the rear wheel lifts off again, brake presssure on the rear wheel is reduced to zero, the wheel lifts off and is no longer slowed down by the brake, wheel speed thus remaining almost unchanged.

The speed of the front wheel continues to drop in accordance with the rate of deceleration, the system once again "sees" that the rear wheel has lifted off (**Fig. 10.8**) and brake pressure at the front is reduced. The rate of deceleration is reduced accordingly, the load acting on the rear wheel increases, the rear wheel once again establishes ground contact and is slowed down by the brakes. With load now acting on the rear wheel again, brake pressure may be increased.

Fig. 10.8: The BMW R1150 RT shortly before lift-off control

10.4.5 Residual Brake Function

The control valves are designed to ensure a residual brake function even with the ignition switched off and the system not activated or in the event of Integral ABS failing to operate altogether.

While the rider has to move the brake lever further and with greater force in order to obtain the residual brake function, the brake force provided in this way still guarantees that the motorcycle can be brought safely to a standstill at any time. This also makes application of the brakes when manoeuvring a simple, unproblematic exercise.

Fig. 10.9 presents the residual brake function in two steps: If the hydraulic pump is not active, the system acting on the control plunger and the control pin will press the ball into the ball seat until it is surrounded by brake fluid in the wheel brake circuit. Then the residual brake plunger is moved, the reflow valve downstream from the pump is closed, and the system can no longer build up any brake pressure on the wheel. Since the hydraulic pump no longer delivers the brake fluid volume required, the residual brake plunger has to compensate this volume, requiring a corresponding increase in brake lever travel.

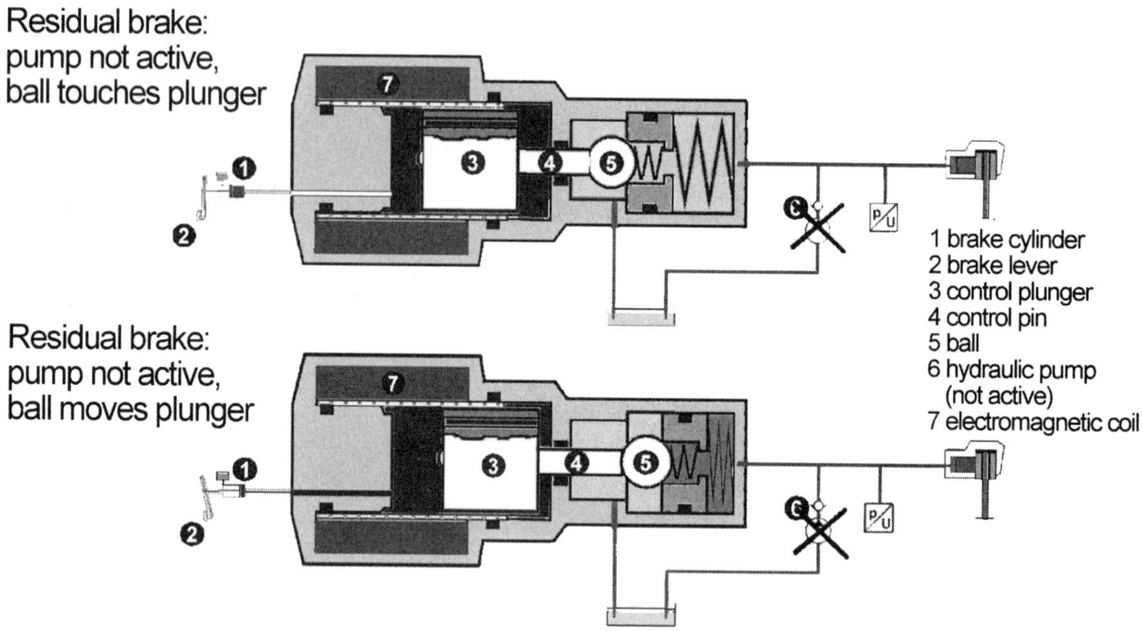

Fig. 10.9: Residual brake function

10.4.6 Diagnosis

Integral ABS fulfills the high safety standards applied by BMW. The system conducts a self-diagnosis from the start, as soon as the rider switches on the ignition. In this process two lamps are activated first, the general warning lamp (with a triangular symbol) coming on for about three seconds and going off if there is no defect in the system. The ABS lamp flashes on and off quickly for about two seconds during self-diagnosis, then switches over to a slower flashing pace until the proper function of the wheel sensors has been checked and verified at a start-up speed of approximately 4 km/h. Then this lamp also goes off and the system is fully available.

BMW Integral ABS constantly monitors itself while the motorcycle is on the road, immediately informing the rider of any malfunction.

10.5 Knowledge Gained under Practical Riding Conditions and Response from Motorcycle Riders

Introduction of a technology as innovative as BMW's Integral ABS is a very substantial challenge. The crucial point is to ensure that this technology is accepted by sensitive customers who wish to really understand "their machine" and the way it works.

Precisely this is why BMW keeps a very close eye on customer and press response.

Elaborate comparative tests and measurements conducted by MOTORRAD, Germany's leading motorcycle journal, reveal a very positive overall result: "Regardless of whether they were risky or normal riders, all testers were able to achieve a brake force in the region of 10 m/sec^2 in brake tests from a speed of 100 km/h. And thanks to Integral ABS, they were able to repeat this performance time and again, without having to be afraid of stoppies or locking wheels." [10-5]

The following three diagrams (**Fig. 10.10 - Fig. 10.12**) published in MOTORRAD [10-8] confirm this higher standard of performance and present the physical features and characteristics of Integral ABS versus ABS II:

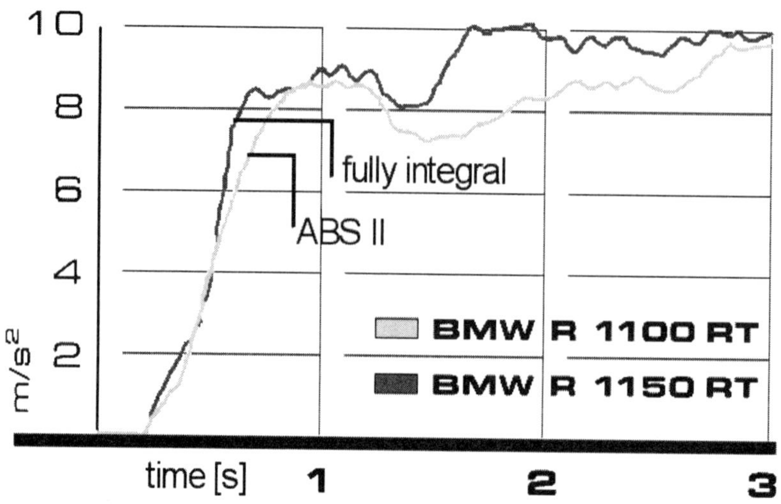

Fig. 10.10: Comparison of the response of ABS II and Integral ABS [10-8]

Responding more quickly to operation of the brakes, as shown in **Fig. 10.10**, Integral ABS starts the deceleration process at an earlier point in time. This is ensured by the hydraulic pumps in the ABS modulator building up pressure more quickly. At the same time the EVO brake system speeds up the response of the brake discs/pads rubbing against one another.

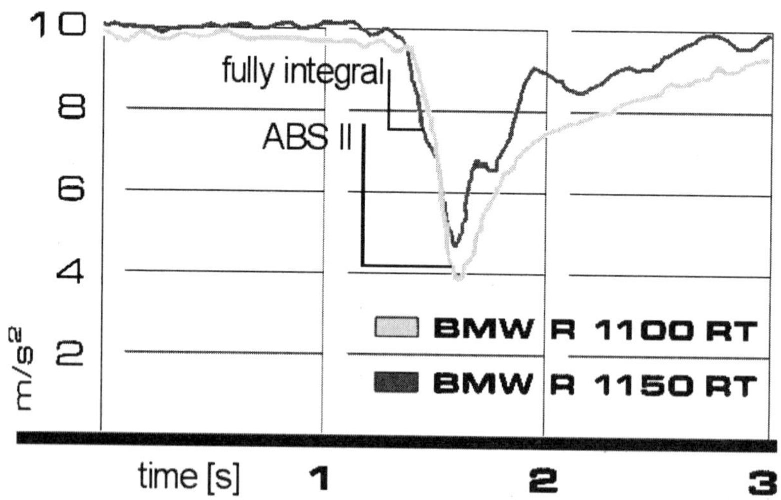

Fig. 10.11: Comparison of deceleration with ABS II and Integral ABS in a sudden change of frictional coefficient on the road surface

The new control principle incorporating lighter components allows the pressure in the system to go down and build up more quickly. The results of this more efficient process are presented in **Fig. 10.11** showing the smaller and shorter interruption in the deceleration curve of the R 1150 RT fitted with a fully integral brake system.

The faster response, better use of the frictional coefficient and simultaneous operation of both wheel brakes explain the **shorter brake distance** achieved by test riders using Integral ABS in the tests conducted by MOTORRAD, as shown in **Fig. 10.12**.

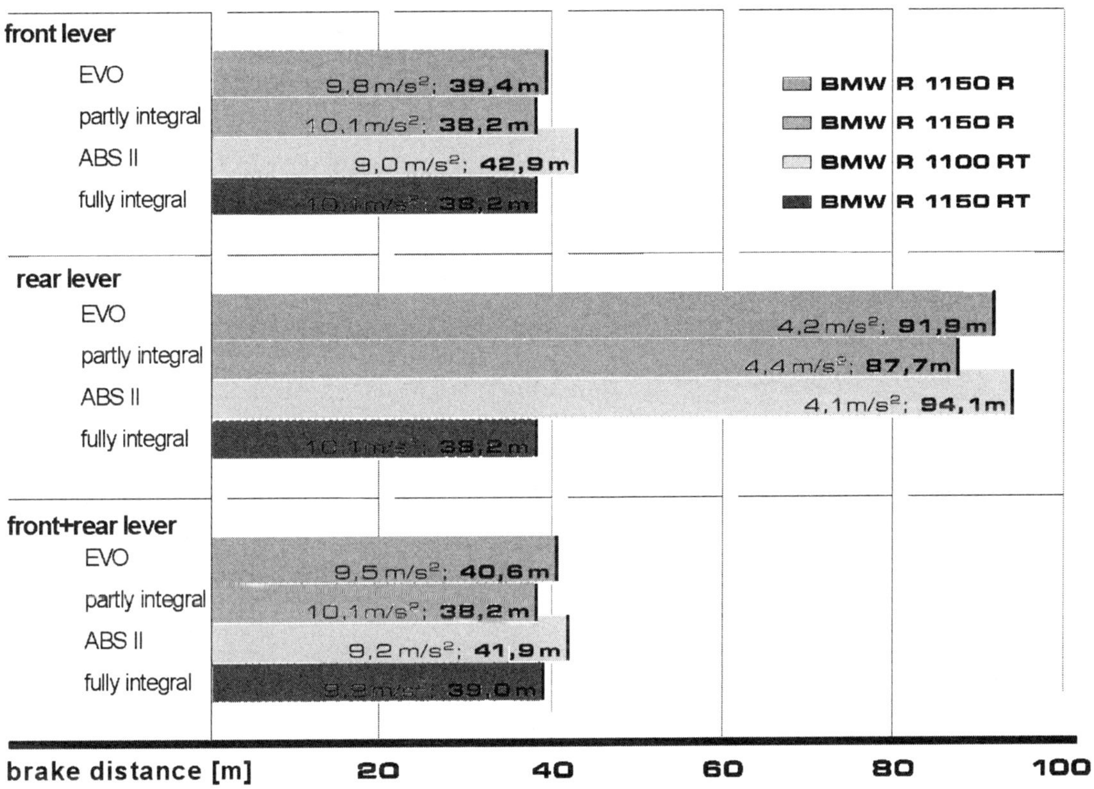

Fig. 10.12: Comparison of brake distance EVO-Brake, ABS II, Integral ABS

Customer reports in the internet show the same positive response:

"Yesterday I had the opportunity to try out partly integral ABS on a BMW K1200 RS. It was simply wonderful ...". "... true, you have to get used to it a bit at the beginning, but then you wouldn't want to miss it." "I must admit I was a bit unsure to start with because of the press. But the test ride was simply wonderful ...". "... you really get used to the brake servo and its response very quickly. And the integral system is great!". "... the only thing BMW might have got a bit better in my opinion is the way you dose the footbrake." "At the very most, I needed the first three days to get used to the system. Then everything was easy." "I have the new K with the partly integrated brake system myself - and I wouldn't take anything else any more!" "Dosing the brakes was no problem for me, I can really recommend this new brake system." "... it really is more fun now to apply the brakes than it is to accelerate!".

Dosing the brakes properly is obviously regarded as an exercise requiring some practice, particularly at very low speeds with the rapid increase in pressure within the system boosted by the brake servo. Precisely this is where BMW had to find a compromise between comfort and safety (short stopping distances).

With the footbrake on the fully integrated system generating exactly the same high level of deceleration as the handbrake, riders have to change their braking habits. Indeed, this is not surprising considering that so far the footbrake (with the exception of Moto Guzzi and the Honda Pan European, which also have a compound system) was regarded by motorcyclists more as an additional or ancillary brake due to its rather moderate effect. Now the fully integral brake system lifts the footbrake to the same function level as the handbrake. The fully integral function is in fact absolutely vital in practice for the rider who, due to a lack of experience or because he is conditioned by the way he drives his car, fails to use the front-wheel brake. In this case the comparative measurements conducted by MOTORRAD show a decrease in stopping distance from 91.9 to 39.0 metres when applying the brakes from a speed of 100 km/h, that is an improvement by 57 per cent!

10.6 Conclusions and Outlook

Introduction of BMW's new Integral ABS with the two partly integral and fully integral versions may be regarded, judging by initial customer response, as an outstanding success.

Despite criticism raised on the issue of brake dosage at low speeds, both the press and customers were very happy to accept the new system and acknowledge its benefits. Boosting brake power, introducing adaptive load distribution with the integral function, and ensuring far-reaching diagnostic options, BMW Motorrad has more than closed the technological gap to brake systems in the passenger car. And introducing this brake system step-by-step on all multi-cylinder models, BMW is furthermore showing quite clearly that safety of the highest standard should be offered not just to the purchasers of top models.

10.7 References

[10-1] B. Spiegel: Die obere Hälfte des Motorrads. Vom Gebrauch des Werkzeuges als künstliches Organ, Stuttgart 1999

[10-2] Biker bremsen mangelhaft – Gebremstes Bremsen, München 21.09.2000 (bho/tde)

[10-3] H. Eberspächer, Verhaltenswissenschaftliche Aspekte der Motorradbremsung, Auftaktsymposium besser bremsen, Essen 21.05.2001

[10-4] M. Braunsperger; S. Beulich; H.-A. Wagner: Das neue Integral ABS von BMW Motorrad, ATZ 3/2001

[10-5] Jörg Thomas, R-frischend neu, Zeitschrift Motorrad 8/2001

[10-6] Waldemar Schwarz, Verzögerungstaktik, Zeitschrift Motorrad 10/2001

[10-7] Jürgen Stoffregen: Motorradtechnik, Grundlagen und Konzepte von Motor, Antrieb und Fahrwerk, Vieweg Technik, Braunschweig/Wiesbaden 1999

[10-8] Zeitschrift Motorrad: "Verzögerungstaktik", Heft 10/01

11 SBC – The Electro-Hydraulic Brake System from Mercedes-Benz

Dr. U. Stoll, DaimlerChrysler AG

Sindelfingen, Germany

Summary

The electro-hydraulic brake (EHB) is the first generation of future „Brake-by-Wire“ systems for cars. EHB is a new mechatronic brake system, which links adaptive, digital control technology with integrated pressure sensors and high-precision hydraulic valves which are, however, fit for volume production. It permits highly dynamic, individual wheel brake pressure control which provides excellent stability and optimum short braking distances. The „Brake-by-Wire“ interface of the system permits driver-independent brake applications to improve vehicle safety.

The system is developed by DaimlerChrysler in co-operation with the system supplier Robert Bosch under the brand name SBC (Sensotronic Brake Control) to series manufacture.

11.1 Introduction

Like other components in the car the brake can be seen to be in a period of technological change. Purely mechanical or hydraulic systems are replaced or supplemented by electronic systems. The driver is gradually being supported and relieved by electronic assistance systems, which take over control and vehicle guidance tasks.

The hydraulic transmission of the brake force from the brake pedal to the four wheel brakes goes back to the twenties of the last century.

It was only from 1978 that the linkage of conventional brake technology to electronics brought further decisive leaps in the progress of development.

Electronically-superimposed assistance and control systems such as ABS (Antilock Braking System) and ASR (Acceleration Skid Regulation) for the first time took over automated partial tasks in the control of vehicle behaviour via the brakes.

The series introduction of the ESP safety system (Electronic Stability Program) in 1995 and the BAS (Brake Assist) in 1996 represent the peak in active safety with the equipment available today.

The decoupling of the mechanical-hydraulic link between the brake pedal (driver's wish to brake) and the wheel brakes represents the next important step in brake technology.

Operation of the brake pedal is thereby measured by sensors and transmitted to an electronic control unit, which then triggers the corresponding actuators (valves, electric motors) („brake–by-wire“).

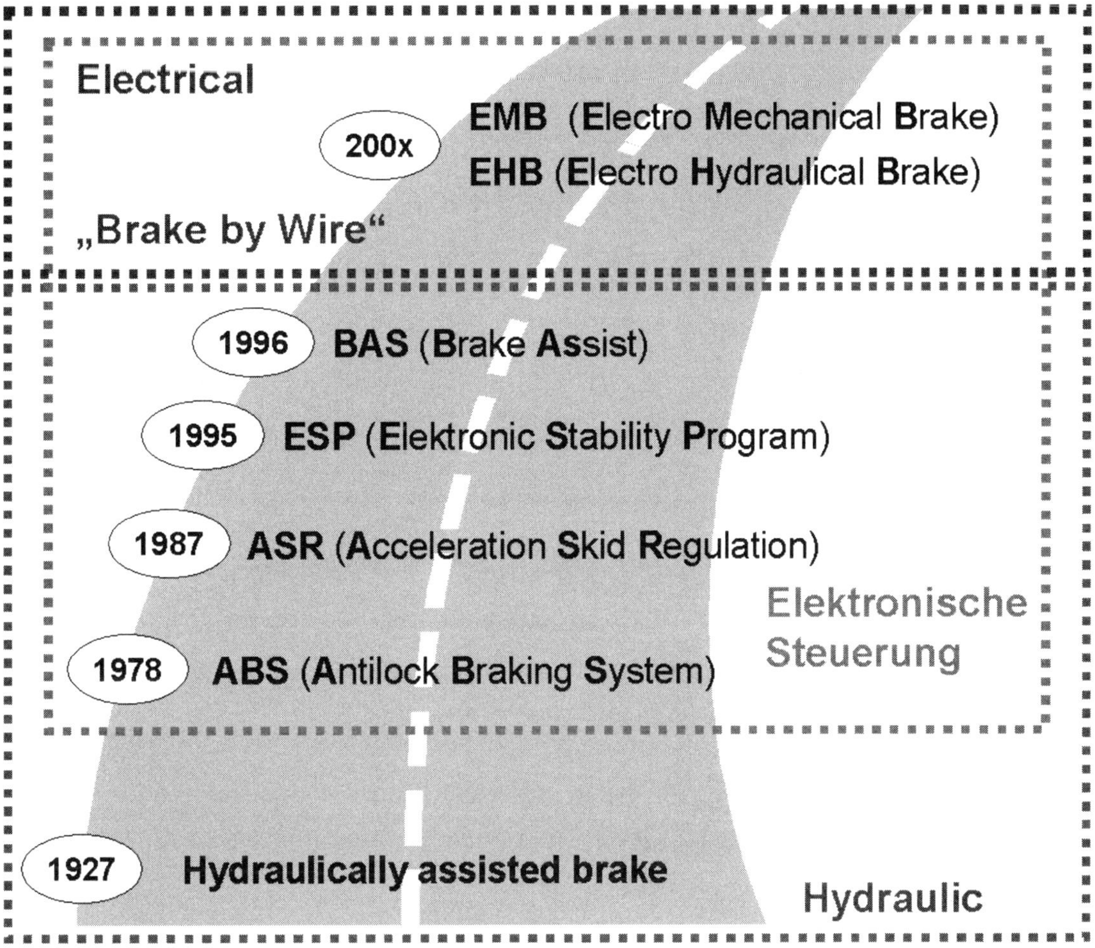

Fig. 11.1: Development of the brake system

Thereby electrical-hydraulical concepts as well as pure electrical concepts are possible:

The electro-hydraulic brake, which continues to use the well-proven hydraulic wheel brakes. Here the brake pressure is applied to the wheels from a high-pressure storage vessel, which is regenerated by an electrical pump.

Or the electromechanical brake, which operates completely without any brake fluid. The necessary brake pressure is applied to the individual wheel brakes by electrical servomotors.

11.2 Requirements of future brake systems

Apart from satisfying the legal requirements, such as fulfilling the statutory minimum deceleration if the power fails with corresponding display and warning regulations, a brake system must also satisfy the customer's needs and the needs of the vehicle manufacturer.

Above all the customer desires safety, performance and comfort from the brake system. Here neither noises (squeal) nor perceptible reactive effects (rubbing, knocking, uneven braking) are wanted. This is true for the driver braking as well as in the event of the activation of the control systems.

The desire of the customers for higher performance, low pedal forces and spontaneous response, but also high stability during sequenced braking, must be met just as much as his demand for the highest possible dynamics and stability in emergency situations.

From the vehicle manufacturer‘s point of view there are additional, - partially contradictory requirements to be added to this.

The increasing pressure on costs forces the developer towards integrated solutions, which if possible can easily be transferred to the most varied vehicles. Low air resistance and low vehicle weight to reduce fuel consumption require ever-smaller and lighter components, which for crash and design reasons must also be capable of being installed as variably as possible in the available internal areas. Thereby installation-friendly manufacturing is just as important to bear in mind as is good accessibility for repairs and servicing in connection with expanded diagnosis functions at system and component levels.

Beyond these aspects the newly developed brake systems and brake components must naturally have a guaranteed future. These systems form the basis for all future systems, which take over the tasks of longitudinal and lateral vehicle guidance.

Through the electronic networking the linkage to further vehicle systems is secured. As examples of this given here are the connection of future navigation or camera systems as well as the possibility of energy recovery (recuperation) during braking.

The brake systems are thus an elementary component of the infrastructure of future vehicles.

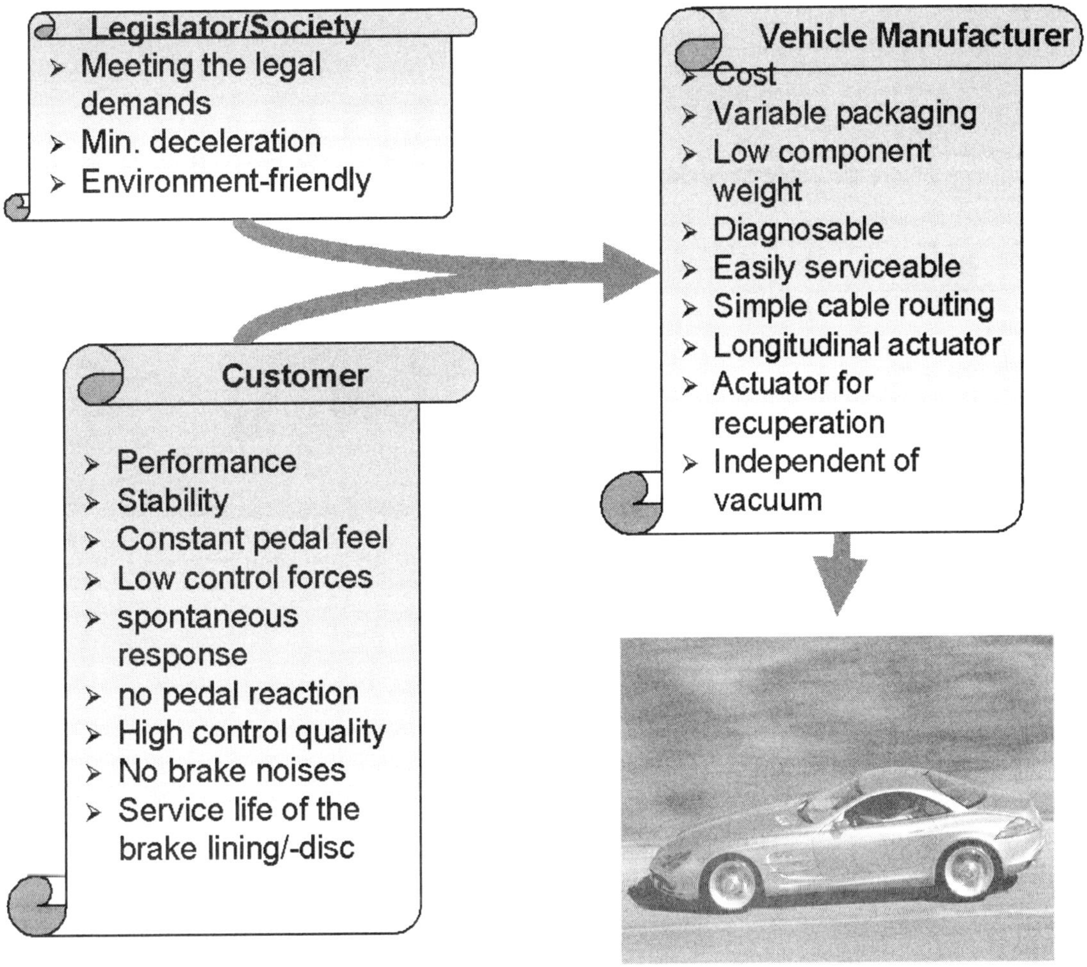

Fig. 11.2: Requirements of future brake systems

11.3 Evaluation of the system

The next development step in brake systems must create the possibility of being able to set the brake pressure in each wheel brake individually and independently of the position of the brake pedal and therefore independently of the influence and assistance of the driver. Thereby of course the vehicle deceleration during partial braking must follow accurately the braking desire of the driver and provide him with the familiar pedal feel.

The following systems have been evaluated with respect to these criteria:

- EHB (Electro Hydraulic Brakes) with central hydraulics
- EMB (Electro Mechanical Brakes) with decentralized wheel modules

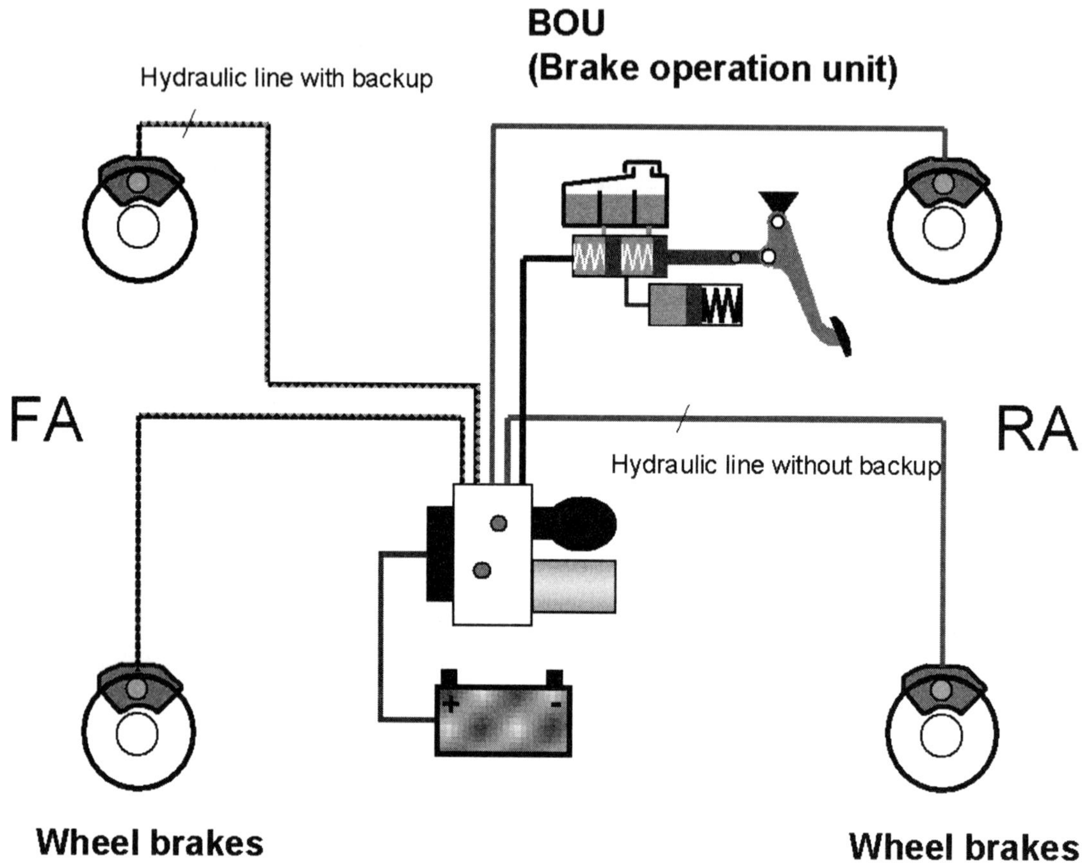

Fig. 11.3: EHB (Electro Hydraulic Brakes)

In the case of EHB the well-known and well-tried hydraulic brake components continue to be used. Unlike conventional brakes the pressure is however not applied by the driver and amplified with a low-pressure brake booster.

The pressure feed to the wheels comes from a high pressure storage vessel, which is kept constantly at operating pressure by an electric pump. The exact dosing of each wheel individually is carried out by pressure modulators assigned to each wheel via integrated pressure sensors.

The driver himself is normally decoupled from the brake by isolating valves. His braking wishes are interpreted from his actuation of the brake pedal by measurement of the path and the pressure build-up in the actuating unit. In order to be able to provide the driver with his familiar pedal feel, an hydraulic, spring-loaded pedal path simulator is connected up.

This capability of separating the hydraulic link between the driver and the wheel brake offers the advantage that if the electrical supply fails, the hydraulic separation is automatically released. The design of the hydraulic transmission in the main brake cylinder in connection with the brake pedal transmission, must thereby be selected in a way that the legally-required minimum deceleration can be achieved by the driver without amplified support.

The expenditure of guaranteeing the electrical energy supply increases for purely electromechanical brakes. In the case of these completely „dry“ brakes the hydraulics are omitted and electrical servomotors are attached in modular form directly onto the brake calipers.

However, with respect to the size and capacity of the brakes – due to the necessary high electrical power of an EMB – it will only be possible to meaningfully show the design together with the future on-board networks (e.g. 42 V). The necessary failure safety must be guaranteed by means of adequate on-board network concepts with separate energy supply and storage. Here too the familiar pedal feel during braking must be created by means of a simulator.

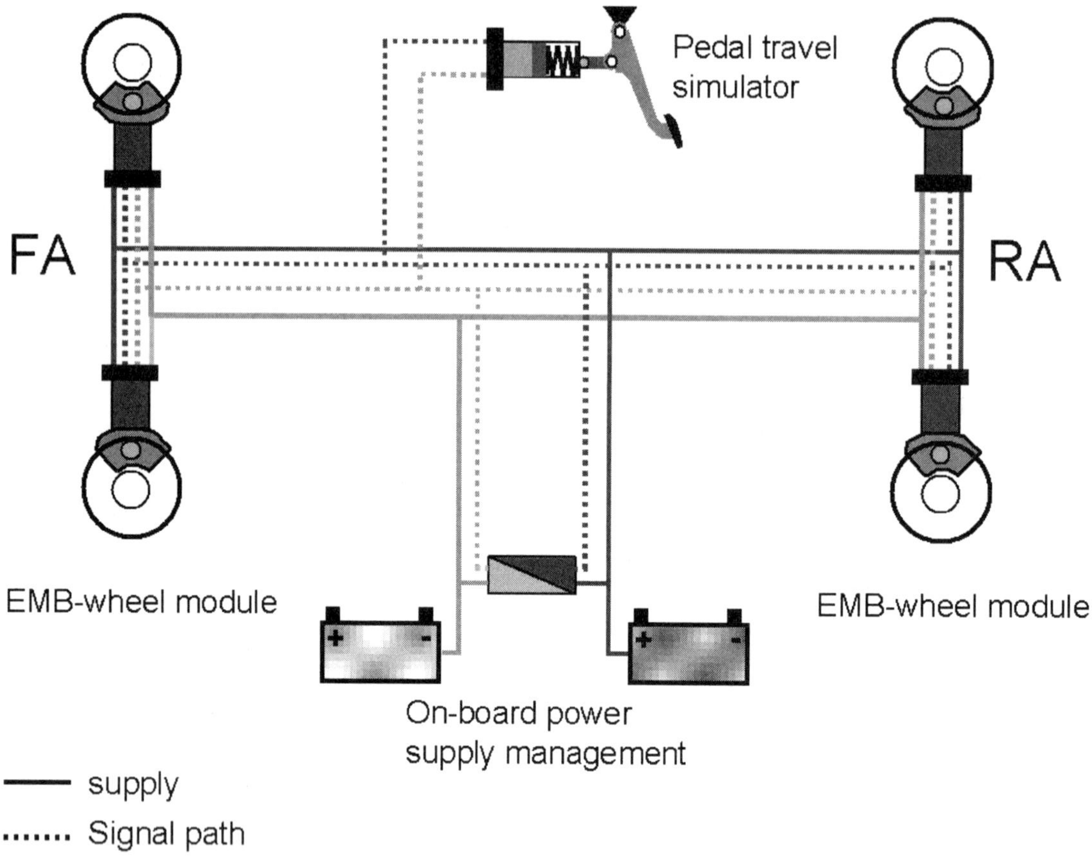

Fig. 11.4: EMB (Electro Mechanical Brakes)

The concept comparison of future „brake-by-wire" systems with conventional brakes clearly shows the limits of today's systems.

The main weaknesses of the conventional brake lie in the poor representation of a reactionless pedal feel, which is not dependent on the size of the brake booster at low actuation forces.

The strengths of the new brake systems are the high dynamics of the pressure build-up, the ability to apply sensitively the brake pressure in stages and the distribution of the braking force individually to each wheel. Herewith both new and higher performance safety systems as well as easy-to-use, automated, driver-independent longitudinal control as well as - for example - recuperation can be represented as well as providing the possibility of recuperation.

The strengths of today's conventional brake systems are in the costs of the components which have been optimized over the years through many ratio steps.

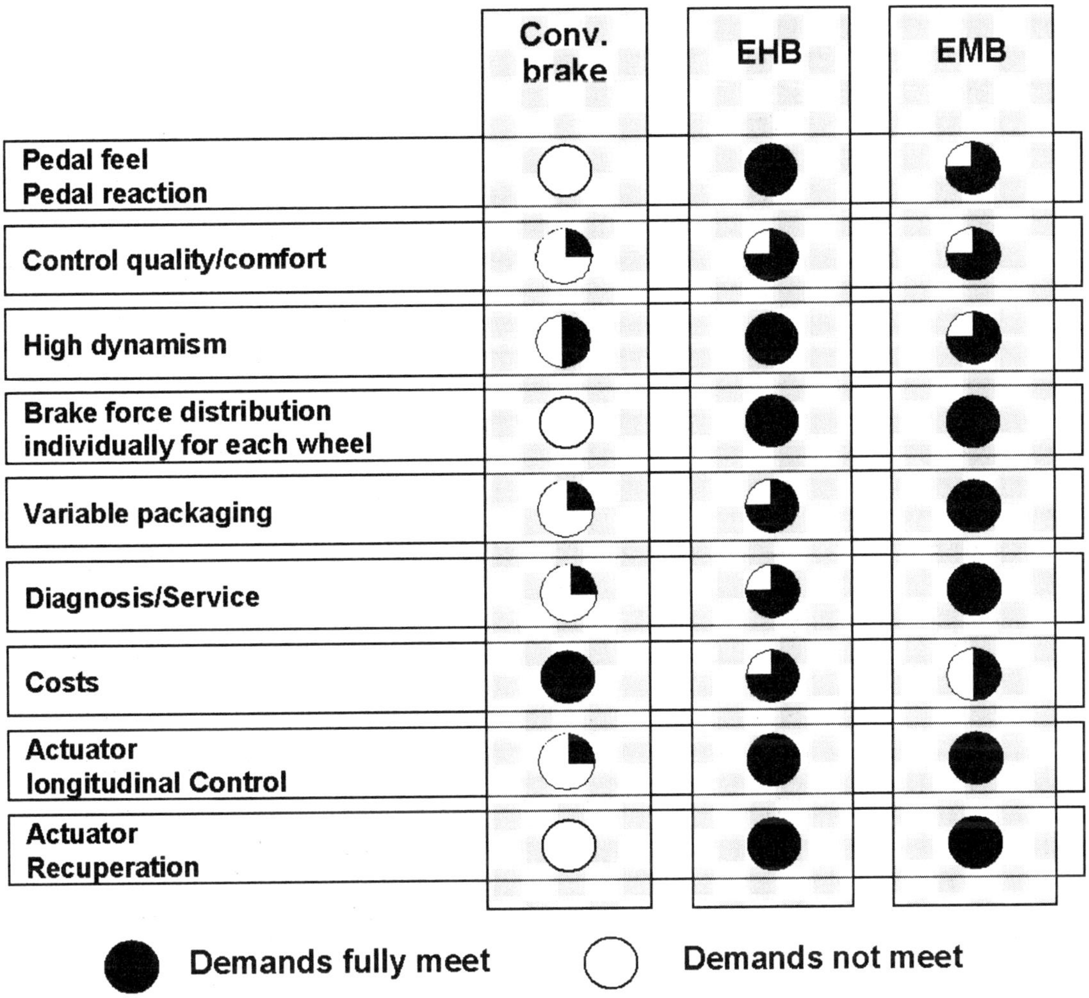

Fig. 11.5: Concept comparison (fulfilment of requirements)

The requirements of a brake system, which is guaranteed for the future, are fulfilled significantly better by the EHB and EMB „brake-by-wire“ systems overall. They only differ slightly in their individual aspects due to the different system-specific prerequirements.

If, however, this concept is evaluated with a view to the technical prerequirements in today's vehicles, then the choice clearly falls in favour of the EHB with central hydraulics and hydraulic back-up level.

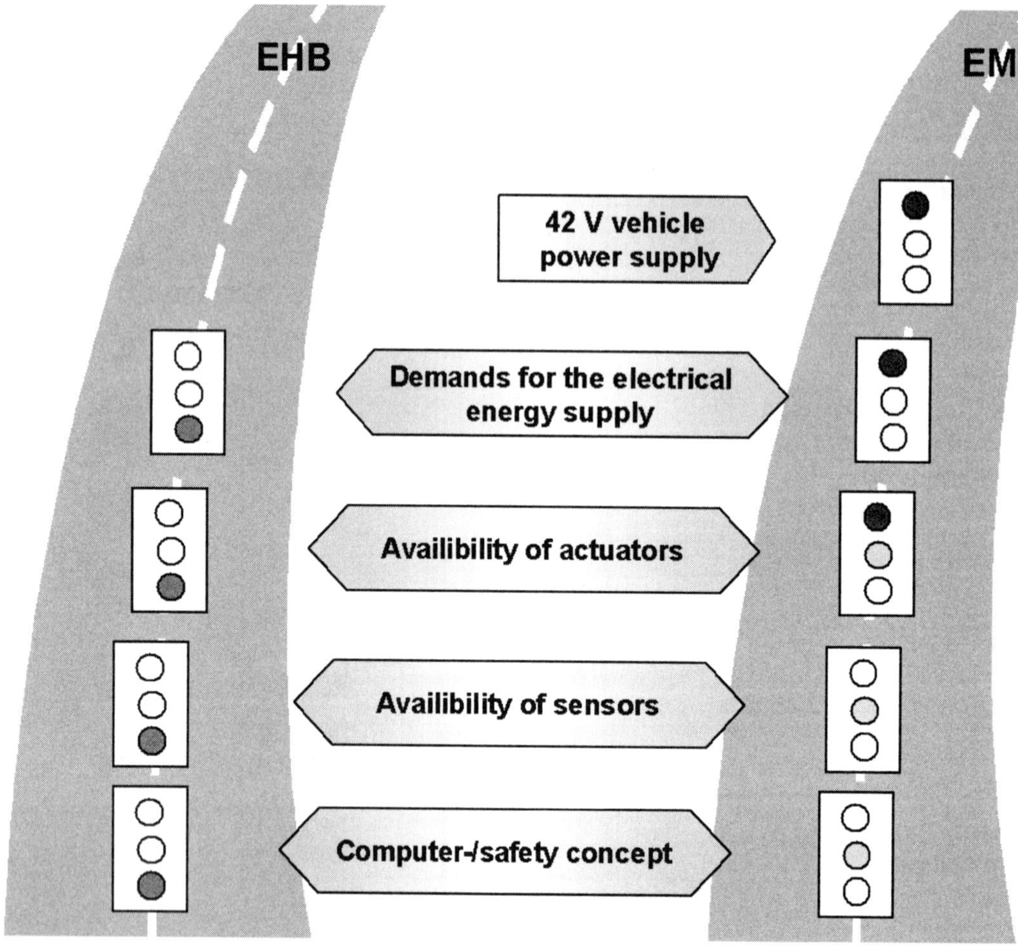

Fig. 11.6: Concept comparison (technical prerequisites)

Since mature 42 V on-board networks with the respective, series-made components such as generators and batteries are still only at the development stage, the use of electromechanical brakes will only be practicable in a few years time.

This was the reason for the decision of the Daimler-Benz company, to initially proceed along the route of the electro-hydraulic brake system.

11.4 Forms of cooperation

The firm of Robert Bosch was selected as a competent development partner and system supplier.

A joint project centre was founded together with Robert Bosch for the purpose of working on the demanding tasks from component development to vehicle integration. The objective of this was to be able to develop the complete EHB mechatronic system efficiently and with high functionality and quality in its connections and interfaces with the brake mechanics, electronic controls, new actuators and sensors as well as intelligent control and monitoring algorithms. Through direct cooperation between vehicle manufacturer and supplier it was possible at early stage to demonstrate and test the full function of the system on the complete vehicle.

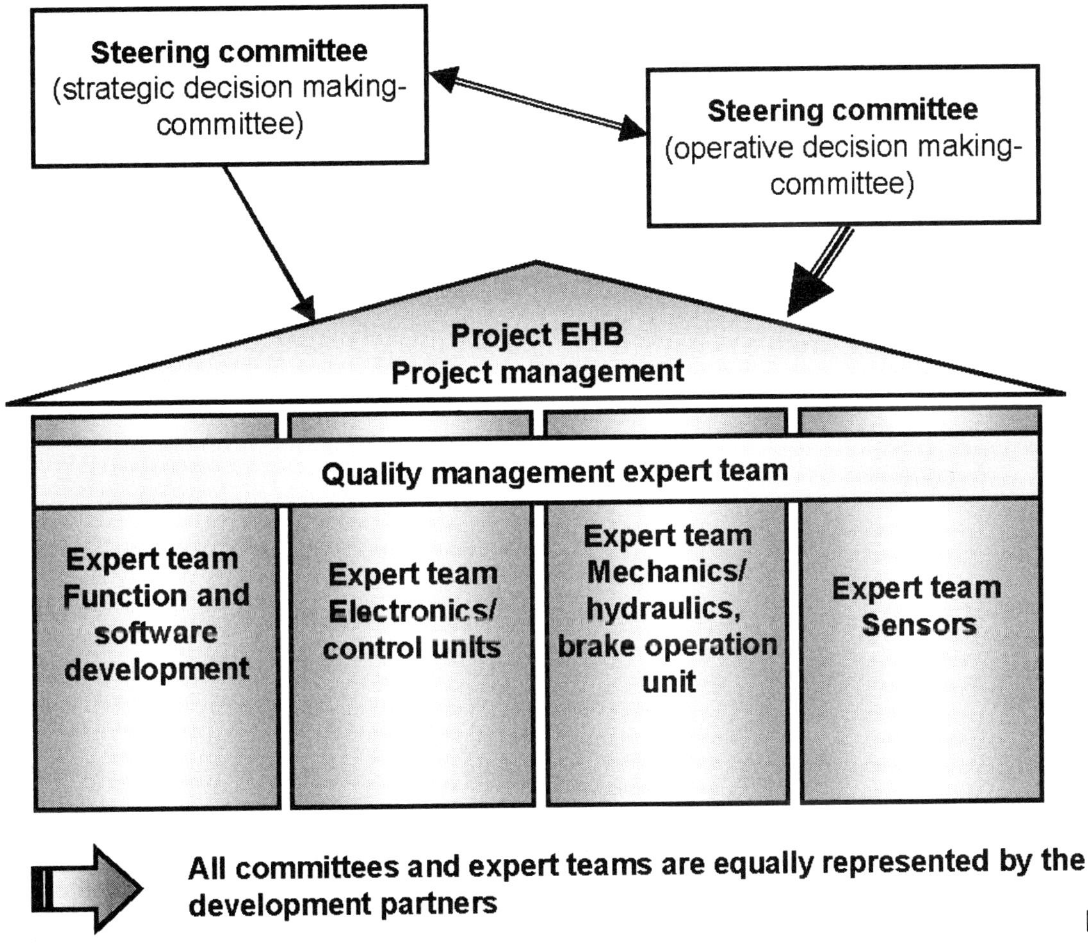

Fig. 11.7: EHB Project Centre

11.5 EHB becomes SBC (Sensotronic Brake Control)

It was only with the use of highly-sensitive sensors, which could be manufactured in a robust, integrated design in major series processes at a reasonable price, that the full and reliable function of the EHB was achievable. Beginning with the creation of the sensing of the driver's braking desires by means of the pedal path and pressure, the basic braking function is fulfilled by means of accurately adjusting the storage vessel and wheel pressures via the sensors integrated into the add-on control unit. The full functionality is achieved with the linking of the wheel speed sensors and the ESP sensors (yaw rate, lateral acceleration and steering angle sensors).

Making use of the concept of brakes, which can be electronically controlled with the help of sensors, the brand name „SBC“ (Sensotronic Brake Control) was derived from the working title „EHB“ at Mercedes-Benz.

11.6 Method of function of the SBC

SBC (Sensotronic Brake Control) is an electronically-controlled brake system („brake–by-wire“) with hydraulically-assisted braking using normal muscular force.

The central control unit of the SBC is a mechatronic device, which contains the hydraulic valves, the pump motor, the brake pressure sensors, the electronics and the software.

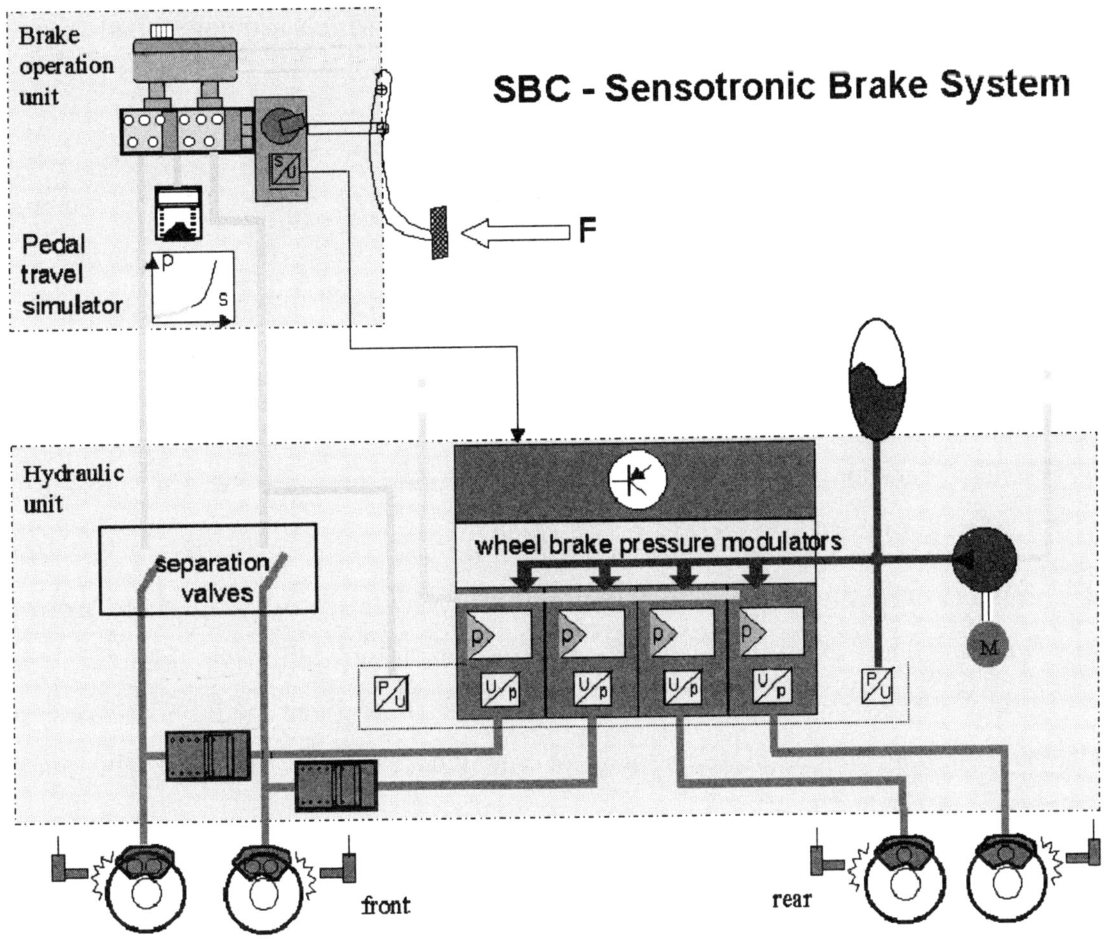

Fig. 11.8: SBC Principle diagram

Stored hydraulic energy is the external power used for the electronically-controlled operating mode. An hydraulic high-pressure reservoir contains the brake fluid volume for several braking operations. It consists of a nitrogen-filled gas diaphragm storage vessel, which is preloaded to a pressure of 140-160 bar. This is attached to the central SBC device.

The integrated control unit controls the pressure in this reservoir and controls the electromotor for driving the pump.

The brake pressure is applied from this reservoir via the wheel pressure modulators to the brake calipers of the four wheels of the vehicle. The pressures in the brake calipers are individually metered from the central SBC unit. The wheel pressure control system is represented as a closed loop circuit and uses the integrated pressure sensors.

For each modulator there is an inlet and outlet valve, which are controlled from the control unit by current-regulated power amplifiers.

The conventional main brake cylinder with vacuum brake booster is replaced in the SBC system by a brake operation unit. This includes the sensors for the sensing of the brake pedal operation, an hydraulic simulator to provide the desired brake pedal feel and a tandem brake cylinder for the manually-actuated auxiliary brake.

When starting braking, the driver is decoupled from supplying the pressure to the wheels by two isolation valves. Thereby the pedal operation is measured in terms of lever path. Together with the main brake cylinder pressure as the measure of operating force a desired deceleration is formed.

The simulator, providing the brake feel, is thereby hydraulically separated from the wheel brakes, the properties of which are normally definitive for the brake operation. Thus a braking force can be assigned to a given pedal force or pedal position – independently of the status of the wheel brakes and also the loading condition of the vehicle.

The separation of the brake fluid in the lines to the front wheel brakes from the fluid in the modulators is carried out by isolating pistons integrated into the SBC unit. In the event of a fault this, therefore, prevents mixing of the hydraulic fluid used for the back-up function with the fluid used in the active part.

11.7 The SBC power modules

11.7.1 Optimum pedal feel

One of the most decisive development objectives in the SBC development project was the representation of an optimum pedal feel for the „normal driver". Thereby the relationship between the pedal force applied by the driver by means of the pedal path and the deceleration thus established was decisive.

In the case of conventional wheel brakes the influences exerted by the brake dimensions, vehicle weight, size and the performance of the vacuum booster as well as the absorption volumes, elasticities and flow regulation of the brake components strongly restrict the design of the pedal characteristics.

By decoupling the driver from the wheel brakes in the case of the SBC, there is a greater degree of freedom for the desired pedal characteristics in the design of the simulator and the adaptation of the deceleration algorithm in the software.

The criteria for the matching of the pedal feel with the required pedal characteristics are mainly:

- **Low response forces at still high recuperation**
- **Slightly increased pedal forces with sufficient pedal travel in the metering range for targeted braking actions and in partial braking actions and during manoeuvering**
- **Harmonius transition in the progression range with a force level which makes it possible also for weaker drivers to exceed the ABS control limit**
- **Maintaining the familiar damping feeling when pressing the pedal rapidly**
- **No pedal reaction during brake control actions**

Fig. 11.9: Criteria for optimum pedal feel

In the solution shown here for a mechanical-hydraulic simulator linked to a braking unit, in which a non-linear spring assembly is compressed hydraulically via a bypass, the above requirements can be met.

Based on fundamental investigations on conventional vehicles with different brake and booster designs, it was possible to achieve an optimum pedal force characteristic by matching the spring assembly in the simulator. The deceleration as a function of the pedal force and travel, actually adjusting at the vehicle, is determined by the deceleration algorithm in electronics. A small deceleration in high pedal counter acting force subjectively leads to a „blunt“ brake, a high deceleration with low pedal forces results in subjectively „sharp“ brake, badly to be dosed. Determination of the deceleration characteristic was done by averaging the scatter range of the subjective evaluation of decision-relevant experiments.

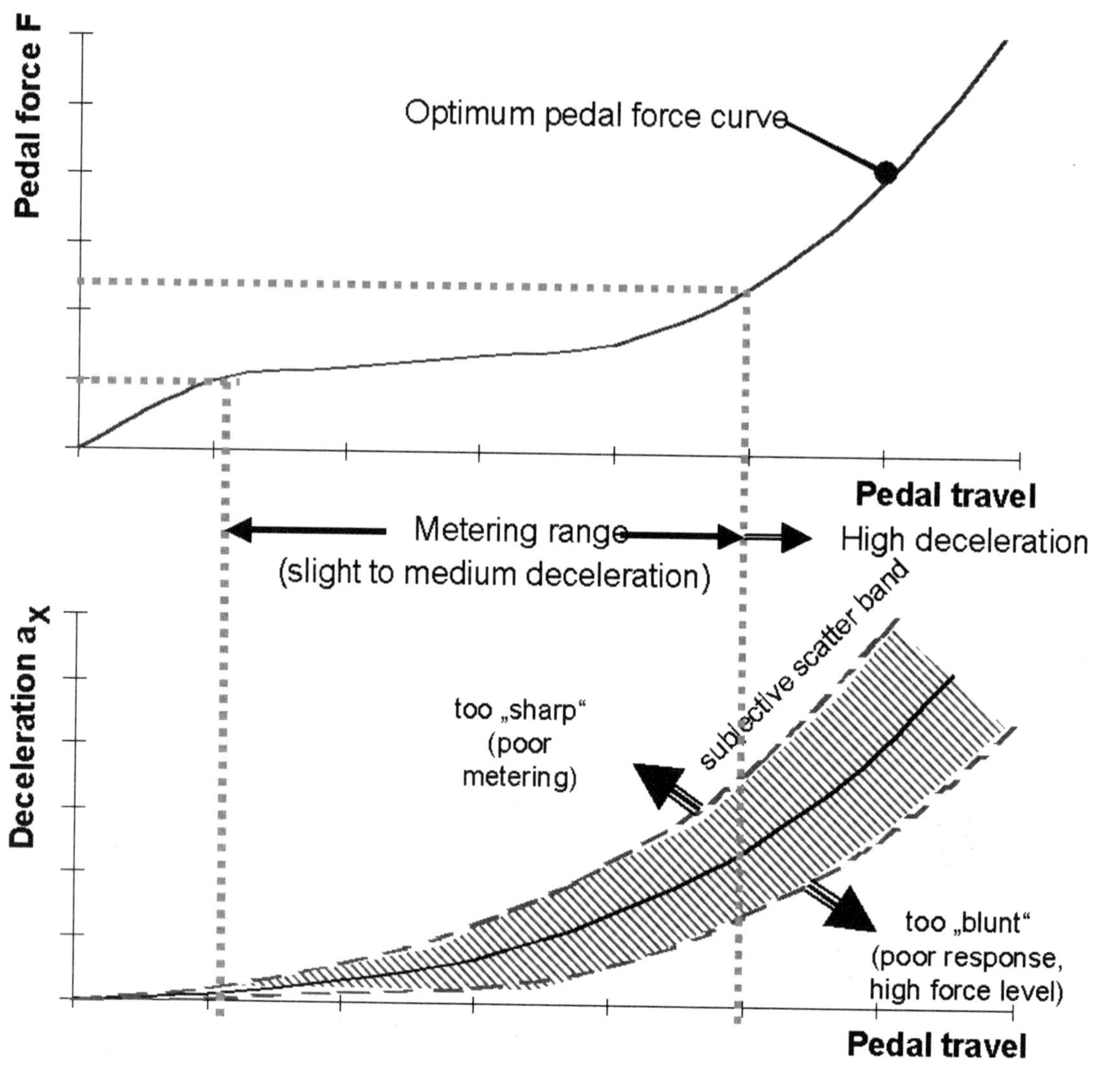

Fig. 11.10: Design range of deceleration characteristic

Due to the very close relationship between the above mentioned characteristic and the deceleration characteristic the reverse conclusion can be drawn as shown in **Fig. 11.11**, that for a given deceleration characteristic of the series software, only a tolerance range of the pedal force characteristic, which must be accurately defined, is also permitted.

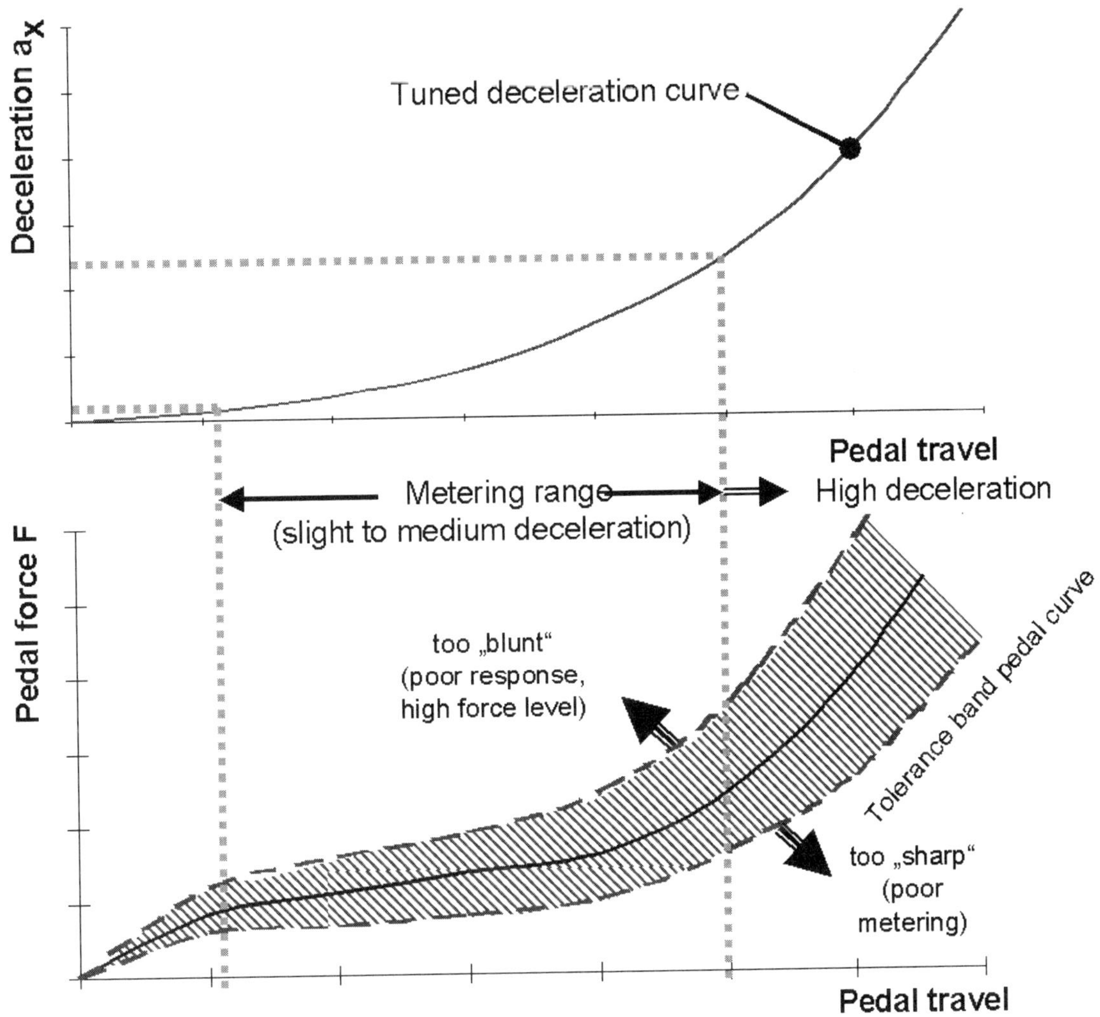

Fig. 11.11: Series scatter range for pedal force characteristic

11.7.2 Pedal reaction during controlled braking

With the decoupling of the driver from the pressure-modulating part of the brake system it is possible for the first time with the SBC to eliminate the ABS pulsing, which disturbs many drivers. Investigations in the Mercedes-Benz driving simulator show that a high percentage of drivers, in spite of experience with ABS, do not increase the pressure further once they feel the ABS pulsation starting, or they may even sometimes reduce the pressure. Since not all the wheels can thereby be kept at the optimum slip limit, valuable braking distance is thus lost.

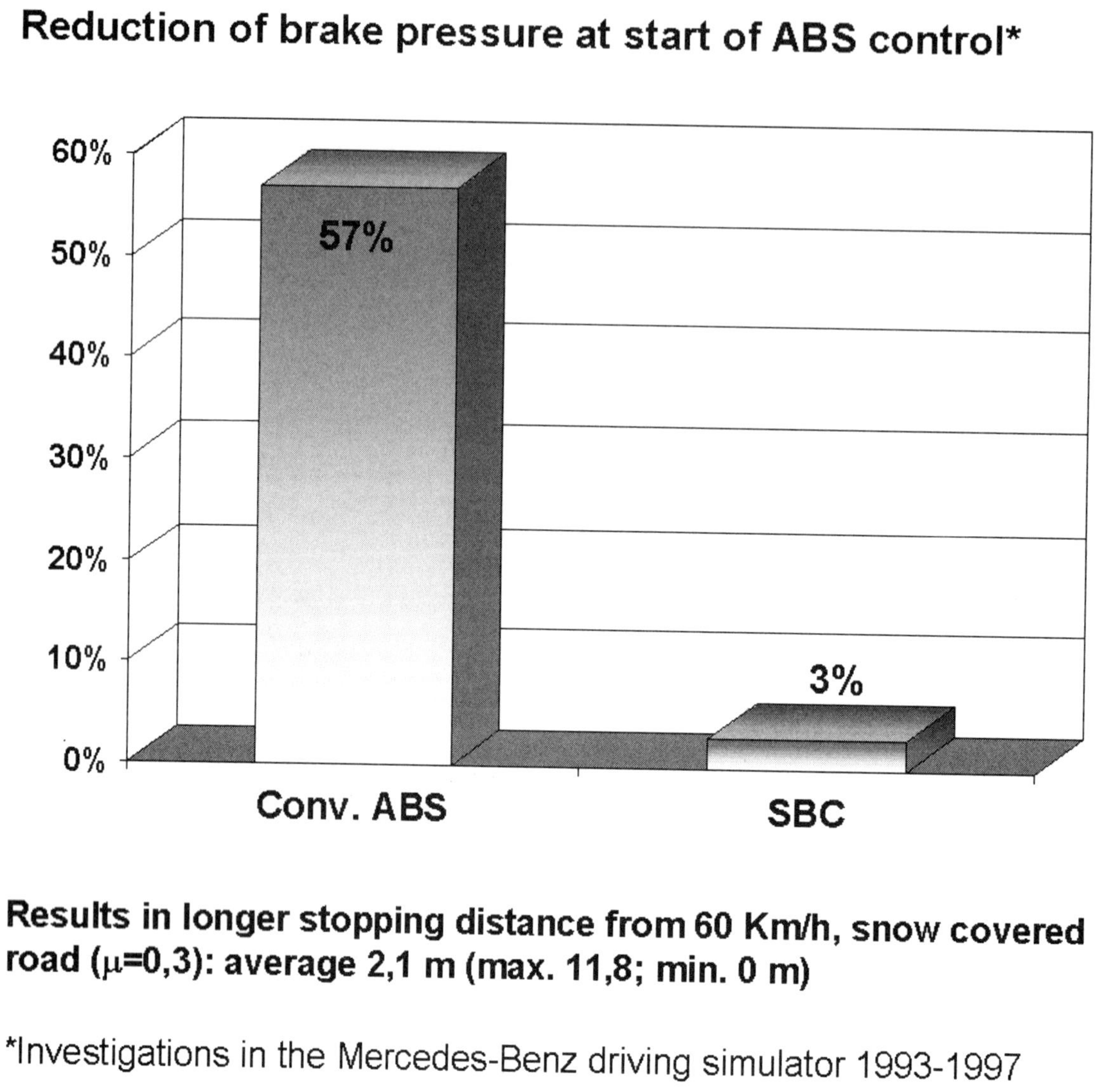

Fig. 11.12: Safety-gain by omitting the ABS reaction on the brake pedal

11.7.3 Improved pressure build-up dynamics

In the case of conventional systems the pressure build-up gradient in the event of an emergency brake application – even when supported by braking assistants – is limited by the damping and flow restriction in the brake booster and in the ABS or ESP hydraulic unit.

On the other hand the SBC in this situation can feed both the necessary volumes as well as the desired pressure to the wheel calipers from the high-pressure reservoir in a very short time. In an emergency braking operation introduced by the driver this leads to a faster build-up in deceleration, which results especially when braking from high speed in measurably shorter stopping distances. In the case of automatically-triggered ESP or ASR control, SBC is significantly superior to today's systems due to its higher dynamics. Through fast, sensitively dosed brake applications an out-of-control vehicle can easily and quickly be stabilized.

11.7.4 Ideal braking force distribution

In today‘s car braking systems overbraking of the rear axle must be avoided for driving stability reasons. Critical driving manoeuvres in this area involve braking from high speed, in which the rear axle slips due to a too low power take-off. Just as critical is when braking is carried out on curves, in which the adhesion of the rear axle can decrease due to the slippage of the inner wheel on the curve with a high lateral acceleration and possibly additional alternating load reactions. To avoid this in the design work the brake system for the front axle is given an overproportionally-high percentage. This fundamental design is also useful for SBC.

With SBC, however, due to the option of being able to distribute the brake forces as desired to the four wheels, there is the advantage - in uncritical partial braking ranges or at low speeds – of using the braking force reserves on the back axle by raising the back axle proportion. This results in uniform wear of the brake pads and improves the response of the full braking system.

A significant gain in safety is achieved during braking in curves. By specific distribution of the brake forces to the outer curve wheels, which due to the wheel load bearing arrangement can carry higher loads, and with a reduction on the inner curve wheels, a stable braking behaviour is possible with optimum deceleration values.

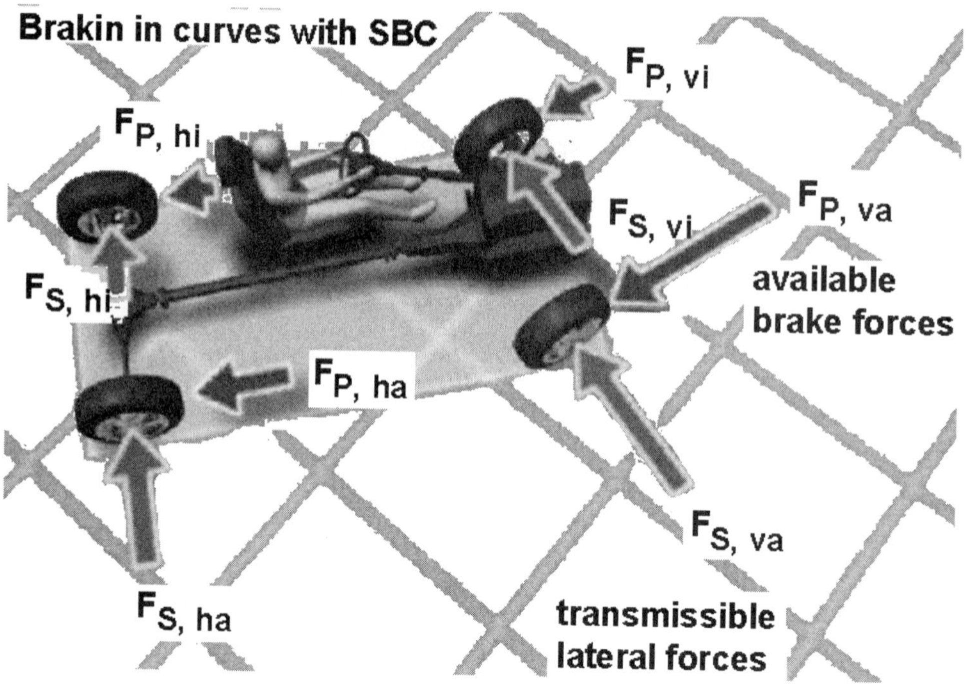

Fig. 11.13: Ideal braking force distribution

11.8 Outlook

The change in brake system technology is in full swing. On the path to mechatronic systems the SBC can be recommended as the successor to the well-proven hydraulic brake components.

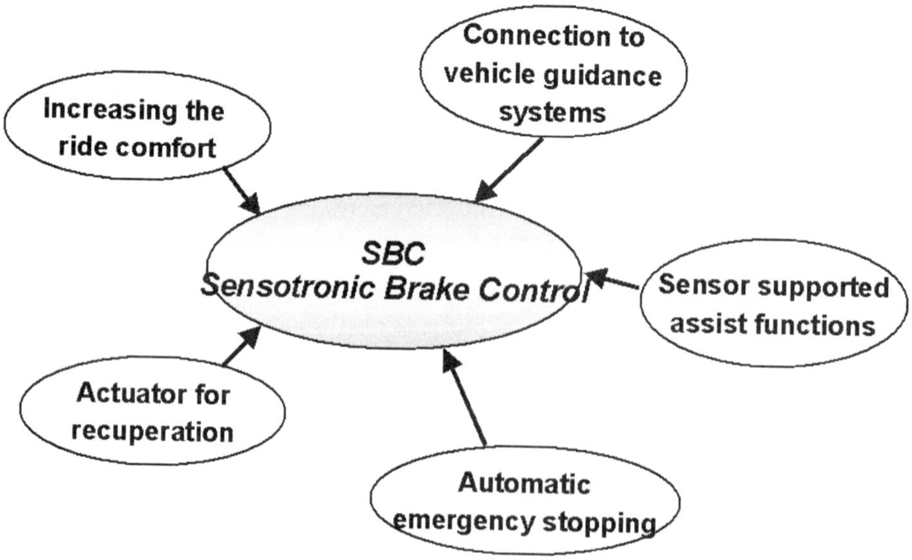

Fig. 11.14: Future options for SBC

With this technology and the option thus provided of being able to stabilize or brake a car independently of the driver, far-reaching perspectives are opened up through the linking with or the connection of other intelligent systems inside or outside the vehicle.

11.9 References

[11-1] Schunck, E.; Stoll, U.: Das Sicherheitskonzept einer Elektrohydraulischen Bremse; VDA-Vortrag, 1999

[11-2] DaimlerChrysler: Die intelligente Bremse; in: HIGHTECH Report'99, S. 52-53, 1999

[11-3] Jonner, W.-D., Winner, H., Dreilich, L., Schunck, E.: Electrohydraulic Brake System – The First Approach to Brake-by-Wire Technology; SAE Technical Paper Series, No. 960991, 1996

[11-4] Eglmeier, P.: Untersuchung des Zusammenhangs Pedalkraft/weg und Fahrzeugverzögerung auf der Basis von Feldversuchen; Dipl.-Arbeit; Daimler-Benz AG, 1997

[11-5] Stellner, U.: Meßsystem- und Verfahrensentwicklung zur meßtechnischen Beurteilung des Bedienkomforts einer Fahrzeugbremse; Dipl.-Arbeit, Daimler-Benz AG, 1995

[11-6] Kuhn, F., Greiner, M., Grezlikowski, H.: Alternative ABS-Rückmeldung; Untersuchung im Daimler-Benz Fahrsimulator, Berlin, 1997

[11-7] Faulhaber, Grezlikowski: Einfluß der ABS-Rückwirkung am Bremspedal auf das Bremsverhalten von Normalfahrern; Untersuchung im Daimler-Benz Fahrsimulator, Berlin, 1993

12 New mechatronic part disc brake

Dipl.-Ing. M. Semsch, Darmstadt University of Technology

Darmstadt, Germany

Abstract

Brake-by-Wire systems will be gaining in importance in future vehicle concepts since they simplify the wheel-selective brake application. However, the economically reasonable realisation of these braking systems will be depending on the system's benefits, such as dynamic behaviour, energy and power demand, space required, mass, and reliability. Especially the connection between actuator and friction brake through a suitable transmission system, which meets the requirements of a vehicle brake, appears as a problem that is not yet solved adequately. For this reason, a mechatronic wheel brake is being developed at the Automotive Engineering Department Darmstadt (*fzd*) within Special Research Department 241 IMES, supported by the Deutsche Research Association [12-12]. Current research topics are both the improvement of the efficiency of several components, and the use of the vehicle's kinetic energy to actuate the brakes. Either shows potential for reducing the absorption of electrical energy of the actuator, controlling the self-amplification effect with the help of the mechatronics.

Fig. 12.1 shows the history of electromechanically actuated wheel brakes. The author does not claim that this history is necessarily shown in its entirety.

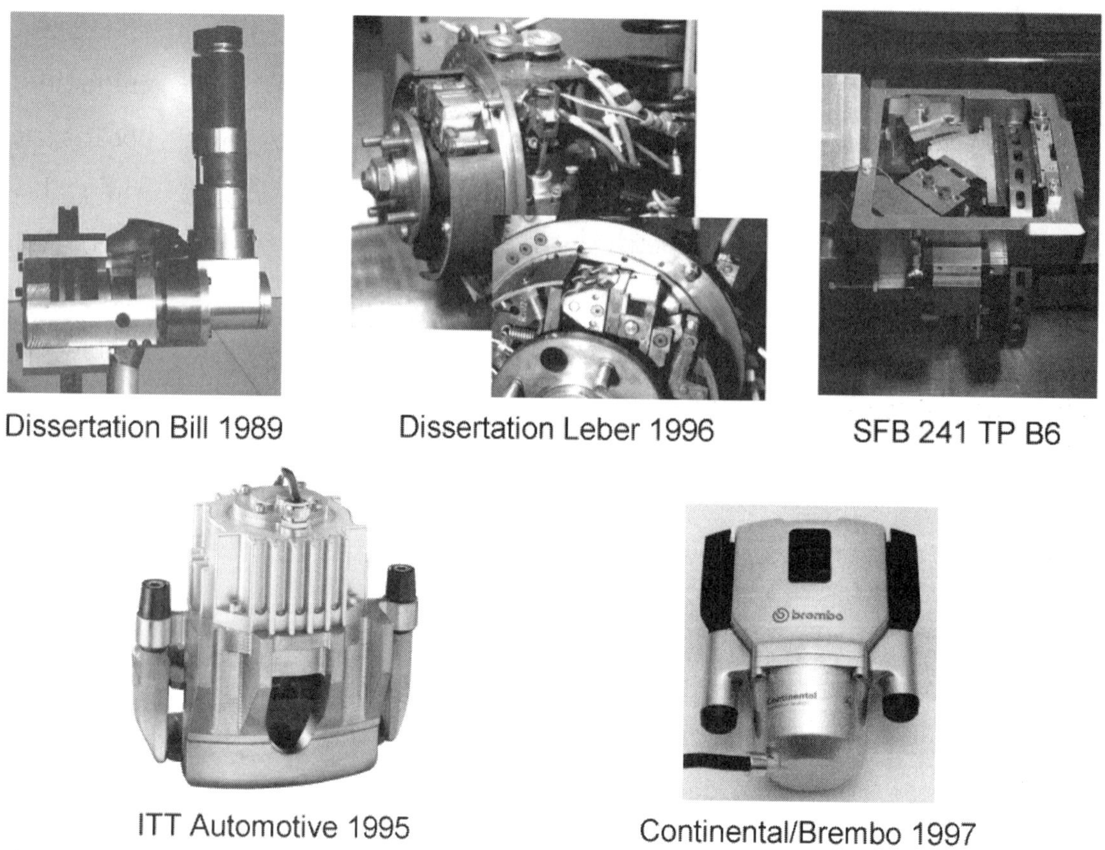

Fig. 12.1: Mechatronic wheel brakes

12.1 Introduction

The increasing function potential of current braking systems and the omission of beneficial auxiliary energy (vacuum), due to new types of engine concepts requires new methods in braking technology. Future vehicle concepts will require braking systems, which operate without complex auxiliary energy components and which have a suitable interface to higher level vehicle systems. If one considers the efforts made towards supporting the driver through new electronic assistant systems, the necessity for individual vehicle sub-systems being able to communicate with each other is clear. As an example, we are quoting the function of the distance controller (ICC) with easy external intervention in the braking system. Such a Brake-by-Wire system, in conjunction with the tire and chassis sub-systems, can achieve optimum comfort, safety and driving dynamics.

Electrically actuated braking systems, due to their vacuum independence, are particularly suitable for use in modern vehicles with optimised combustion or electric drives. Another advantage is the dynamics, since the brake booster, as a transmission component in the chain, with its specific contact behaviour, has a negative effect on the dynamics. The man/machine interface, due to the "by-Wire" connection, is independent on the mechanical components. This creates the opportunity to realise a better adaptation to the "man" element, thereby increasing both active and passive safety [12-4].

Fig. 12.2 shows a general wiring diagram of an electron-mechanical braking system. This type of modern braking system with maximum function potential can be structured by modules using standardised hardware. By omitting the hydraulic brake fluid, the system – with the exception of the brake lining – becomes maintenance free and therefore essentially more eco-friendly.

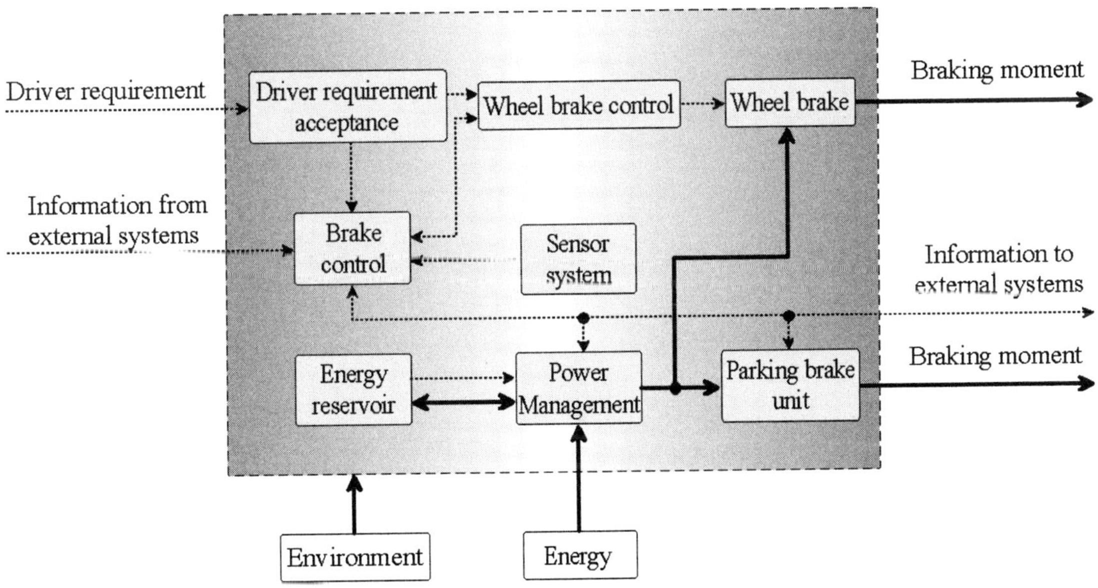

Fig. 12.2: Electromechanical brake system

12.2 General structure of an electrically actuated friction brake

In the widest sense of the word, an electrically actuated vehicle brake represents an electromechanical transducer with lengths and strengths, which vary over time.

Input variables are the electrical energy and the data flow to control the system; the output variable is the required braking moment, **Fig. 12.3**. For a more precise description of the mechatronic system, the function and energy flow inside the wheel brake must be further divided:

- ACTUATOR: Converts electrical energy into mechanical energy.
- GEARS: Creates the compatibility between the actutaor and the braking mechanism.
- BRAKE: The brake represents the friction brake, whereby the tension force is provided, the braking force created, the braking moment produced on the rotating friction surface and the braking energy deflected.

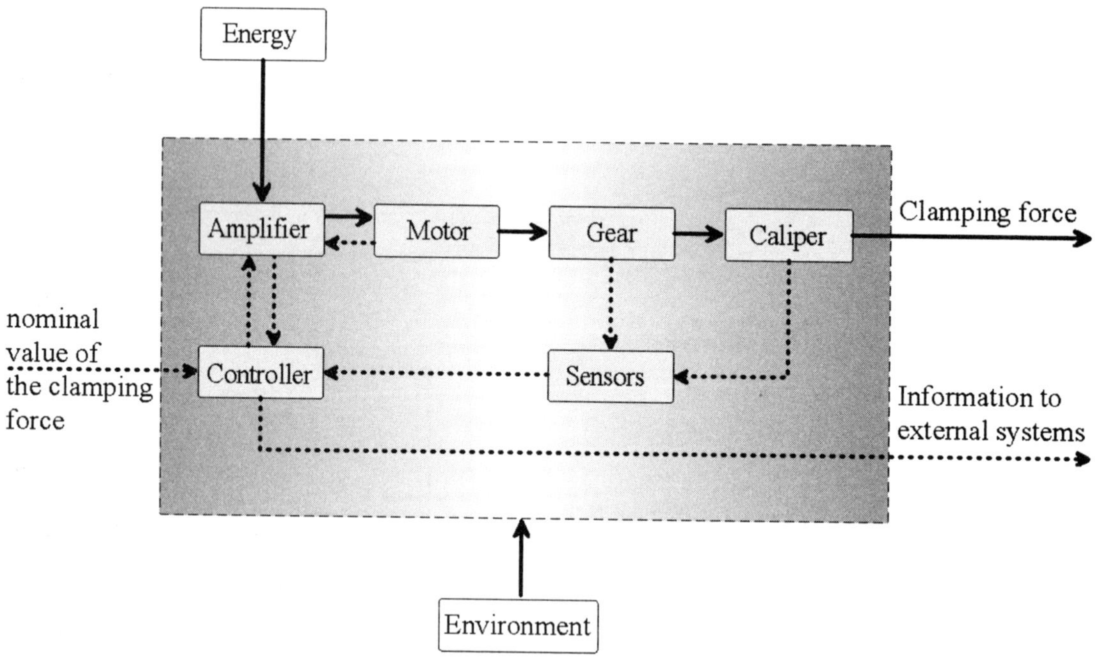

Fig. 12.3: Electromechanical wheel brake

Several options exist for realising the electrically actuated wheel brakes in terms of selecting the structural main components – actuator, gears, brake. The components may be chosen in accordance with the specified structural objectives, whereby particular application-related requirements (temperature, contamination, acceleration, time behaviour...) are to be taken into consideration.

12.2.1 Problems with electro-mechanically actuated wheel brakes

In contrast to conventional brakes, where the auxiliary energy is converted to hydraulic energy through additional components (vacuum brake booster, pumps) in the engine compartment, the electromechanical wheel brake must convert the auxiliary or external energy where the wheel is fitted into mechanical tension energy. Since the energy and performance density of the hydraulics is higher than that of the electro-mechanics, the required fitting area and the weight of the previous prototypes is far higher than that of conventional hydraulic wheel brakes.

The efficiency of the electro-mechanical transducer and gears of such wheel brakes, which is lower than that of the hydraulic system, has too high a capacity and energy take-up for current vehicle on-board networks. Furthermore, the costs of known solutions are unacceptable as a result of using very high value components for subsequent series use. The issue of integrating with other vehicle sub-systems which relates to the conversion of auxiliary force brake systems to external force brake systems, also requires further clarification, along with failure safety.

12.3 Solution

Fig. 12.4 shows the mathematical derivation of the actuation energy to tension a part disc brake, whereby the basic structure of a mechatronic wheel brake, consisting of actuator, gears and friction brake are considered in general. The result shows the "bolts" which can be turned to minimise the actuation energy and therefore the actuation efficiency. The required peripheral force is provided by the vehicle and can be only slightly influenced by the brake construction. Likewise, the tension path can be only slightly influenced by the necessary play and the rigidity present in the system.

Thus, as a solution for total mechanical efficiency, the product of the individual efficiencies for all used components and the braking value C * are required. These two "bolts" are used to provide a more detailed description.

12.3.1 Consideration of efficiency

From the point of view of the electro-mechanical transducer, an optimum efficiency adjustment to the consumer (brake) is difficult. At the start of braking actuation, the air play must be surmounted. Normally, only slight forces are required here. As the braking force then increases, significantly stronger forces are required for approximately the same actuation paths. Since the creation of the required tension force is of paramount importance, the gear system must be designed accordingly. In terms of air play range, optimum adjustment is not possible for gears with constant translation, which has a negative effect on the brake actuation time. Optimum adjustment to the consumer can take place only in connection with non-constant translated gears or two actuators responsible for each operating area of the brake.

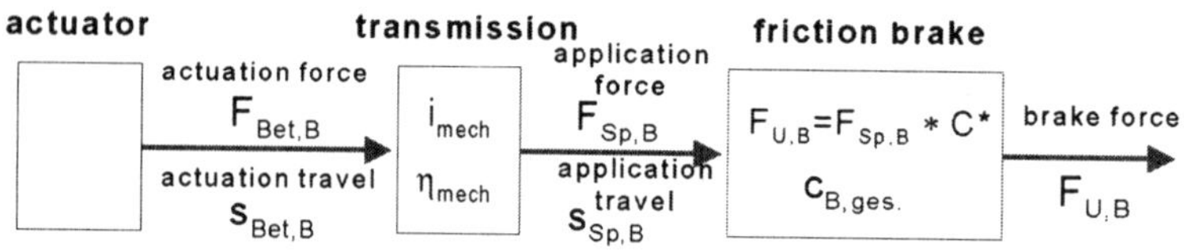

$$W_{Bet,B} = \int_S F_{Bet,B} ds_{Bet,B}$$

with: $$i_{mech} = \frac{s_{Bet,B}}{s_{Sp,B}} = \frac{ds_{Bet,B}}{ds_{Sp,B}}$$

$$\eta_{mech} = \frac{F_{Sp,B} \cdot s_{Sp,B}}{F_{Bet,B} \cdot s_{Bet,B}}$$

$$\rightarrow F_{Bet,B} = \frac{F_{Sp,B}}{\eta_{mech} \cdot i_{mech}}$$

$$F_{Sp,B} = c_{B,ges} \cdot s_{Sp,B}$$

$$W_{Bet,B} = \int_S \frac{c_{B,ges} \cdot s_{Sp,B}}{\eta_{mech}} ds$$

$$W_{Bet,B} = \frac{c_{B,ges} \cdot s_{Sp,B}^2}{2 \cdot \eta_{mech}} = \frac{F_{Sp,B} \cdot s_{Sp,B}}{2 \cdot \eta_{mech}}$$

and $F_{U,B} = F_{Sp,B} \cdot C^*$ with: $C^* = f(\mu)$

$$\boxed{W_{Bet,B} = \frac{F_{U,B} \cdot s_{Sp,B}}{2 \cdot \eta_{mech} \cdot C^*}}$$

Fig. 12.4: Actuating energy

12.3.1.1 Transducer and brake coupling

For purposes of simplification, the wheel brake is a spring, whose strength c and length s is non-constant over the operating time due to wear and thermal load. The dynamic self-behaviour of the transducer, in connection with the requirements of the brake actuator shown in [12-2], in an ideal situation, requires a very complex and varying adjustment of the gear translation to consider the non-constant rigidity and geometric relationships of the brake (model of non-variant spring).

The continuous line in **Fig. 12.5** shows one possible course of the actuation force in relation to the actuation path. This shows the progressive force increase during the transition area from air play to tension force. As wear to the lining increases, this transition shifts to longer actuation paths. The course of the force increase changes over temperature and is dependent upon the mechanical brake rigidity and also compressibility of the brake lining (dotted line).

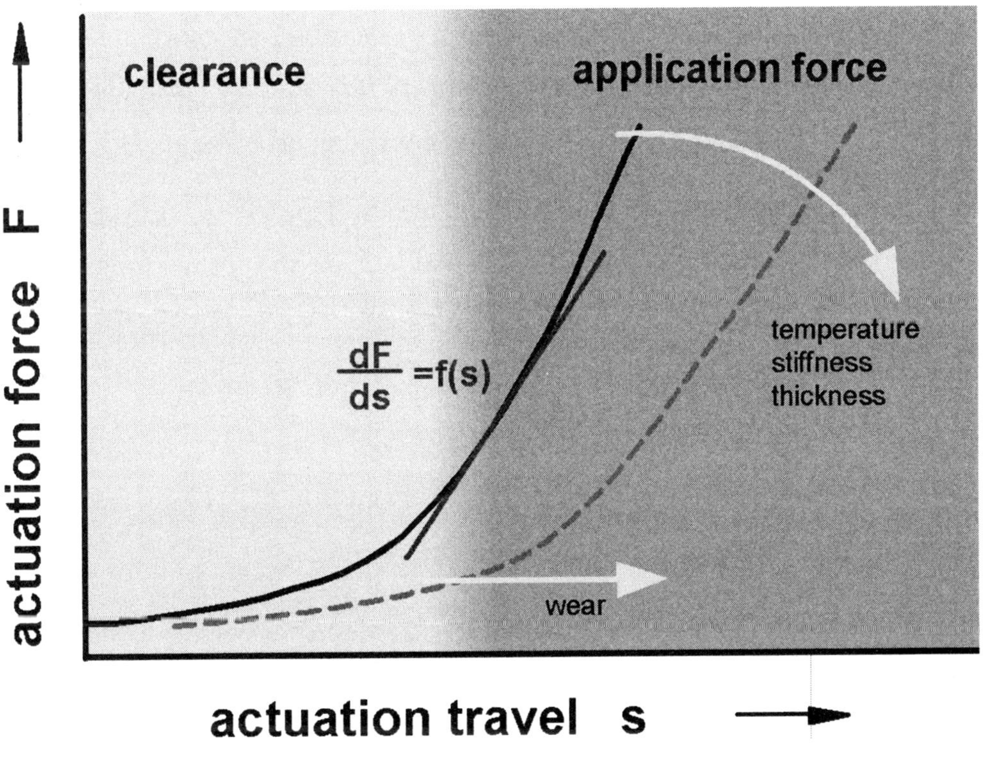

Fig. 12.5: Divergence of the static specifications (function: brake clamping)

The systems for wheel brake actuation must be small and light, in order to take account of the limited installation area in the wheel housing, the requirement for low un-sprung masses and low energy requirement. The further requirement of a high mechanical efficiency is derived from the need for high braking actuation dynamics, retaining the ability to self-release over a wide temperature range and a low energy consumption. The efficiency depends on the type of gear (front wheel, screw...), the bearing arrangement and the lubrication.

In consideration of the framework conditions (requirements) it is clear that the gear 'linking component' has special value for the realisation of electrically actuated wheel brakes. Efforts in respect of new on-board network architectures [12-14] in the car under more favourable operating conditions for high energy consumers ($U > 12$ V) may even lead to an increase in the degree of electro-mechanical efficiency, this has no direct effect however on the mechanics and size of the gear components.

12.3.1.2 Rotatory drive

In terms of the high working capacity of a rotating drive, the use of electro-mechanical transducers with rotary output is described in more detail below, as shown in [12-1] for example. Both the rotary/translatory transducer and also support of the brake tension force is taken up by the gear system.

The theoretical efficiency of ball and roller drives is dependent on degree of increase *p*, the screw diameter d_0 and a coefficient dependent on the screw design *k* [12-13]:

- Ball drive: $0.018 \geq k \geq 0.02$
- Roller drive: $0.025 \geq k \geq 0.0455$

The theoretical screw efficiency η, according to Equation (1) [12-13] becomes greater with increasing and/or smaller screw diameters.

$$\eta = \frac{1}{1 + \frac{k \cdot d_s}{p}} \qquad (1)$$

The dependency of the efficiency on the screw increase at constant screw diameter is shown by **Fig. 12.6** as an example of a roller and ball drive at optimum initial force (no lateral forces). Particularly when using screws for brake lining actuation, it must be remembered that the screw efficiency is not adversely affected by lateral forces or bending moments, which are introduced into the screw gear.

Due to the minimal limit diameter of the balls in this type of drive, smaller screw increases can be realised. Also, the supportability of the roller drives is higher than ball drives with comparable dimensions. However, these characteristics, including efficiency losses are connected between screw and rolling element due to the more severe sliding actions.

Depending on the required screw axial force and the available moment of drive on the screw nut, the screw diameter and the screw increase must be chosen. The latter depends on the required proportionability of the braking tension force at a specified positioning capability of the electro-mechanical transducer.

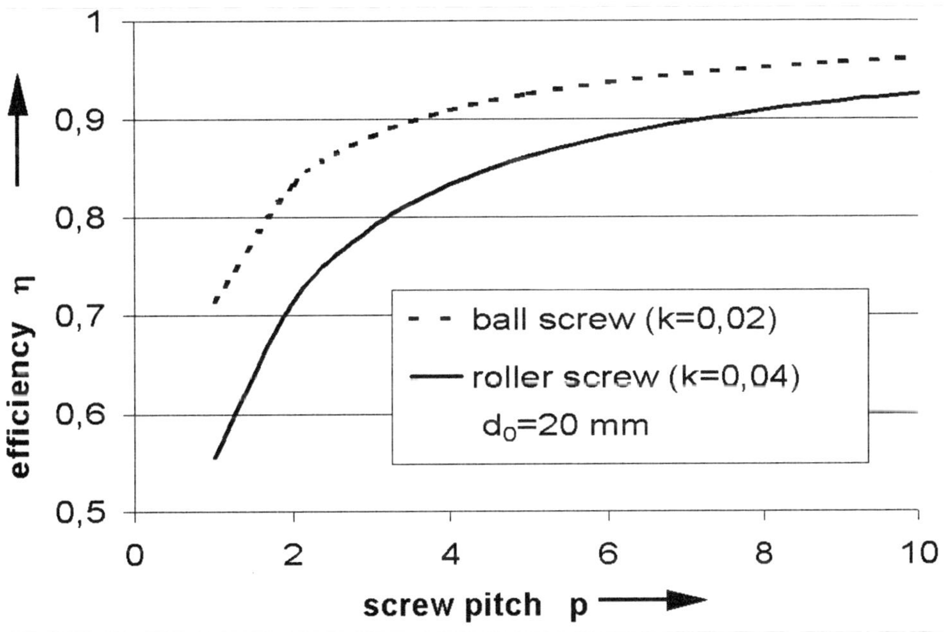

Fig. 12.6: Screw efficiency

An adequately dimensioned axial bearing, which has a severe effect on the overall gear efficiency, supports the axial force provided by the screw. The friction moment of the bearing can be described qualitatively according to Equation (2) can be described qualitatively from the sum of a bearing moment part M_0 and a part M_1 [12-7]. M_0 depends essentially on structure and lubricant, M_1 is dependent on load and increases progressively with the linear increase in load.

$$M = M_0(\vartheta,\nu,n,d_L,f_0) + M_1(\mu,F,n,d_L,f_1) \quad (2)$$

Predominantly the parameters speed n, diameter d_L and load F can be slightly influenced. The operating viscosity of the lubricant ν is largely dependent on the required strength and the range of temperature applications. The bearing coefficients f_1, f_2 and the bearing coefficient μ cannot be directly influenced constructively and depend only on the definitive bearing design (including the rolling element geometry).

Fig. 12.7 shows the principal correlation between the two bearing friction instants M_0 and M_1 as a function of the bearing load.

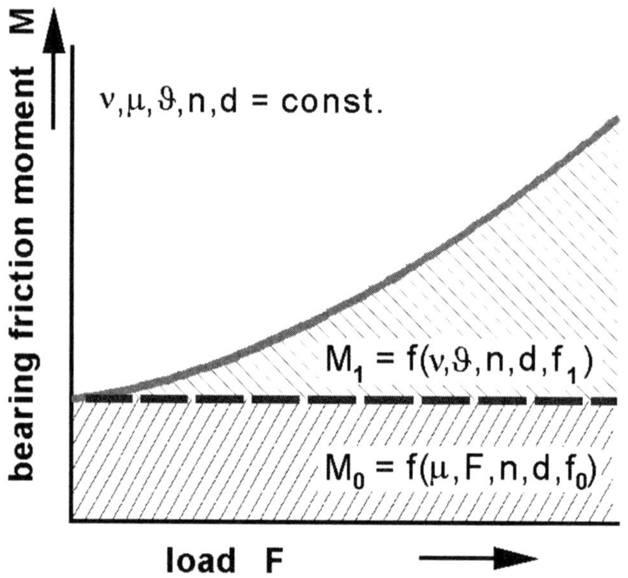

Fig. 12.7: Moment of friction of the bearing depending on the stress

The decrease in bearing diameter reduces the friction moment of the bearing. The required bearing supportability C and the diameter of the rotating gear components however limit the constructive ways in which the bearing diameter can usually be influenced.

The bearing moment can be reduced by reducing the bearing osculation (e.g. the transition from line to point contact of the rolling element through ball instead of roller bearing) at the same bearing diameter. The measures for optimising fitting area, efficiency and mass of the gear/bearing combination to actuate the electrical wheel brake on the basis of screw gears is summarised in **Table 12.1**.

Table 12.1. Ways of optimization.

	Measure	Consequence	Remark
screw	F ↓	η↑ d_S↓ p↑	Influence on rolling element geometry:
	d_S ↓	η↑ F↓	Ball: k ↓; Roller: k ↑
	p ↑	η↑ F↓	
	k ↓	η↑ F↓	
	F ↓ (C ↓)	η↑ d_L ↓ f_0 ↓	
Bearing	d_L ↓	η↑ F↓	Influence on roller element geometry:
	f_0 ↓	η↑ F↓	Ball: f_0 ↓; Roller: f_0 ↑
	n ↓	η↑	

It is clear that when optimising the screw/axial bearing system for brake actuation, the level of force plays a significant role, since the basic geometric conditions are largely dependent.

One way of achieving this objective is to improve the efficiency of the individual components and also a variable load adjustment for translation. A force-controlled, two-stage gear for reducing the screw axial force and the normal bearing force is shown in [12-3]. Another way of reducing engine capacity and relieving the gear components lies in the friction brake itself. If the C* value of the brake is increased, some of the kinetic energy stored in the vehicle being used to tension the brake, this energy can be saved on the actuation side.

12.3.2 Using the vehicle's kinetic energy

Drum brakes use the vehicle's kinetic energy by way of the construction principle, for tensioning. Depending on the design, the C* factors of such brakes are between 1.5 and 10. Drum brakes and regulation of the C* value of these brakes have been thoroughly examined [12-11]. The analogy between the drum brake principle and a lever system is shown (**Fig. 12.**).

The resultant friction force of the drum brake impacts the lining at a distance of a, tangential to the rotational movement of the drum. With the lever arm r for the jaw support point, the wedge angle (α) is derived from the ratio a/r. In the case of a part disc brake, this angle occurs again on the support lever. With modern brake discs, the angle $\alpha = 90°$ and thus splits the tension force from the support force. No self-boosting effect therefore occurs.

The desire to use the vehicle's kinetic energy and combine it with the part disc brake, is not new. Many patents from the 50's and 60's related to this subject.

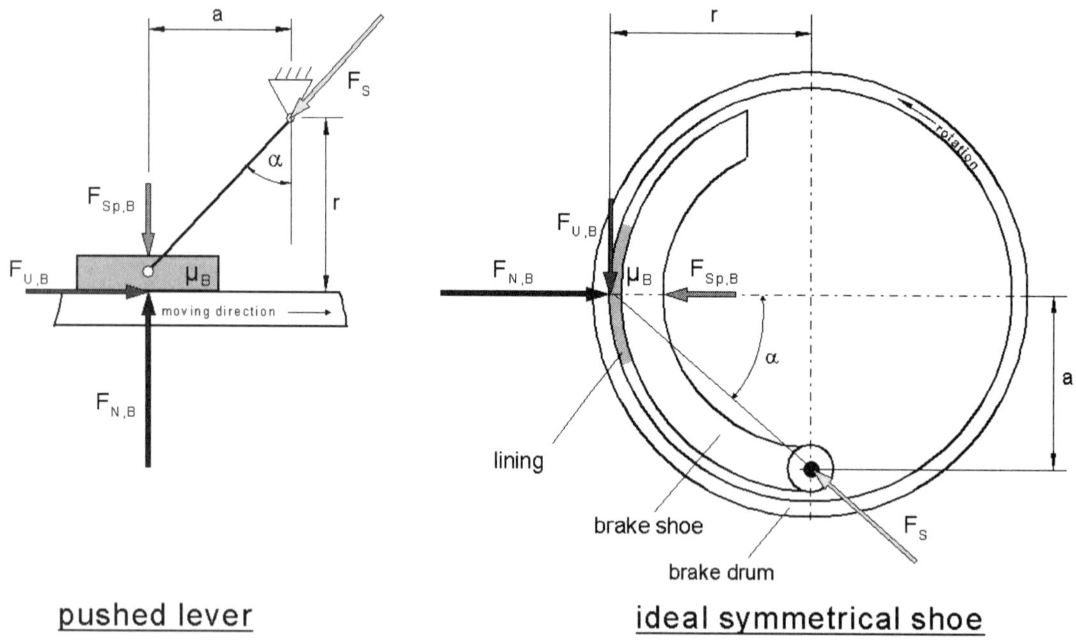

Fig. 12.8: Drum brake in analogy to lever brake

The progressive coding of such brakes in relation to friction value fluctuations and therefore associated braking moment fluctuations however made the safe use of such self-boosting part disc brakes either very difficult or almost impossible. Particularly the wear and subsequent adjustment of the linings posed serious problems for such constructions. The lever system shown above changes its wedge angle with wear and therefore the coding. Such systems are therefore difficult to manage.

One improvement in this respect has been achieved using the wedge system, as shown in **Fig. 12.9**. With this type of system, the wedge angle remains constant, irrespective of the wear to the lining and disc. This provides a positive condition for using the vehicle's kinetic energy.

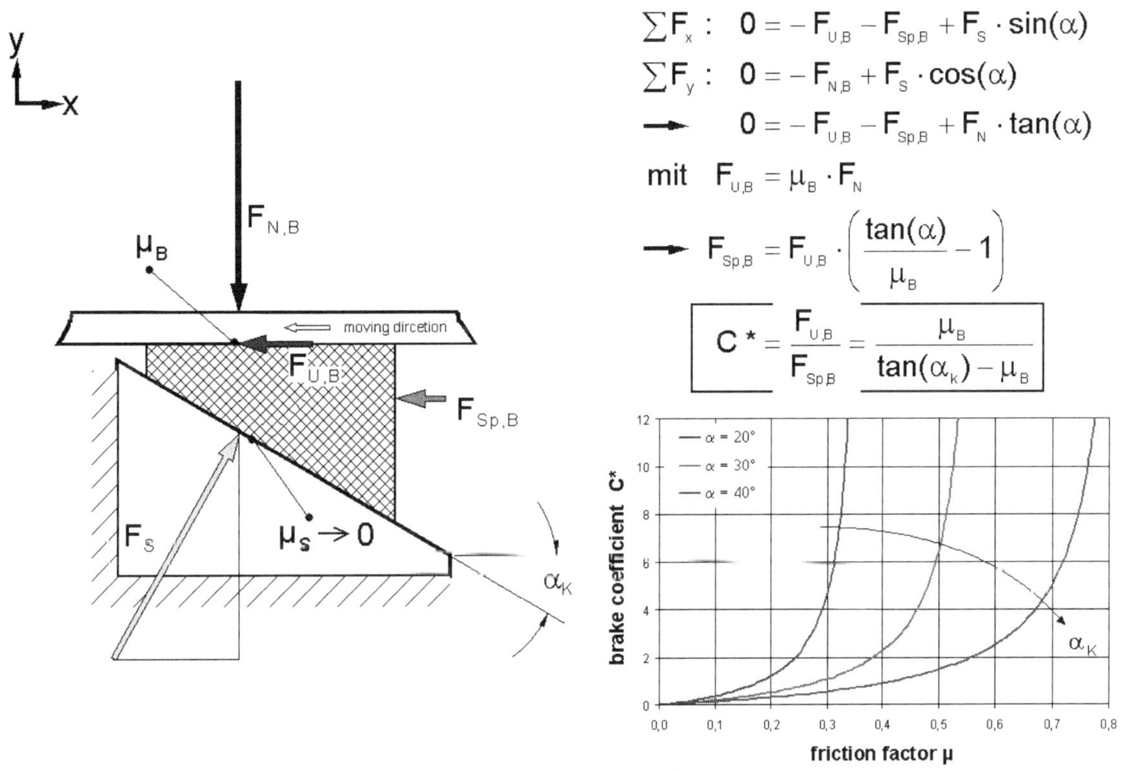

$$\sum F_x : \quad 0 = -F_{U,B} - F_{Sp,B} + F_S \cdot \sin(\alpha)$$

$$\sum F_y : \quad 0 = -F_{N,B} + F_S \cdot \cos(\alpha)$$

$$\rightarrow \quad 0 = -F_{U,B} - F_{Sp,B} + F_N \cdot \tan(\alpha)$$

$$\text{mit} \quad F_{U,B} = \mu_B \cdot F_N$$

$$\rightarrow \quad F_{Sp,B} = F_{U,B} \cdot \left(\frac{\tan(\alpha)}{\mu_B} - 1 \right)$$

$$C^* = \frac{F_{U,B}}{F_{Sp,B}} = \frac{\mu_B}{\tan(\alpha_k) - \mu_B}$$

Fig. 12.9: Wedge brake

This principle, in combination with current options for mechatronics, has given us the impetus to describe such a system in more detail and build a suitable prototype brake. Particularly the option of being able to regulate the brake to the tension force or alternatively the braking moment, using modern micro-controllers, within a loop time of just a few milliseconds, in combination with the option of being able to actively withdraw the self-boosting lining by means of an actuator, creates new methods for using the kinetic energy stored in the vehicle to actuate the brakes.

12.4 Total solution for realised wheel brake

Fig. 12.10 shows the general drawing of the realised wheel brake. The main component groups of engine, gears and friction brakes are constructed in modules. The electro-motor follows an initial rotary translation stage. This then works on a screw drive. The screw introduces a translation movement into the friction component of the self-boosting part disc brake.

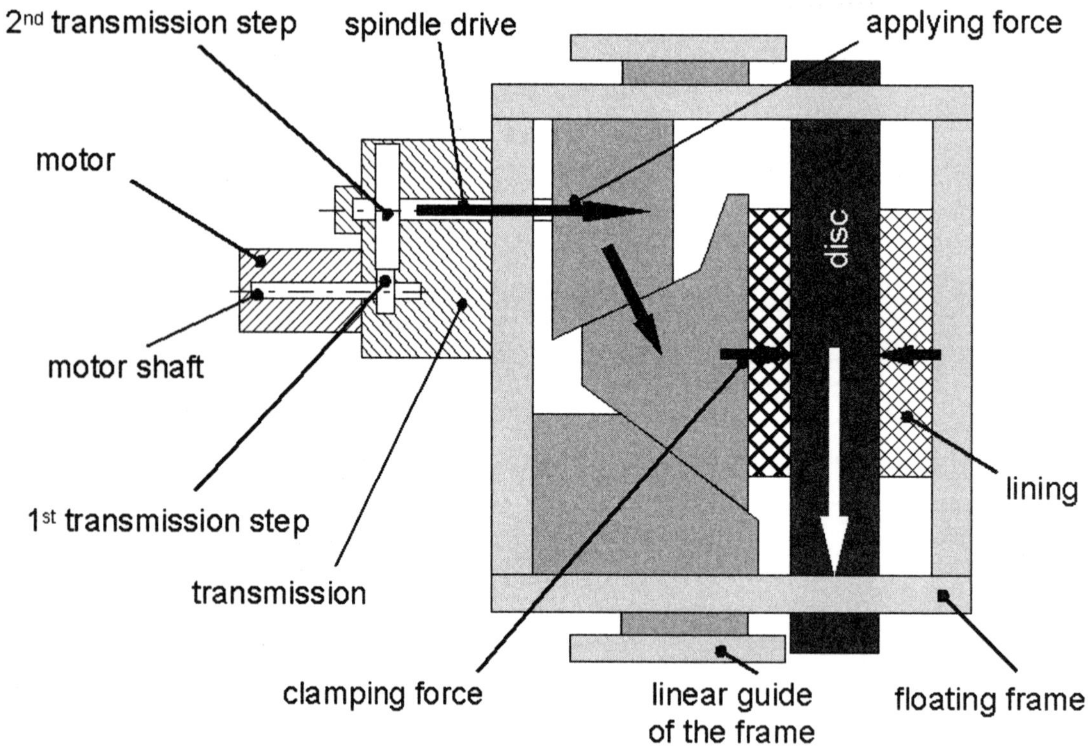

Fig. 12.10: Sketch of the realized wheel brake

Fig. 12.11 shows the realised friction brake in 3D. A floating frame disc brake with a self-boosting lining (Lining 1) and a standard lining (Lining 2) are realised. The friction brake is designed such that the support unit and the force actuation unit can be interchanged, in order to vary the self-boost within appropriate limits. The wheel brake also contains the necessary sensor technique for experimental verification of the calculated and simulated results.

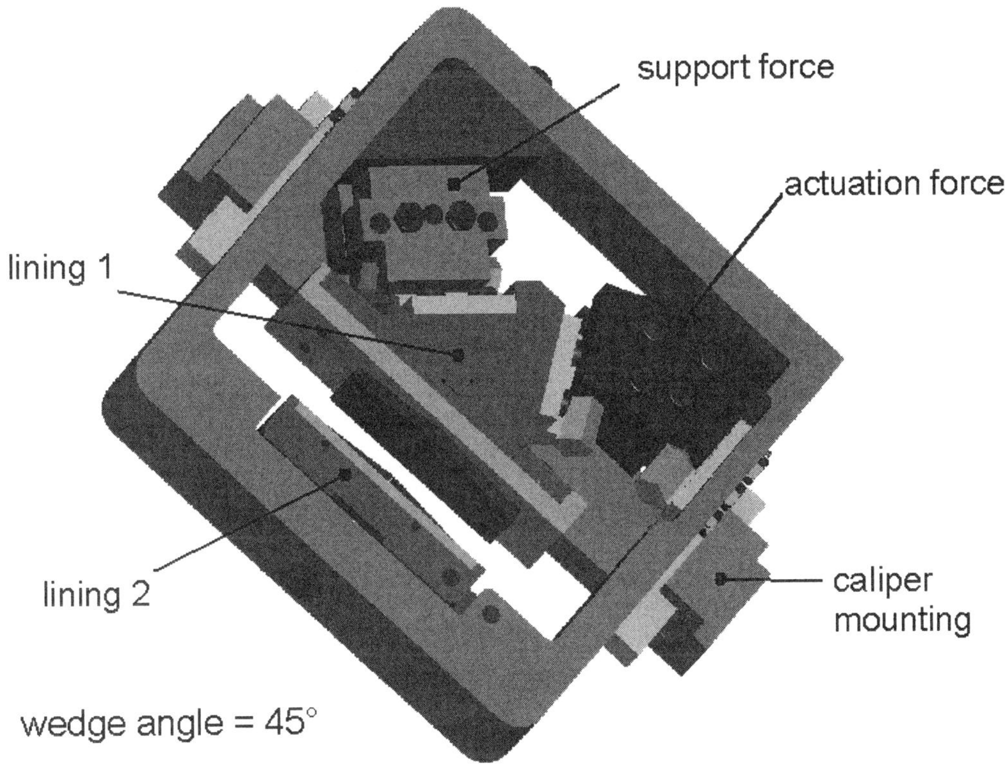

Fig. 12.11: Self-amplifying disc brake

Fig. 12.12 shows the total solution for the wheel brake. The individual components – actuator, gears and friction brake are designed such that the individual components can easily be replaced and modified. The brake was built for a middle class of vehicle and uses the series internally ventilated brake disc.

Fig. 12.13 shows the realised prototype brake and how it is fitted into the test stand in a series produced 17" rim.

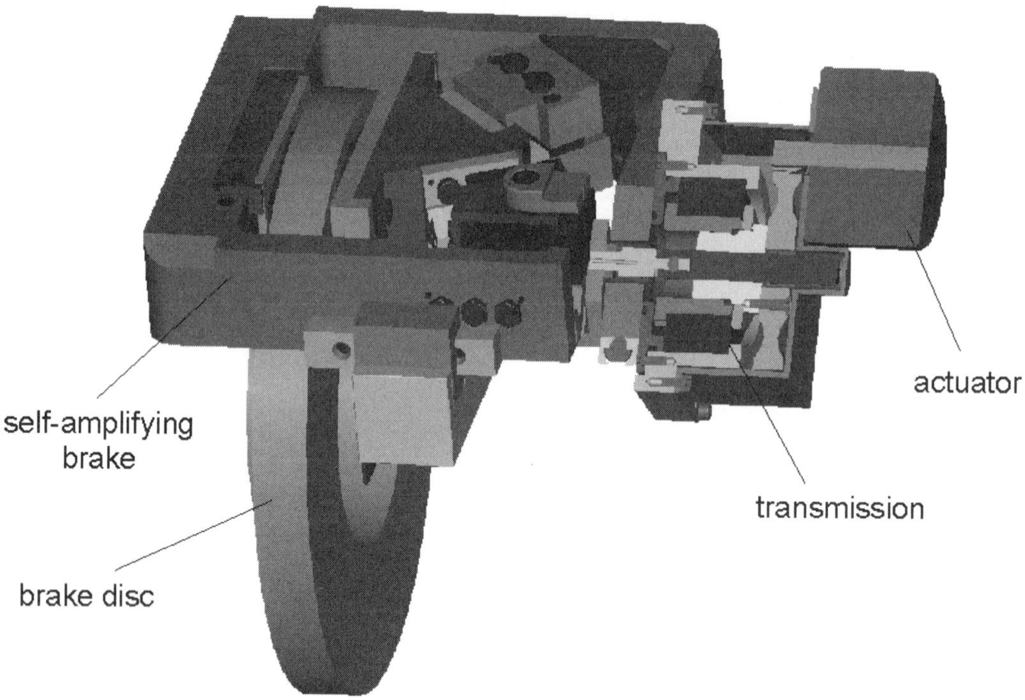

Fig. 12.12: View of the brake prototype

Fig. 12.13: Brake prototype in the 17" rim

12.5 Test stand tests

To test the mechatronic wheel brake on the test stand and subsequently illustrate the intelligent vehicle corner (section 12.6), a quarter vehicle has been designed and built. This (**Fig. 12.14**) is operated on the internal rolling test stand. A series produced complete wheel is positioned on the test stand by means of a special wheel suspension system. The wheel runs in the gap of a roller on the test stand. The wheel load is produced through a pneumatic cylinder, which is infinitely adjustable during operation. The braking moment produced by the prototype brake can be recorded using a measuring hub.

The wheel suspension system in this case is a longitudinal swivel arm in a triangular design. The dynamics of this system (retracted longitudinal arm) increases the wheel load under the influence of a braking moment. This effect is used to transfer high braking moments, which means that the braking behaviour of heavy vehicles can be simulated.

Fig. 12.14: Quarter-vehicle test stand for proving the mechatronic wheel brake

The test stand has the following key values:

- max. shearing force of the pneumatic cylinder: 6.3 kN
- Amplification factor through the braking moment: variable from: 1...2

- max. wheel load: 13.6 kN
- max. transferrable braking moment: 4280 Nm
- vehicle mass capable of being simulated: $\leq$2400 kg

12.5.1 Test results

In addition to calculating and simulating the wedge principle, tests must also be carried out to validate the theoretically derived results. At the time of producing this report, the first focused tests had been performed on the above test stand. Two measurements are indicated and described briefly below.

Fig. 12.15 shows braking to a standstill from 100 km/h. With this test, the standard screw force is used as a feedback for the control. For the braking shown, it amounted to approx. 2200 N. The resultant normal lining force was around 12 kN. This means a relief of the axial bearing and the screw by a factor of approx. 5 against a direct lining actuation with the screw. Furthermore, the electrical drive may also decrease by this factor. The electrical output during this braking action was around 15 Watts. The peak output, which was required to overcome the air play, was around 70 Watts. The braking moment was approx. 1000 Nm. The brake disc temperature at the start of braking was around 100°C and at idle, around 140°C.

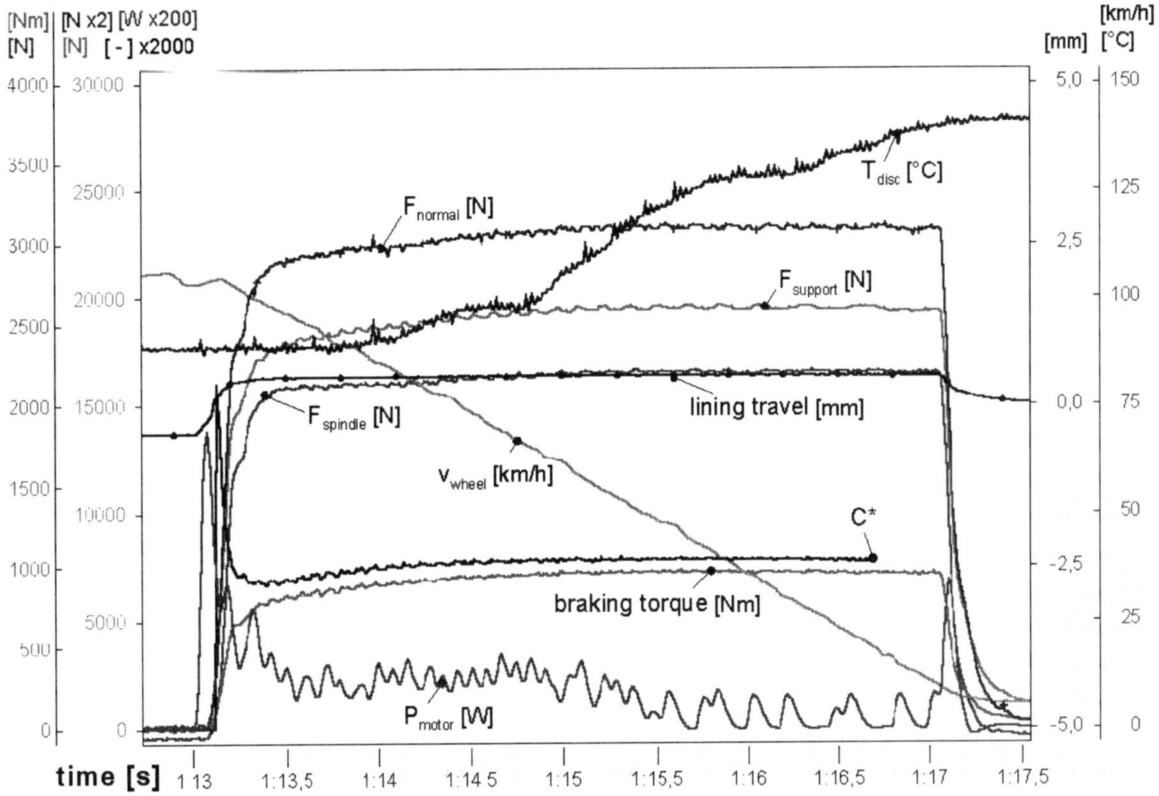

Fig. 12.15: Braking to a standstill with 1.000 Nm

Fig. 12.16 shows braking to a standstill with approx. 1200 Nm braking moment. The standard screw force is just less than 3 kN. The resultant standard lining force was around 15 kN. A construction-related design of 5 is derived in this case. The electrical output during this test was 25 Watt, the peak output for overcoming the air play was around 125 Watts.

From this and other focused measurements, no scientific knowledge can be drawn, however, confirmation of the theoretical considerations may be assumed. It is our view that great potential exists here, which could be exploited to use the electromechanical brake.

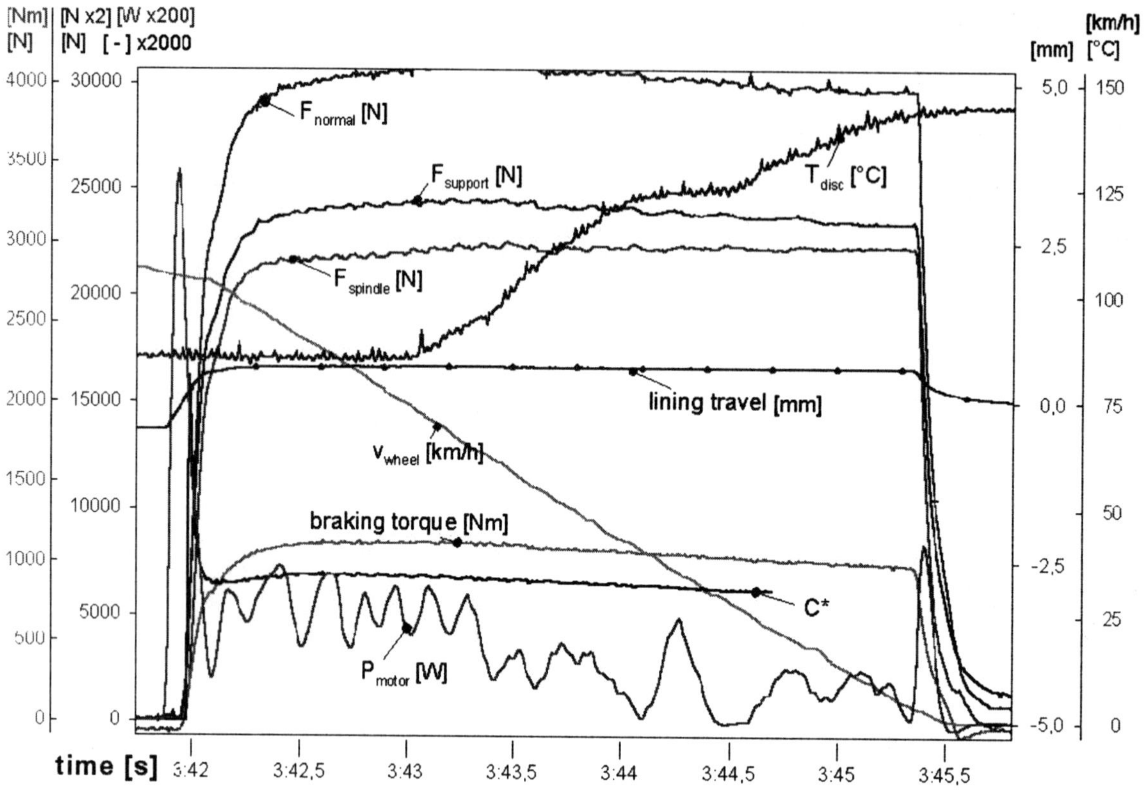

Fig. 12.16: Braking to a standstill at 1.200 Nm

12.6 Prospects

The continuous development of mechatronic vehicle sub-systems and driver assistance systems require skilful management of these systems. In this way, the active and passive safety of today's vehicles can be improved, thus achieving genuine use by the customers.

In the Special Research Report 241 IMES (New Integrated Mechanic Electronic Systems For Machine Building) [12-12], in addition to sub-project B6 (Mechatronic Wheel Brake), other sub-projects are also being worked on. In the area of vehicle technology, in addition to B6, sub-projects A3 and A7 were researched. Sub-project A3 (Darmstadt tire sensor) deals with the subject of tire sensor systems. This sensor, which is now used in the 4th generation, is able to record all relevant physical conditions in the tire contact area, i.e. the vehicle/road interface and provides the current braking moment for the wheel brake regulator. Sub-project A7 researches mechatronic actuators for use in modern vehicles. The combination of these sub-projects for the mechatronic vehicle, and the joint regulator, should indicate the advantages of the interaction between the individual sub-systems. Networking individual sub-systems in future vehicles will contribute to the safety and comfort levels, but requires the appropriate interfaces. Mechatronic systems have such interfaces and will therefore become more significant in the future.

Brake-by-Wire or X-by-Wire systems for heavy goods vehicles are appropriate for use in future vehicles and necessitate research and development in this area, in order to be able to offer competitive products.

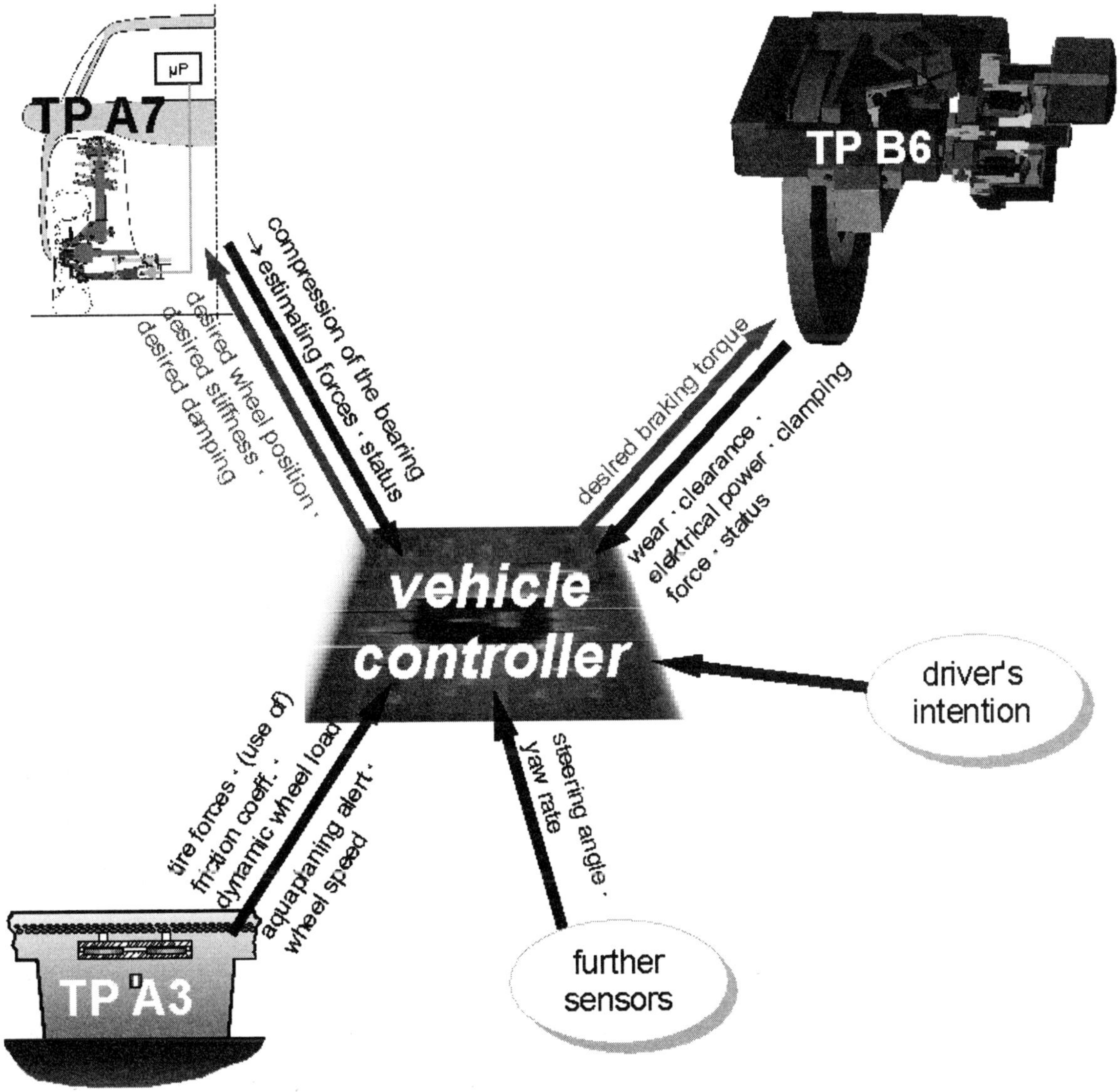

Fig. 12.17: Mechatronic vehicle corner

12.7 Appendix

12.7.1 Symbols

C*	brake factor
c	spring constant
D_S	screw diameter
D_L	searing diameter
F	force
I	transmission
f_0	bearing coefficient (design, lubrication)
f_1	bearing coefficient (load direction)
K	screw coefficient
M	moment
M_0	bearing friction moment
M_1	bearing friction moment (depending on load)
p	sScrew pitch
μ	friction coefficient
n	revolutions
ϑ	temperature
U	voltage
X	shift
ν	operating viscosity of lubricant

12.7.2 Indices

G	gGear box
N	iIndex
S	screw
Sp	tension
W	transducer

12.7.3 References

[12-1] Balz, J.; Bill, K.-H.; Böhm, J.; Scheerer, P.; Semsch, M.: Concept for an electro-mechanical vehicle brake. ATZ Automotive Publication 98 (1996), S. 328 ff.

[12-2] Bill, K.-H.: Basic examinations for using electrical wheel brakes in cars. Wheel brakes in cars. Fortschritt-Berichte VDI Reihe 12 No. 166. Düsseldorf: VDI-Verlag 1992, ISBN 3-18-146612-3

[12-3] Bill, K.-H.; Semsch, M.: Translation gears for electrically actuated vehicle brakes, ATZ Automotive Publication 100 (1998) 1, S. 36 ff.

[12-4] Bill, K.-H.; Semsch, M.; Breuer, B.: A New Approach to Investigate the Vehicle Interface Driver / Brake Pedal Under Real Road Conditions in View of Oncoming Brake-by-wire-systems. SAE 1999-01-2949

[12-5] Breuer, B.: Script for Lecture on Commercial Vehicles I, Technical College Darmstadt, Vehicle Technology, 1997

[12-6] Dieckmann, Th.: EMB – An electro-mechanical braking system. XVII. – Symposium 1997, Progress Reports 340, Series 12, VDI GmbH Düsseldorf, 1998, ISBN 3-18-334012-7

[12-7] Eschmann, P.; Hasbargen, L.; Weigand, K.: Roller bearing in practice, Manual for calculating and designing bearings, ISBN 3-486-31102-6, 1978

[12-8] Jahn, M.; Jante, A.: Brake systems in busses, Automotive Manual, Volume 2, 18th Edition, Section. 3.37, Technical Publishers Herbert Cram, Berlin 1965

[12-9] Klein, H.: Patent P 3718944.1, The Archimedes Gear System

[12-10] Koeßler, P.: Manual for Truck Engineers, Section 7: Brake Systems Deutsche Verlagsanstalt GmbH, Stuttgart, 1973

[12-11] Leber, M.: Wheel brake with mechatronic key value regulation – Examination of operating behaviour and driver linking problems in respect of Brake-by-Wire-Systems. Progress Reports 358, Series 12, VDI-Verlag GmbH Düsseldorf, 1998, ISBN 3-18-335812-3

[12-12] N.N.: DFG-Special research 241, New integrated mechatronic systems for machine building (IMES). Working and Results Report - July 1996 – March 1999

[12-13] N.N.: Transrol Selection Table for Ball and Roller Gears, SKF, Schweinfurt, Copy Ref: 3501T, 1984

[12-14] Schöttle, R.; Schramm, D.: Future power networks in vehicles Kraftfahrzeug, VDI-Year Book 1997, VDI-Verlag GmbH, Düsseldorf 1997, ISBN 3-18-401608-0

[12-15] Semsch, M.; Bill, K.-H.; Dausend, U.; Breuer, B.: Concept and design of the gears for a mechatronic part disk brake. Progress Report 743, Series 8, VDI-Verlag GmbH Düsseldorf, 1998, ISBN 3-18-374308-3

13 Introduction into the Technology of Airvehicle Braking Systems

Dipl.-Ing. B. Ohly, formerly MBB, DASA (retired)

Ottobrunn, Germany

Summary

Dealing with the subject of aircraft brakes is somewhat new in a conference for automotive systems. In the following presentation, specific differences in the layout of airvehicle braking systems and associated equipment shall be introduced to the auditorium.

The classical big commonality amongst brake specialists of all technical disciplines (except railway applications) remains in the fact that by means of a set of pedals a braking signal will be created to activate a more or less sophisticated control loop, in order to induce a braking force which has to transfer via some square inch of rubber a retardation force from a ground surface back to the associated vehicle in a way that a safe standstill can be guaranteed within a broad spectrum of friction coefficients.

It will be matter of this presentation to introduce the enormous effort necessary to safely decelerate an aircraft after its transformation into a "landcraft" with a rapidly shrinking runway length in front of the pilots eyes.

13.1 Introduction

The evaluation of a braking system for an airvehicle (aircraft, helicopter, spacecraft etc.) is highly depending from detailed specification requirements of the customer and is basically to be understood as an integrated part of an overall deceleration system. Military projects, in particular fighters might require substantially different deceleration features than civil aircraft. Whereas brake parachutes and arrestor hooks on most fighters are standard equipment additionally to the wheel brakes, civil projects and military transport aircraft absorb most of the kinetic energy by the application of thrust reversers (jet engines resp. propellers). Additionally many aircraft utilize airbrakes in the high landing speed regime after touch down. In order to achieve sufficient wheel loads, the still existing aerodynamic lift, in particular with delta wings requires rapid lift dump, which only can be achieved by either applying spoilers or wing profile adjustment control.

An efficient utilization of the available friction coefficient between tire footprint area and airfield surface, has been an issue since the fifties of the past century. By means of simple hydro mechanical on-off antiskid facilities (i.e. Maxaret) the number of tire bursts could be partially reduced. Technological progress over a period of some decades finally resulted in the development of analogue and later digital control systems of at least 95% efficiency, which can be offered to the airframers.

This presentation in particular deals with the more complex and different design criteria which have to be considered for the layout of a modern aircraft braking system in comparison to the automotive demands. Neglecting conventional braking systems, this paper shall inform the audience about the state of the art configuration of a "brake by wire system" and also to the complex build up of the functional wheel and brake unit.

Finally a forecast to future projects forms part of this presentation, whereas its verifications are highly dependant from the enormous cost pressure in the aviation industry with its relatively small production rates in comparison to the automotive industry.

13.2 General functional requirement spectrum for aircraft wheel brakes

- energy absorption spectrum with and without the assistance of deceleration aids (thrust reversers, parachutes, etc.)

- differential braking capability to assist nose wheel steering
- hold the airplane when parked
- hold the airplane while running up the engines
- control speed while taxiing
- automatic braking capability with selctive deceleration rates
- automatic braking during landing gear retraction

13.3 Main differences in design criteria for aircraft braking systems in comparison to automotive demands

- centre of gravity positions
  aircraft (12% : 88% (wheel load distribution between nose wheels/mainwheels))
  motor car (50% : 50%)
- generally only main wheels will be braked
- wheel loading dependent from lift dump behaviour
- protection against landings with braked wheels
- pilot and co-pilot each having two brake pedals available (differential braking capability)
- automatic switchover from normal to emergency braking system
- temperature monitoring avoiding landing gear retraction with overheated brakes
- cockpit system status display(ECAM wheel page)
- use of high performance multi disc brakes
- complex wheel design with integrated brake drive keys, heat shields and fusible plugs
- application of fan cooling for critical missions (i.e. short operational turnaround)
- extensive qualification and flight clearance procedures for all system related components and finally for the integrated system (confidence testing)
- extensive reliability requirements (i.e. no single failure shall cause a hazard; sufficient protection against increased exposure to electro magnetic interference (EMI))

13.4 Typical hybrid multi-wheel brake/antiskid system diagram

The systems as shown in **Fig. 13.1** provides brake and skid control for four wheels. Hydraulic pressure is provided by means of two independent pedal command systems. While the normal brake control loop is based on a brake by wire technology, the backup system will be operated conventionally by means of pedal connected master cylinders and dedicated antiskid valves. The brake design of this particular system is based on dual cavity brakes, where via two independent hydraulic ports half the number of brake pistons in each case will be operated. Whereas the servo valves of the brake by wire portion of the system receive a valve current value as calculated by the digital control unit proportional to the pedal transducer signal, the backup system is controlled by the pedal connected master cylinders in conjunction with the metering valves. Electronic control becomes only active when an incipient skid situation occurs and the free pressure passage through servo valves has to be reduced accordingly by an analogue control signal from the associated control circuit.

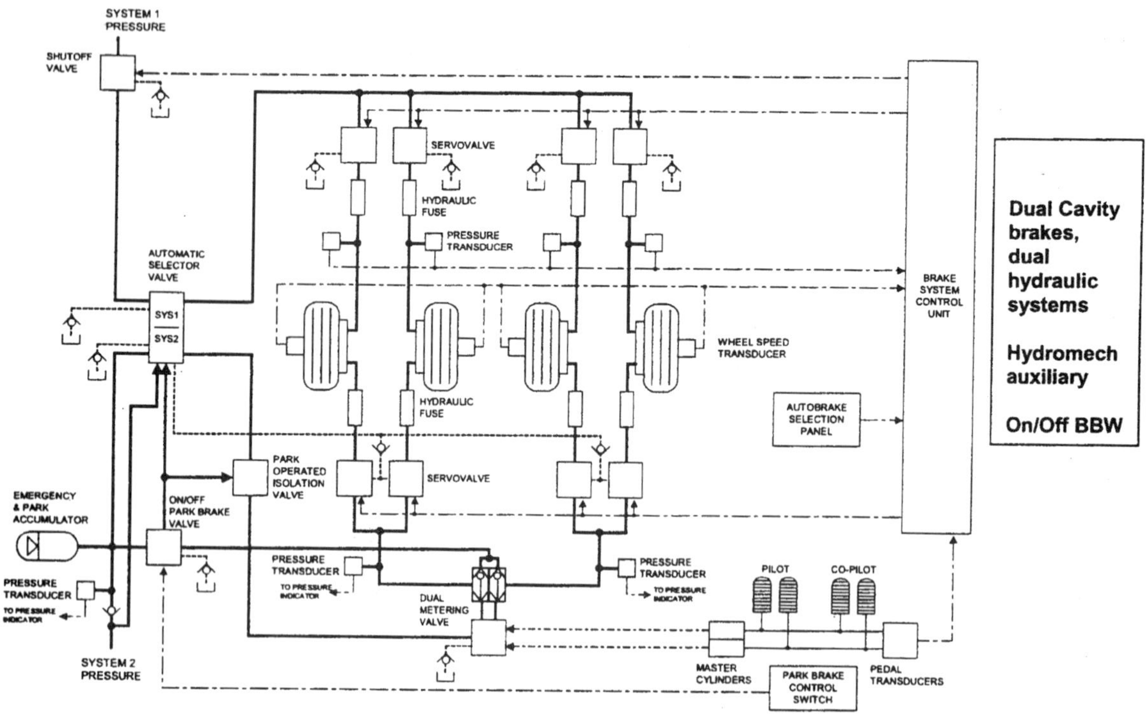

Fig. 13.1: Typical Brake by Wire layout with Hydromechanical Backup System

13.5 Typical two wheel brake by wire diagram

Fig. 13.2 is showing a simplified system architecture for a braking system for two wheels. The system is fully duplexed – hydraulically and electronically. The brakes are of a single cavity type interconnected by means of a shuttle value with both hydraulic pressure lines. A hardwired, manually selectable motor valve can provide unmetered parking pressure from an accumulator. The associated control loop is shown in **Fig. 13.3** in a simplified version.

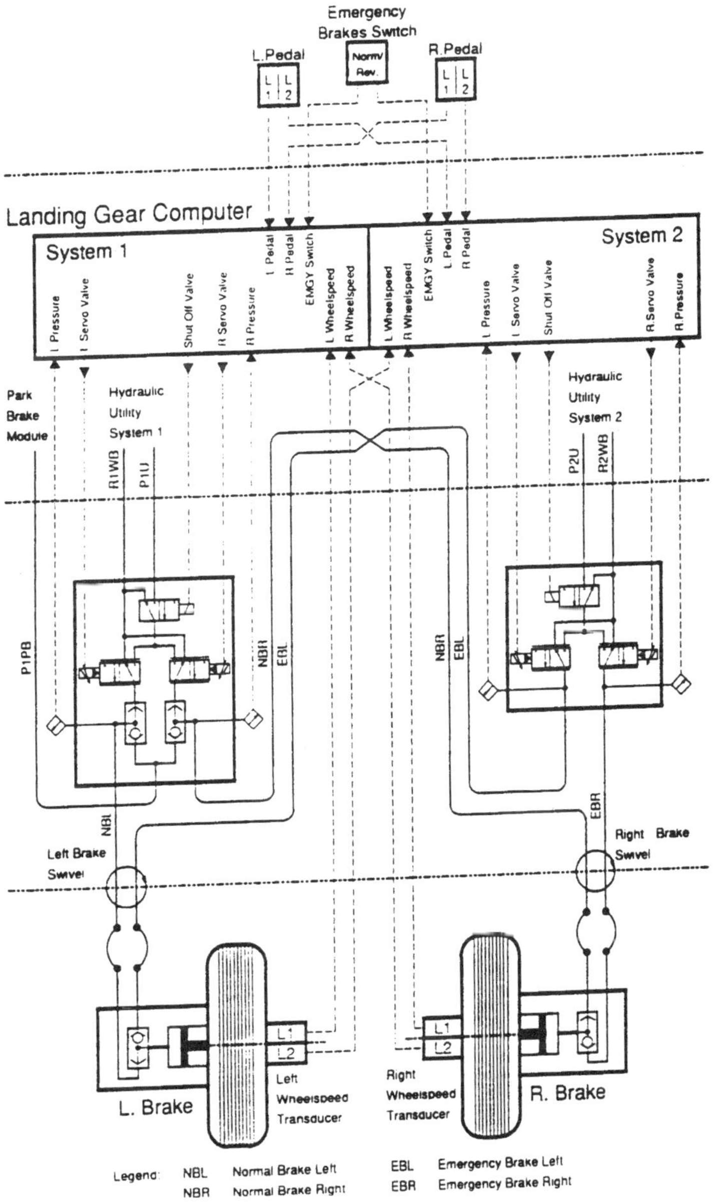

Fig. 13.2: Typical Brake by Wire System Architecture for Two Wheel Application

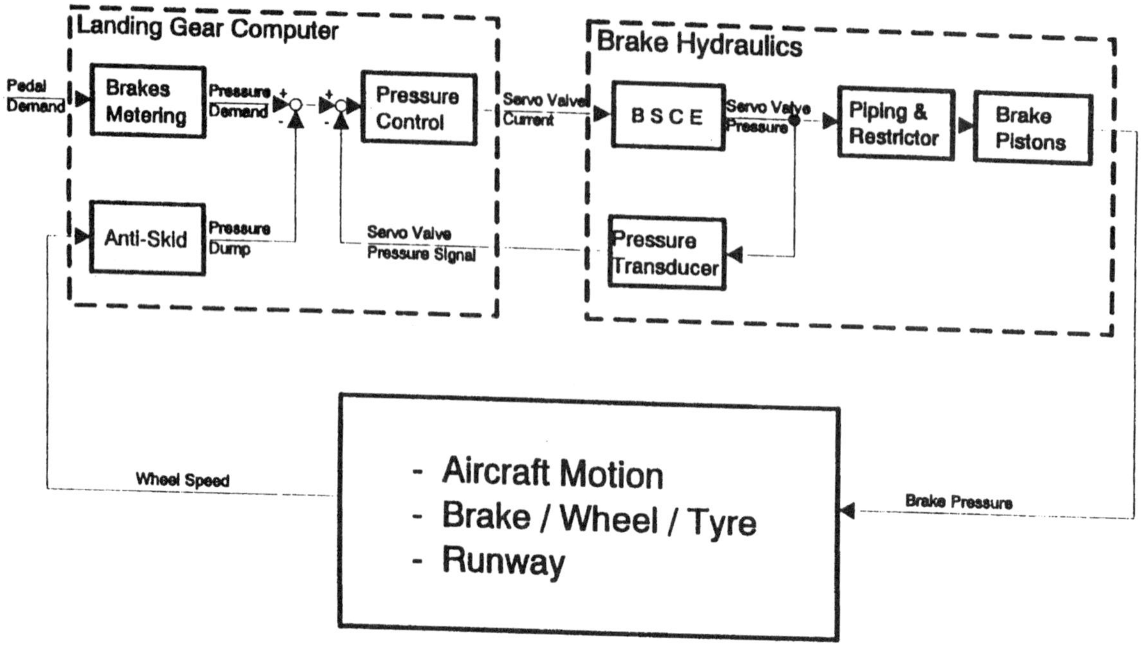

Fig. 13.3: Brake by Wire System - simplified Control Loop

13.6 Design criteria of a typical high performance multidisc carbon brake

Independent from the layout of a heat pack either in steel or carbon the main design criteria of either brake are partly similar. The most obvious difference between a steel and carbon brake is the need of dedicated friction liners of the stators, while stators and rotors of carbon brakes are in direct friction contact to each other. Carbon brakes are about 40% lighter than steel brakes but require a substantial larger installation volume which requires special attention in the early design stage. Torque tubes for both types of brakes are made either from high strength steel or titanium in order to transfer the braking moments under extreme temperature conditions into the landing gear and finally into the aircraft. Integrated temperature sensors transfer the thermal status to a dedicated cockpit display. Feedback of torque data to the control algorithms will be utilized in latest system improvements in order to balance the wide spectrum of carbon brake torque characteristics.

Fig. 13.4 + **Fig. 13.5** show the general build up of a carbon and steel brake assembly.

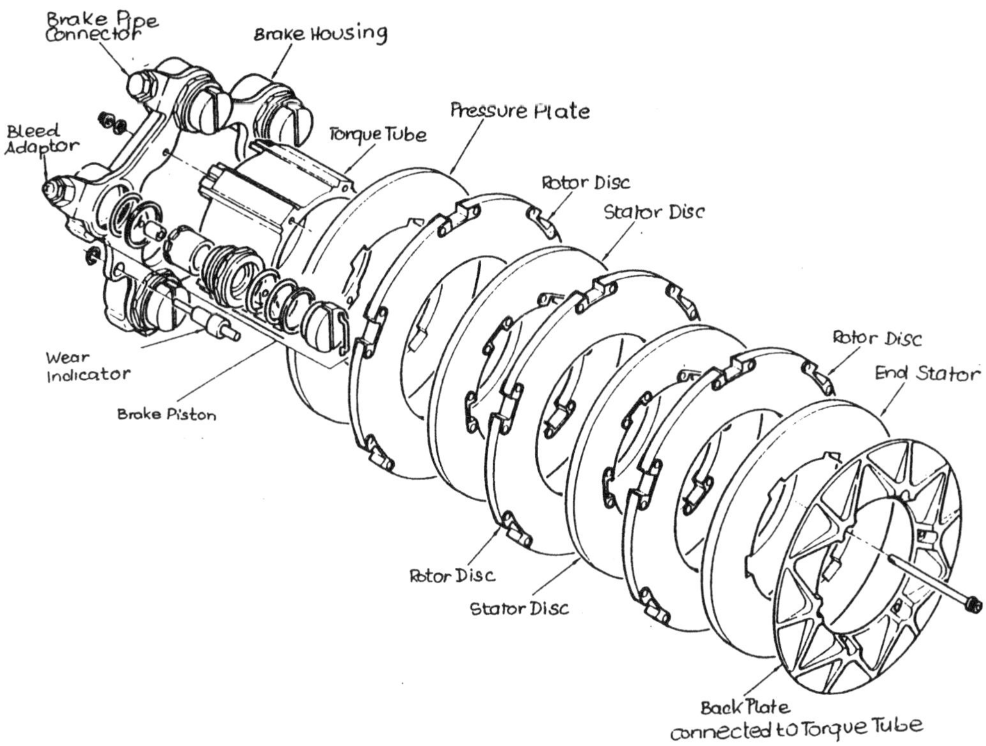

Fig. 13.4: Multidisc Carbonbrake Assembly

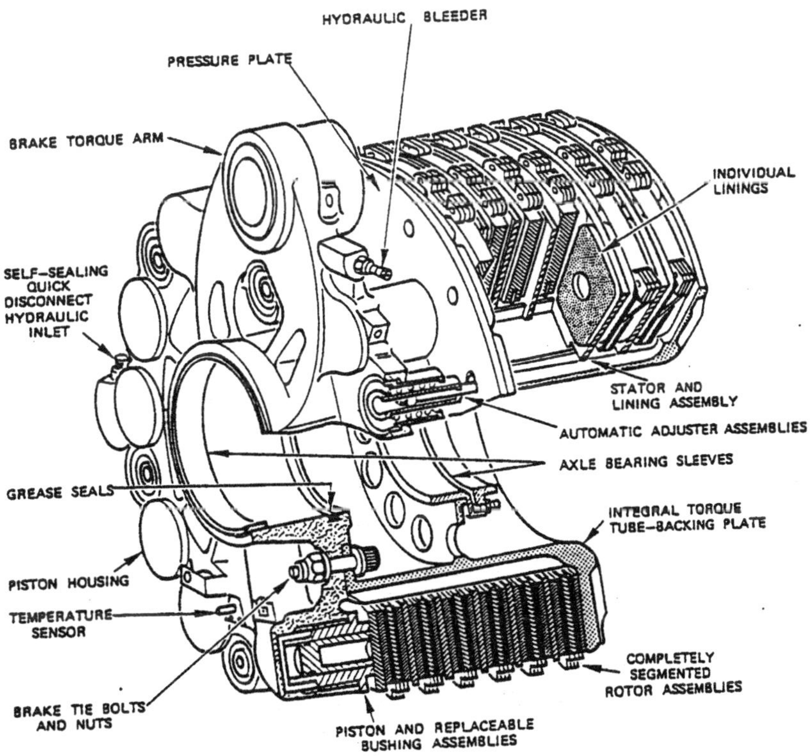

Fig. 13.5: Steelbrake Sectional View

13.7 Design criteria of a typical main wheel assembly in connection with the associated steel brake unit

Aircraft wheels in general comprise of an outer and an inner wheel half, which are bolted together and sealed by an O-ring. Wheels require short intervals of crack inspection, because of their high stress levels and are equipped with tire change counters. The outer wheel half is equipped with all necessary maintenance features (inflation valve, pressure gauge, high-pressure valve etc.). The wheel cap incorporates a drive shaft clutch for the wheel speed generator. The wheel spoke has to provide cooling holes for sufficient brake heat flow. A system of fusible plugs is integrated into the wheel bead in order to avoid tire bursts in case of excessive brake heat radiation. The inner wheel half is equipped with drive keys for the brake rotors. Between the drive keys and the inner wheel diameter a heat shield is circumferentially located. A typical wheel design is shown in **Fig. 13.6**.

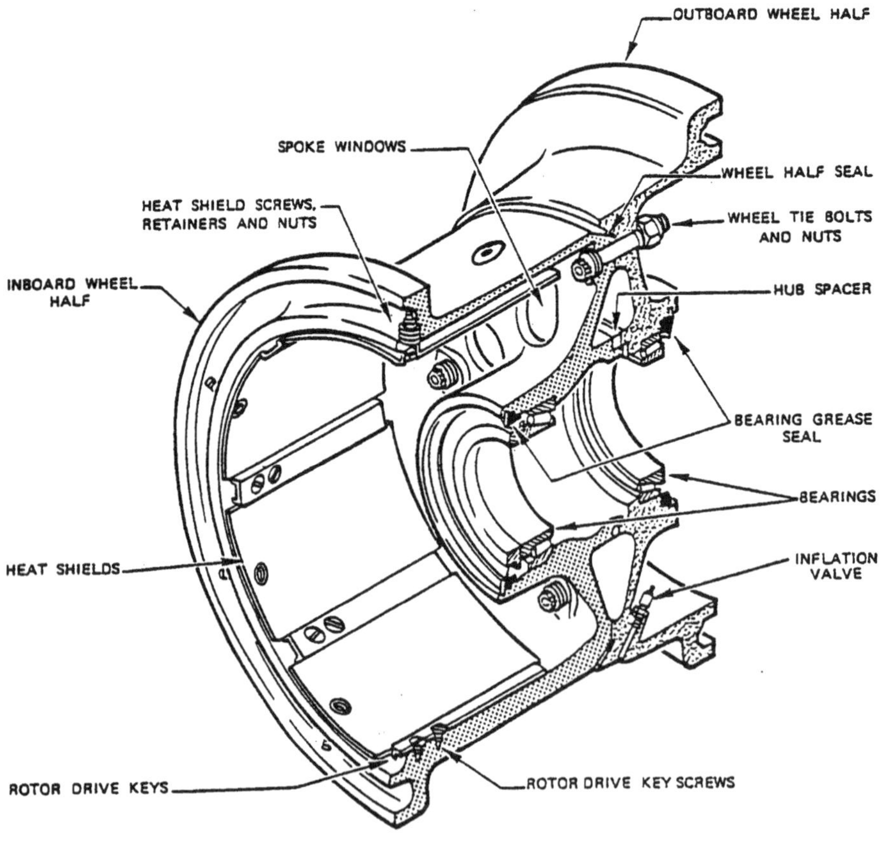

Fig. 13.6: Main Wheel Unit - Sectional View

13.8 Future trends for aircraft braking systems

The majority of currently operating military and civil aircraft are equipped with conventional, hydro-mechanical braking systems and analogue skid control devices. It shall be noted however that there is an obvious trend towards the introduction of brake by wire systems with digital skid control circuits (see **Table 13.1**).

Despite the well known weak points of a carbon brake, this material in its latest generation offers too many technical and commercial advantages which don't justify an outruling in the next future. Partly sponsored by the U.S. Airforce, extensive investigations into the development of electrical actuated brakes took place in the past decade (see **Fig. 13.7** + **Fig. 13.8**). The control loop of such systems received feed back by fibre optic wheel speed transducers as well as from brake torque sensors. Simulation with hardware in the loop took place with a digital brake and skidcontrol system in combination with an original F-16 multi disc carbon brake equipped with four electrical brake actuators. The associated brushless DC motor design incorporates a permanent magnet on a cobalt/samarium basis. The associated aircraft and its landing gears were represented by means of a detailed mathematical model. The operating current for the brake was adjusted to 50 Vdc.

The antiskid performance even exceeded hydraulic brake performance. With a system of this technology, which followed the trend to a more electric aircraft", all hydraulic components of the braking circuit are eliminated (including tubing) and consequently the problem area of leaking, air trapping, valve problems, brake fire etc.

Almost eliminated is the problem of torque gain variations of carbon brakes, because of the torque feedback capability.

Meanwhile flight test results are available, achieved during flight tests of a F-16/Block 40 aircraft equipped with digitally controlled and electrically actuated carbon brakes.

Table 13.1: Examples of Brake by Wire Applications

A/C MODEL	Manuf	A/C Type	# Wheel / Brake	Max Pressure Brake	Max Pressure System	Valve Type	Brake Type	Pedal Data lbs.	Pedal Data Angle	Systems & Remarks
B-70	North American	Bomber	8 wheels 4 brakes	4000 psi	4000 psi	Flapper nozzle 4 way	Steel			Torque feedback, brakes between co-rotating wheels-oronite fluid
Space Shuttle	Rockwell	Research vehicle (NASA)	4	2,000	3,000	Flapper / Nozzle / 4-way	Carbon - dual cavity			Dual Redundant -
Concorde	Aerospatiale	Transport	8	3000	3000	Flapper nozzle 4 way	Carbon dual cavity		14 deg	BBW Primary, hydromech Aux, on/off BBW park torque Limit control – Oronite fluid
F-16	Lockheed-Martin	Fighter	2	2,300	3,000	Pressure proportional Solenoid	Carbon - dual cavity	15-55 lbf	0- 10 deg	Dual simultaneous BBW 1st production BBW
F-117	Lockheed Martin	Fighter	2	2,800	3,000	Pressure Proportional Solenoid	Carbon			Similar to F-16
IDF	TAS	Fighter	2	2,300	3,000	Pressure Proportional Solenoid	Carbon - dual cavity			Similar to F-16
JAS39A JAS39B	SAAB-Scania	Fighter	3	4,000	4,000	Flapper / Nozzle / 3-way	Carbon	22-105 lbf	0- 15 deg	Dual BBW primary mains + Nose wheel on 1 system - hydromech park auxiliary
F2	Mitsubishi	Fighter	2	2,300	3,000	Pressure Proportional Solenoid	Carbon - dual cavity			Similar to F-16
G-IV	Gulfstream	Executive / business	4	3,000	3,000	Flapper / Nozzle / 3-way	Carbon			BBW primary Hydromech park auxiliary
Euro Fighter	EFA Typhoon	Fighter	2	4,000	4,030	Flapper / Nozzle / 3-way	Carbon	15-80 lbf	0-10 deg	Dual redundant BBW primary, Hydromech park auxiliary
A-300 A-310 A-319 A-320 A-321 A-330 A-340	Airbus	Com'l transport	8 8 4 4 4 8 10/8 *	2,550	3,000	Flapper / Nozzle / 3-way	Carbon - dual cavity		14 deg	Dual BBW primary, Hydromech auxiliary system. On/Off BBW park * center landing gear not braked
LJ45	Learjet	Executive / business	4	3,000	3,000	Flapper / Nozzle / 3-way	Carbon - single cavity			Dual pressure input, Inbd/Otbd BBW primary hydromech park auxiliary
EMB145/140/135	EMBRAER	Regional transport	4	3,000	3,000	Flapper / Nozzle / 3-way	Carbon - single cavity	15.4 – 84# (7-38 kg)	0 – 15 °	BBW primary - hydromech park auxiliary
Global Express	Canadair	Executive / business	4	3,000	3,000	Flapper / Nozzle / 3-way	Carbon - single cavity			Dual Redundant BBW primary - hydromech park auxiliary
Rafale	Dassault Aviation	Fighter	2	2,500	5,000	Flapper Nozzle / 3-Way	Carbon - dual cavity		13 deg.	Dual redundant BBW, Brake control, antiskid and decel control
Falcon 2000	Dassault Aviation	Executive Business	4	2,550	3,000	Flapper Nozzle / 3-Way	Carbon - dual cavity			Dual BBW primary /135Simultaneous Application Hydromech Park auxiliary
Falcon 900	Dassault Aviation	Executive / business	4	3000	3000	Flapper nozzle 3 way	Carbon dual cavity			Dual BBW Primary - single hydr-mech auxiliary
F-22	Lockheed Martin	Fighter	2	3,750	4000	Flapper / nozzle 3 way	Carbon – single cavity			Dual redundant BBW, Brake control, antiskid and decel control
Hawker 4000	Raytheon Aircraft	Executive business	4	3000	3000	Flapper nozzle 3 way	Carbon single cavity	15-75 lbf	0-19 deg	Dual BBW primary - dual hydromech park/auxiliary
ERJ170/190	Embraer	Regional Jet	4	3000	3000	Flapper nozzle 3 way	Carbon single cavity			Dual BBW primary - dual hydromech park/auxiliary
FD728	Fairchild Dornier	Regional Jet	4	3000	3000	Flapper nozzle 3 way	Carbon dual cavity			Dual BBW primary - dual hydromech park/auxiliary
F900	Dassault	Executive business	4	3000	3000	Flapper 3 way	Dual cavity carbon			Dual BBQ primary, Single hydomech park/auxiliary

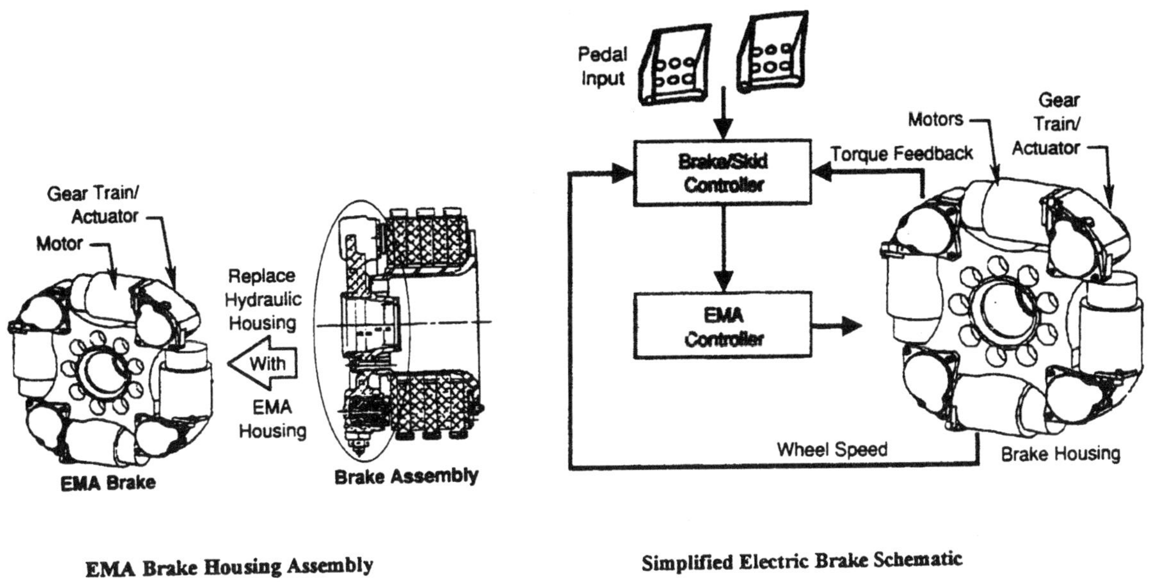

Fig. 13.7: Electrical Brake Arrangement

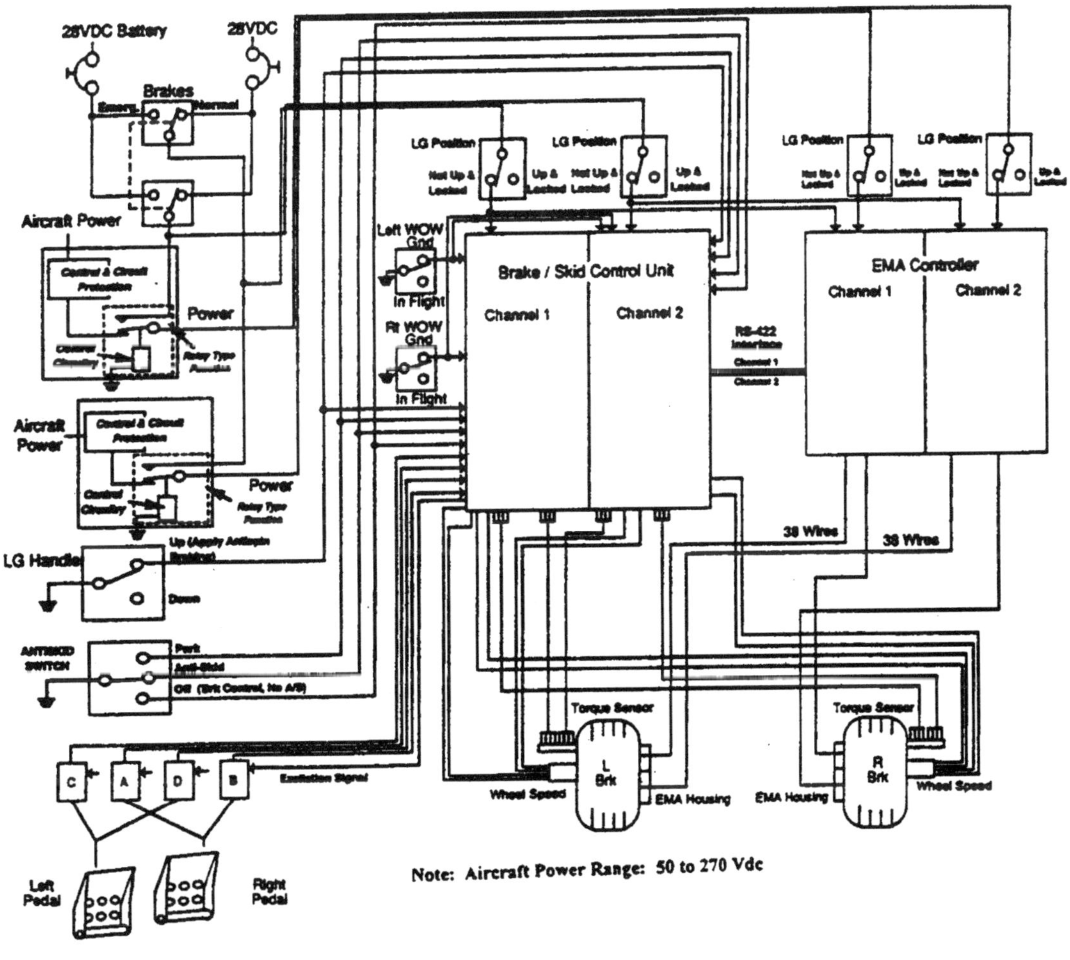

Fig. 13.8: Functional Diagram for Electric Brake Applications

New materials for brake discs are in a process of evaluation sponsored in a magnitude of 50% by the U.S. Airforce together with all major U.S. brake manufacturers.

Evaluation activities are focussed in the development of carbon/silicium materials. The following advantages comparing to carbon are expected to become achievable:

- reduced wear in particular in the low speed regime
- more stable dynamic coefficients of friction
- increased RTO coefficient
- smaller and lighter heatpacks because of 113% more volumetric heat stowage capability

In how far the currently available test results can be verified one day as a serious competitor against the currently used carbon structure is difficult to forecast.

13.9 Conclusion

In contrast to the automotive industry with its tremendous production rates, the relative small quantities of products required by the aviation industry are a certain barrier for the introduction of new technologies, because of the current financial situation of the aircraft market.

However every product which can prove to be technically highly efficient in conjunction with attractive low life cycle costs, can always find an open door as a candidate for new aircraft projects. The challenging slogan of today is to design the "affordable aircraft".

13.10 References

- Archive Ohly
- A-5-SAE-Aerospace Landing Gear Systems Conference Publications
- Airplane Design – Dr. J. Roskam Part IV
- Landing Gear Design Handbook – Norman S. Currey
- Emerging Technologies in Aircraft Landing Gear – John A. Tanner & Co.

Part III

Test and Experiences

14 The Air Disc Brake in Commercial Vehicles - Field Experiences

Dipl.-Ing. A. Siebke, Dipl.-Ing. M. Gruber, Dipl.-Ing. R. Ober, Knorr-Bremse-Systeme für Nutzfahrzeuge GmbH

Munich, Germany

14.1 Development of the market

The air disc brake is today established as standard equipment for heavy commercial vehicles in Europe.

With consideration to the market penetration by this technology since its appearance in the early nineties (**Fig. 14.1**), the drum brake prevailed at the beginning because the disc brake was offered only by a few tractor and bus manufacturers and because its utilisation was initially limited to the front axle. However, since the market introduction of tractor vehicles equipped with disc brakes on all wheels in the middle of the nineties, a consequent U-turn in direction of the disc brake took place.

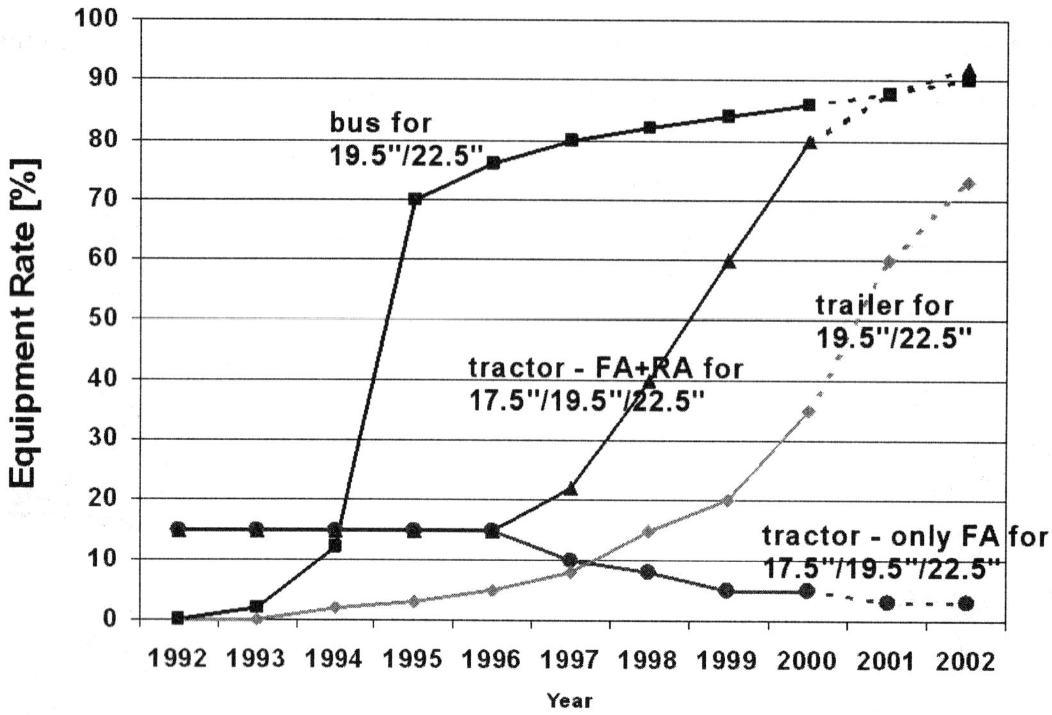

Fig. 14.1: The Air Disk Brake becomes a Standard

In the meantime, 80% of all air-braked tractor vehicles which are new registrations in Europe, are equipped with air disc brakes. For trailer vehicles, an analogue change took place slightly later and today, approx. 40% of all new trailers are equipped with disc brakes. The trend is still rising. In the field of buses, the hydraulically operated disc brake installed on the front axle of touring coaches was first of all replaced by the pneumatic brake which was then used nearly comprehensively on all axles both of touring coaches and city busses.

With reference to the South American market which is strongly oriented to the European technology, the same signs regarding a successive substitution of the drum brake by the disc brake can be registered. In North America and Asia, only a niche position is attributed to the air disc brake, however there is also a trend towards a wider utilisation.

14.2 Utilisation spectrum and applications

The utilisation spectrum of the air disc brake in commercial vehicles is very wide conforming to the variety of the vehicle types offered in the market (**Fig. 14.2**). The same types of brake or identical brakes shall meet completely different requirements with relation to the vehicle type and the use of the vehicle.

Fig. 14.2: An Air Disk Brake for all Types of Vehicles

With consideration of the individual requirement profiles for brakes to be installed on the typical vehicle categories, such as tractor, trailer, touring coach, city bus and construction-site vehicle, it becomes evident that a very wide spectrum of characteristics must be realised to comply with all customer's expectations (**Fig. 14.3**).

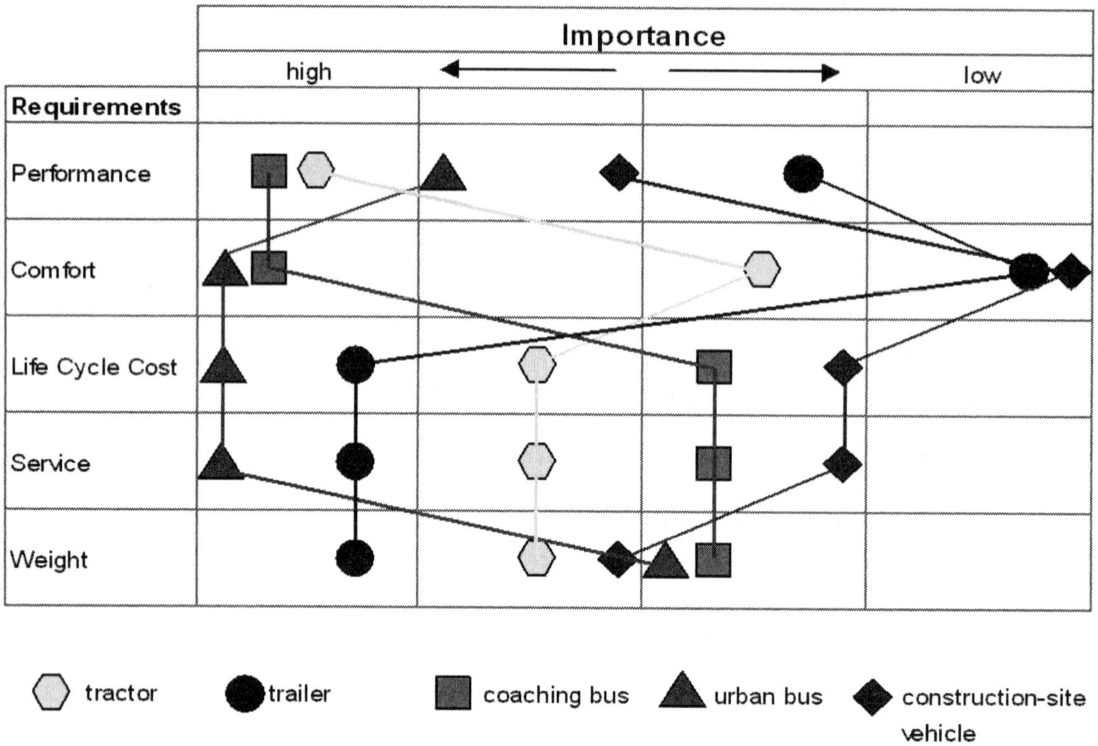

Fig. 14.3: Large Requirement Profiles

Furthermore, there are very different installation situations in the chassis area, which must be taken into account. This is achieved for instance by varying the position of the calliper and by alternative actuator positions. Two typical examples are as follows: the hanging calliper position with axial brake actuator as used in trailers and the upper calliper position with radial brake actuator in the case of rear axles in city busses in the portal design (**Fig. 14.4**, **Fig. 14.5**, **Fig. 14.6**).

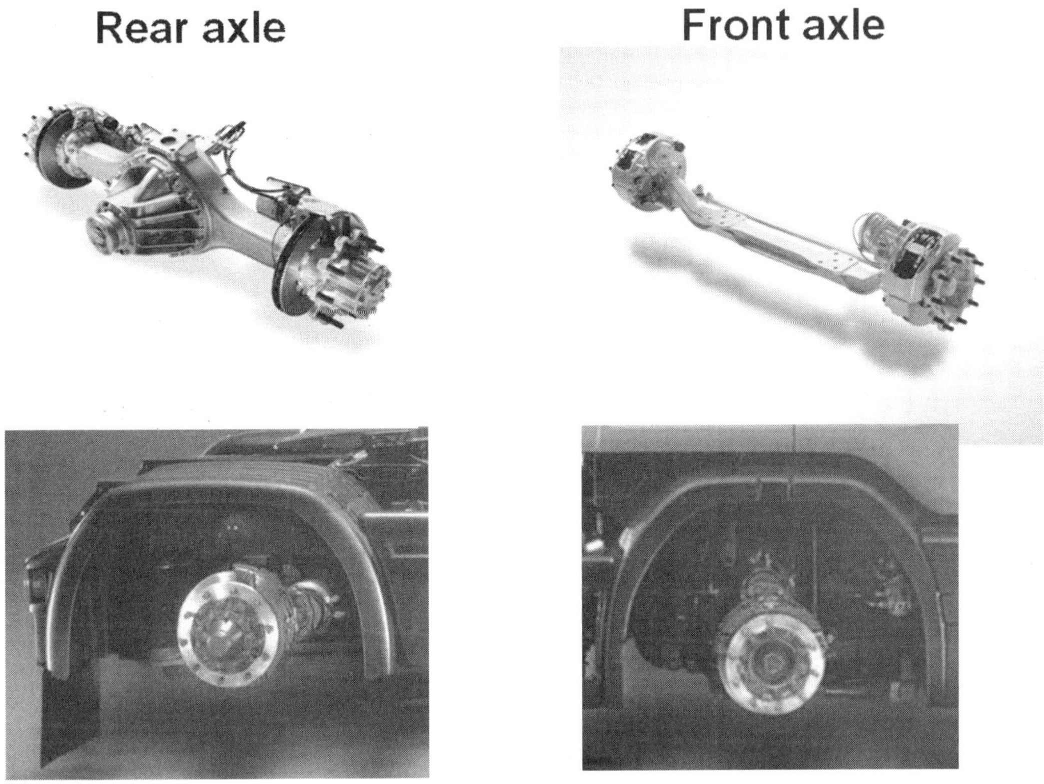

Fig. 14.4: Applications on Tractors

Fig. 14.5: Applications on Trailers

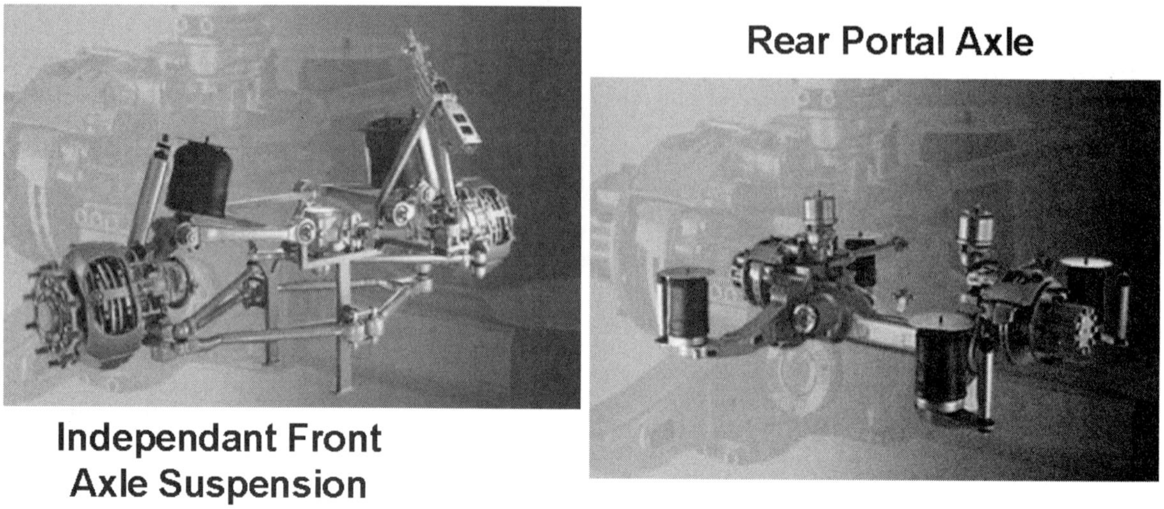

Fig. 14.6: Applications on Buses

Adapted designs are used for off-road or construction-site applications in order to avoid impairment of the brake function due to environmental influences. As an example the splash shields for the pad shaft shall be given which prevent the penetration of small stones into the brake (**Fig. 14.7**).

Fig. 14.7: Protecting Plate for Offroad Application

14.3 Design and function

In the case of air disc brakes, the design of the sliding caliper brake with inner eccentric lever application of the roller bearing type became generally established (**Fig. 14.8**). This type of power transmission is characterised by a high efficiency degree (≥95%) and a small hysterises (≤10%) and therefore, provides best conditions for the use of electronic brake and dynamic drive control systems. The main difference in the design of the three usual sizes for 17.5", 19.5" and 22.5" wheels is that the smallest size has only one piston whereas the larger types mainly correspond to the design with two pistons. This allows for a uniform force application and therefore a reduced tangential oblique wear in the case of the increasing pad length in relation to the brake dimensions.

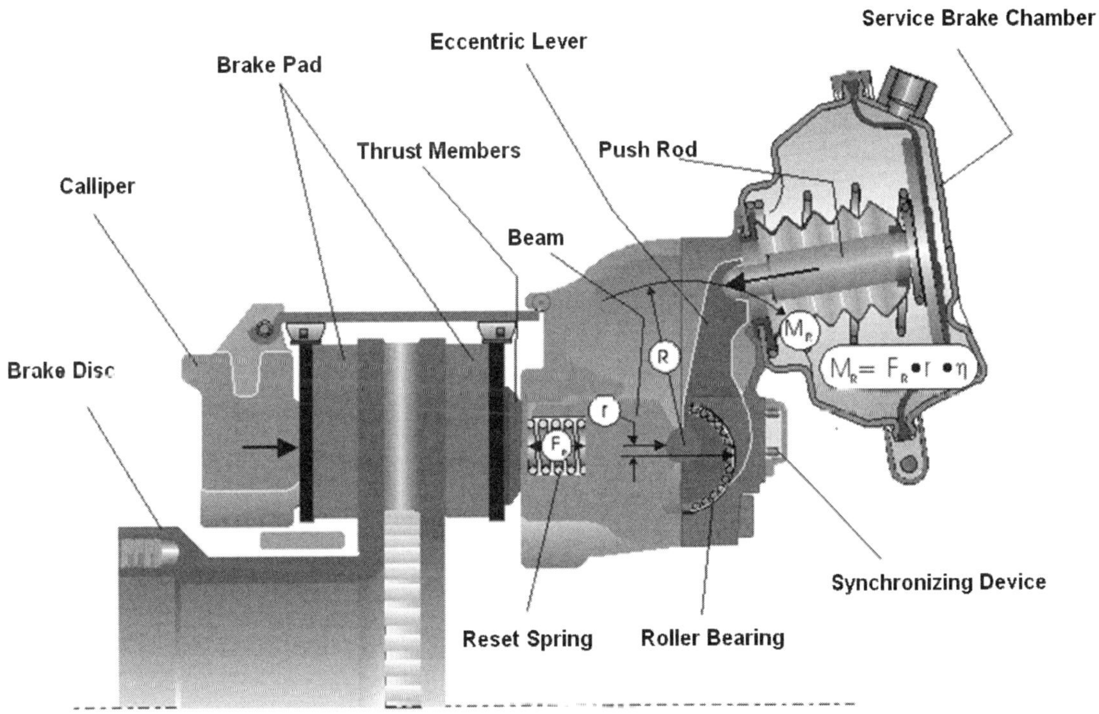

Fig. 14.8: Cross Sectional View

The ratio of the force application amounts between 10:1 and 22:1 according to the brake size. The actuation lever which is operated by the brake actuator flanged on to the calliper, actuates a beam with the help of an eccentric shaft. The pad on the piston side is pressed on to the brake disc by this beam with the help of threaded screw and piston. The reaction force is transferred on to the brake pad mounted on the wheel side with the help of the sliding calliper.

Pad and disc wear compensation is carried-out by means of turning the threaded screw with the help of an integrated automatic wear adjuster which is actuated by the actuation lever. In the case of the design with two pistons, the turning movement of the threaded screws is synchronised for a uniform advance of the contact pieces. The reliability of the adjuster mechanism, even under severe load, in particular its ability to withstand shocks and vibrations, is of great importance for the brake's function. Re-adjustment of the brake during servicing as well as continuous sensing of the brake pad wear (optional) are also carried-out by the adjuster unit mechanism.

14.4 Field monitoring and data evaluation

The air disc brake as described, represents a relatively young product. Therefore, it was and is important for the purposeful further development to monitor its proving under the mentioned, wide operational conditions with the help of active field monitoring.

Feedback from the field regarding e.g. service life, wear behaviour or failures is made available to the product R&D in two ways. One consists in the Quality Assurance with the related customer service organisation and the other is the systematic field monitoring by a team directly belonging to the R&D department. Whereas the customer service carries-out a product related analysis and is mainly active during the warranty period, the field monitoring observes the behaviour of the product within the vehicle system over the entire vehicle's life (**Fig. 14.9**).

Fig. 14.9: Field monitoring influences Development

The continuous field monitoring of the air disc brake over more than five years resulted in a variety of experiences and therefore provided initiatives for the further development of the product. The analysis of vast field data revealed special parameters influencing the function and service life of the brake and of the friction pairs. Within the scope of these activities, approx. 2000 brakes of all sizes with a different service life and from all relevant scopes of application have been analysed up to now. The extreme utilisation on Finnish or Brazilian tracks, but also the operation under difficult conditions in Germany shall be given as examples (**Fig. 14.10**, **Fig. 14.11**).

Fig. 14.10: Extreme Utilisation Conditions: Germany, Bus Application

Fig. 14.11: Extreme Utilisation Conditions: Germany, Bus Application

The given number of inspected brakes corresponds to approx. 1 thousandth of all brakes delivered up to the middle of the year 2000 and therefore, allows to draw conclusions about all products in the field (**Fig. 14.12**).

Basis:

- approx. 2000 brakes
- all vehicle categories > 6 t GVWR
- all fields of application
- various service time (also after warranty period)

	17,5 "	19,5 "	22,5 "
Tractor	60	190	800
Trailer		200	400
Bus			350

Fig. 14.12: Field Monitoring in all Fields of Application

Regarding the wear of pads and discs, "Weibull"-evaluations representing the cumulative failure rate over the brake service life in form of a compensating curve in a logarithmic grid, were carried-out. This allows to make a statement regarding the life expectations of the brake in the above mentioned specific fields of application:

For tractor vehicles (**Fig. 14.13**, **Fig. 14.14**), the characteristic service life T at which 63.2% of the investigated brakes reach a state of complete wear, amounts to more than 450,000 km for the pads and nearly 1,200,000 km for the brake discs. The parameter b = 2.2 or 2.4 regarding the straight line's gradient is typical for wear or fatigue as failure cause. In general, these values correspond to the expectations of the customers, in particular since the nearly integral factor of 3 between pad and disc service life allows for a disc change together with a pad change at an economically viable moment.

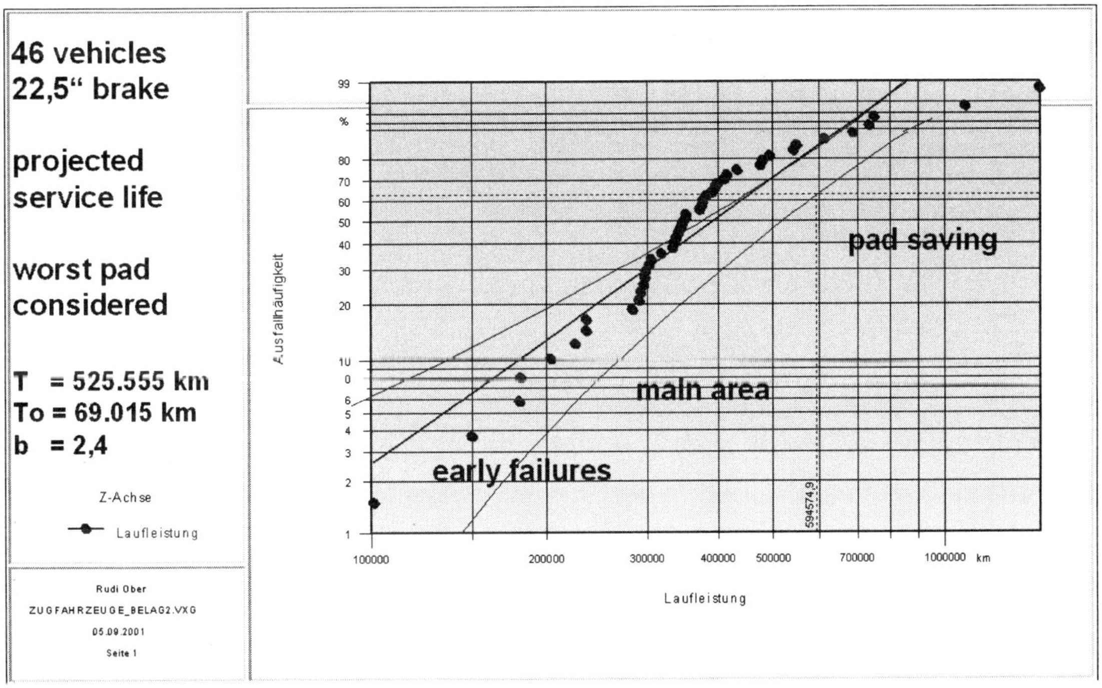

Fig. 14.13: Pad Service Life Tractor

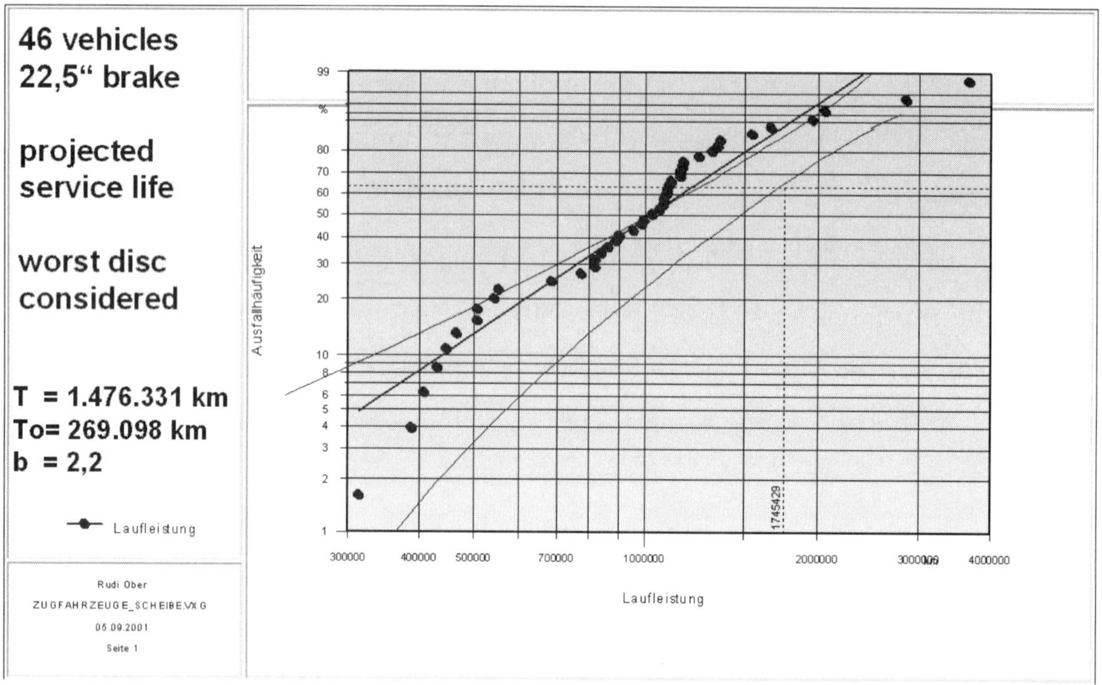

Fig. 14.14: Disc Service Life Tractor

In the trailer sector (**Fig. 14.15**, **Fig. 14.16**), the investigated vehicles give a significantly wider distribution regarding the failure probability which could happen from 110,000 km up to more than 1,500,000 km for the brake pads and from 340,000 km up to 1,500,000 km for the brake discs in the case of a characteristic service life amounting to T> 600,000 km and 1,250,000 km, respectively. This wide dispersion signifies that customer expectations are in some cases not met or are even exceeded. The factor b=1.4 for the pad service life defines an initially stronger, then weaker increasing failure rate. This difference in comparison to the tractor vehicle can be explained first of all by the large influence held by a harmonised vehicle combination when considering the wear of the trailer brakes. For the evaluation of the brake service life in the trailer, the corresponding tractor must also be taken into account (if this allocation is possible).

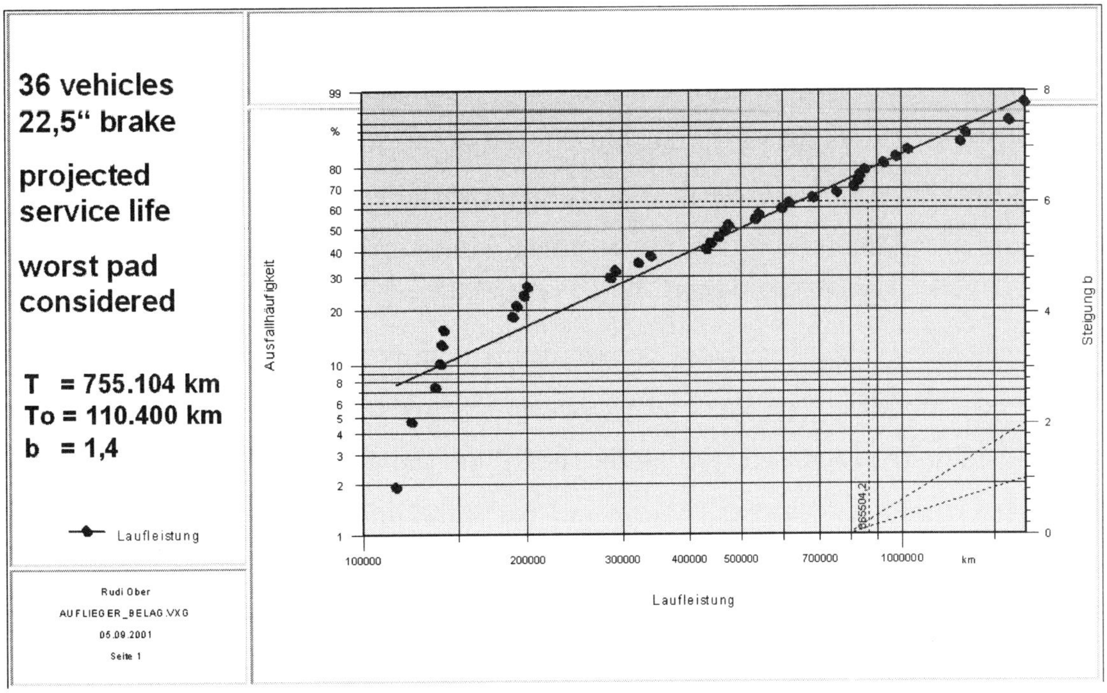

Fig. 14.15: Pad Service Life Semi-Trailer

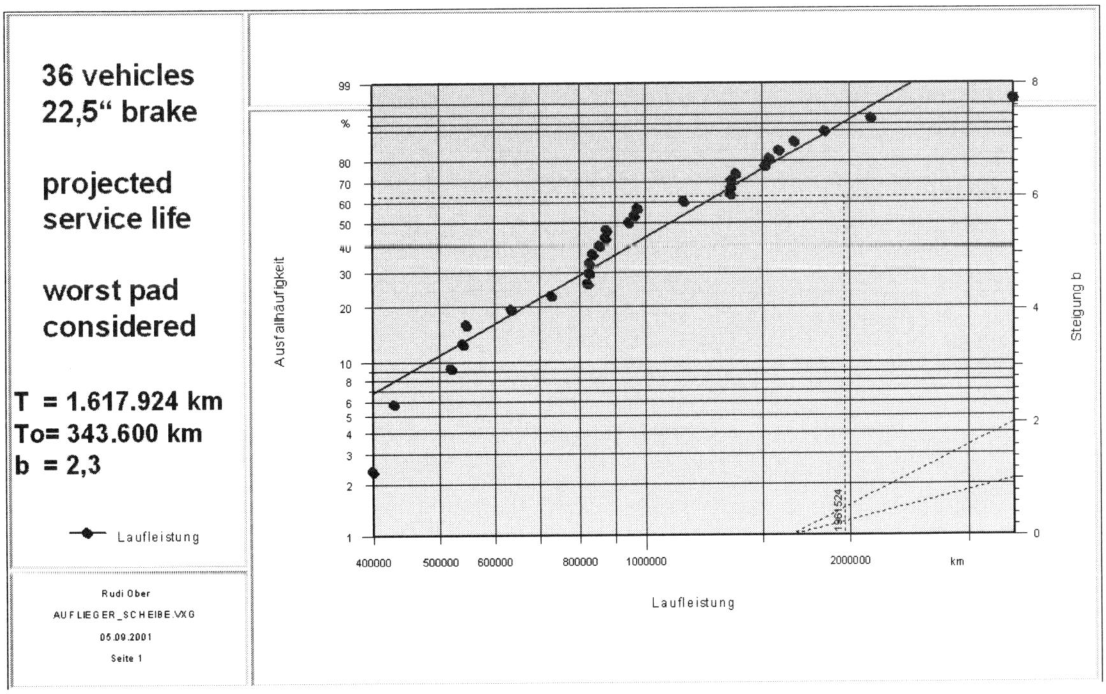

Fig. 14.16: Disc Service Life Semi-Trailer

Finally, in one extreme example of a city bus application (**Fig. 14.17**), the investigated vehicle group showed a characteristic service life of T = 38,000 km for the pads at the drive axle and therefore, does not meet the required performance of at least 100,000 km. The failure frequency b = 11 signifies an extremely progressive characteristic of the failure rate, i.e. the life expectations of the brakes are within a very narrow service life interval. The cause for these results deviating extremely downwards in comparison to the above given values, must be seen first of all in the specific operating conditions of city buses. Frequent stops and limited carrying-off of heat inside tightly enclosed wheel houses result mainly for the drive axles in a high continuous temperature of the brake disc amounting to 400°C up to 500°C. Thus, extremely high requirements are made regarding the wear resistance of the brake pads under high temperatures. Brake and pad manufacturers, in close cooperation with the vehicle manufacturers are urged to realise service life improvements for city bus applications in a short time in order to avoid a strong weakening of the disc brake's reputation.

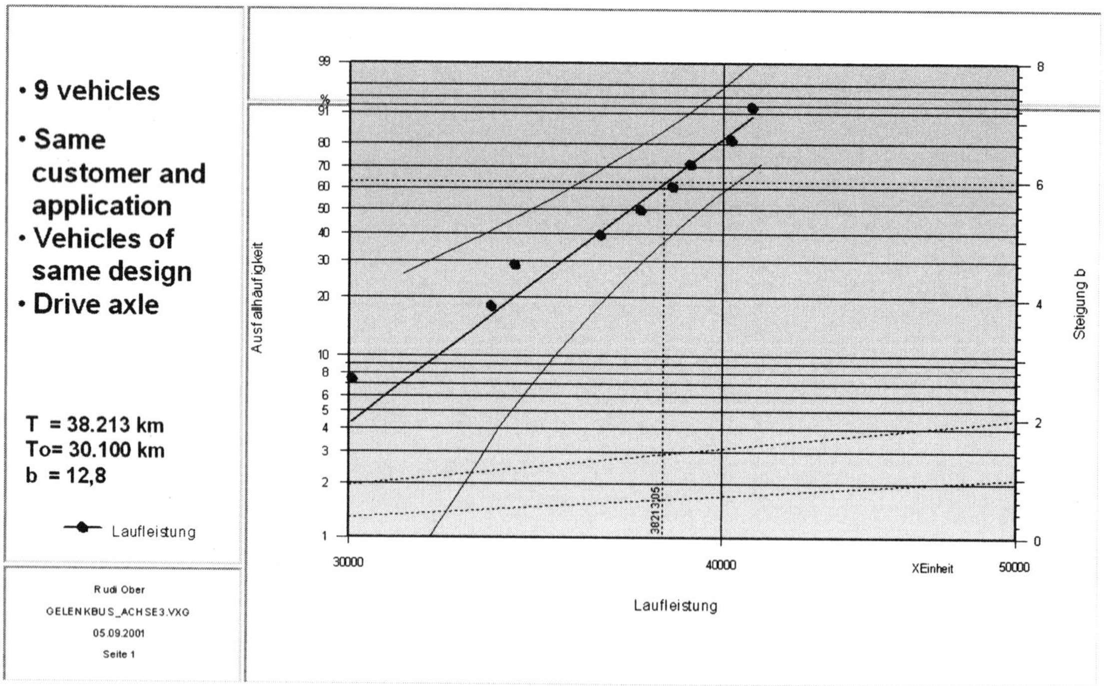

Fig. 14.17: Pad Service Life City Bus, extreme Example

14.5 Technical further developments

In addition to the investigations regarding the wear behaviour of the brake, the proving of the brake under extreme operating conditions was also recorded within the scope of the field monitoring. This allowcd not only the analysis of series products, but also the specific testing of new or modified components under real operating conditions and over longer periods. This resulted for instance in further developments (**Fig. 14.18**) such as

- Optimised calliper guidance under extreme vibrations and abrasive wear
- Replaceable overload protection element at the wear adjuster for easier servicing of the brakes
- Lateral wheel shields against uneven wear of trailer brakes due to dirt accumulation during operation on unmade roads
- Splash shields for the pad shaft against uneven wear on the drive axles of buses due to dirt accumulation
- Hermetical sealing of the calliper interior space in the case of not assembled brake actuator
- Splined disc with spacing elements

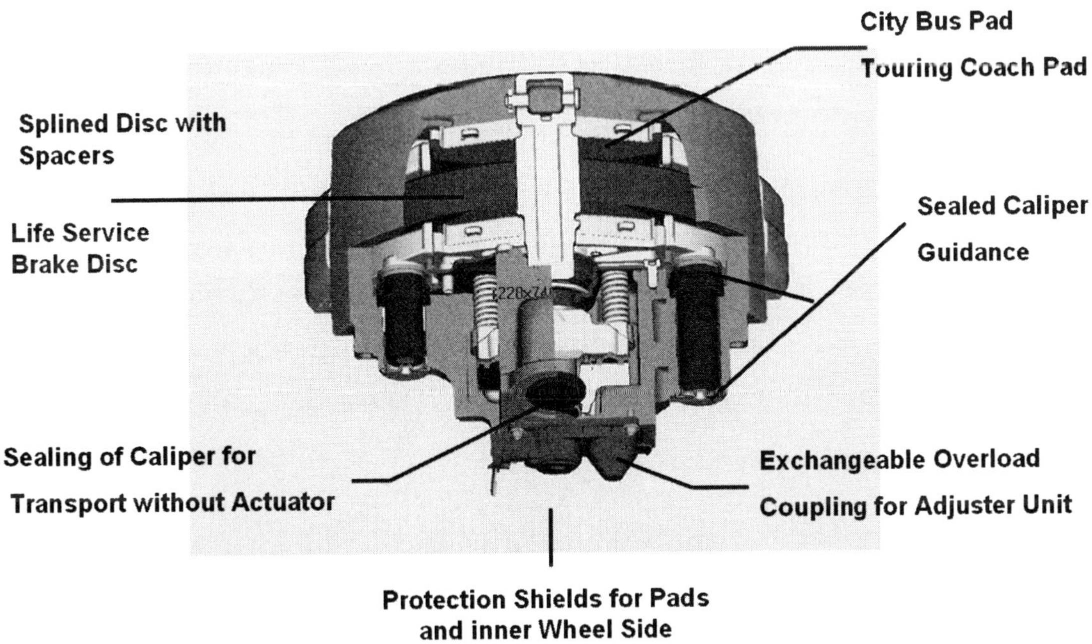

Fig. 14.18: Field Experience influences Product Development

The following projects are still under development such as:

- Specific city bus pad
- Noise-reduced brake pads for touring coaches
- Service life brake disc

14.6 Outlook

Future R&D work regarding disc brakes will take into consideration also the requirements of the young markets such as North America and Asia, but of course also the changing customer requirements in Europe as well as specific market sectors. This is characterised by a further

- Weight reduction
- Installation space reduction
- Increase of the service life for the friction pairs
- Reduction of the life cycle cost
- Substitution of mechanial functions by electrical functions
- Integration of further functions into the brake

Due to the current field experiences which have been made with the air disc brake since its series introduction in commercial vehicles, it is to be expected that this technology will find a further increased distribution in all markets and all fields of the commercial vehicle industry.

The principal advantages of the disc brake in comparison to the drum brake, such as reduced weight, performance, compatibility with electronic brake control systems and ease of servicing, are the basis for this development. In the end however, the proof of reliability and economic viability in the daily operation and the respective customer satisfaction will ensure the permanent success of this product.

15 Effectiveness and Acceptance of Brake Tests in Commercial Vehicles During Periodical Vehicle Monitoring (HU und SP)

Dipl.-Ing. R. Staebler, Dipl.-Ing. H. Erhard, TÜV Verkehr und Fahrzeug GmbH

Munich, Germany

Summary

The outdated intermediate tests for commercial vehicles, such as intermediate tests and special brakes test, were replaced as of December 1, 1999 by the safety tests to take place exactly between the main test dates (HU) – generally referred to as TÜV (German MOT) tests. Together with the car repair garages which used to have sole responsibility for these tests, now the automotive experts responsible for the main TÜV tests can now also carry out the safety test. The safety tests (SP) and main tests (HU) carried out by the TÜV Süddeutschland during 2000 and the experience gained in this procedure provide initial conclusions with regard to the effectiveness and acceptance of the brakes tests as part of the HU and SP.

15.1 Turn two into one

Since the summer of '99, heavy commercial vehicles and all buses in Germany bear another test badge at the back on the left referring to the next date for the safety test performed as of December 1, 1999.

So what has changed?

In 1960, the intermediate and main brakes special tests, or ZU and BSU for short (ZU prescribed as supplement to the main test), were generally adopted as BSU in the Road Traffic Regulations (StVZO) from the "Directive on the Operation of Motorised Companies in Passenger Transport" (or BOKraft for short). On the basis of the former status of vehicle technology and the prevailing accident situation – the statistical average for all technical faults ascertained by the police to be the cause of accidents was 2.8% for all vehicles, compared to 1% today - , the incorporation or introduction of the stated tests in the Road Traffic Regulations prescribed in 1960 and extended considerably in 1971 was seen to be justified. The introduction of the BSU 1960 as supplement to the HU also took account of the fact that in some cases an internal examination of the wheel brakes was necessary, and that this could not be performed by the TÜV testers, because this would involve assembly work. This conforms with the principle: **"The tester may not repair or assemble, and vice versa".** This also avoids any conflict of interests among all those involved, because a neutral expert decides what has to be repaired!

There was always a need for an inner test when demanded explicitly by the vehicle or brakes manufacturer. Such instructions are no longer issued fundamentally today, or only in a very slight scope. This was possible thanks to the improved design of wheel brakes with altogether much better wear and repair characteristics, better rating of brake behaviour to towing vehicle and trailer, wear-free retarding brakes etc. This trend will increase further with the introduction into commercial vehicles of disc brakes, electronic braking systems, retarders and other new aspects, some of which are still in the development stage (ceramic disk brakes). After reviewing the tests prescribed for commercial vehicles (trucks, buses, trailers) together with the corresponding fault statistics and with regard to practical test experience with new vehicle designs, all those involved made the suggestion of combining the two ZU and BSU tests into the Safety Test (**Fig. 15.1**).

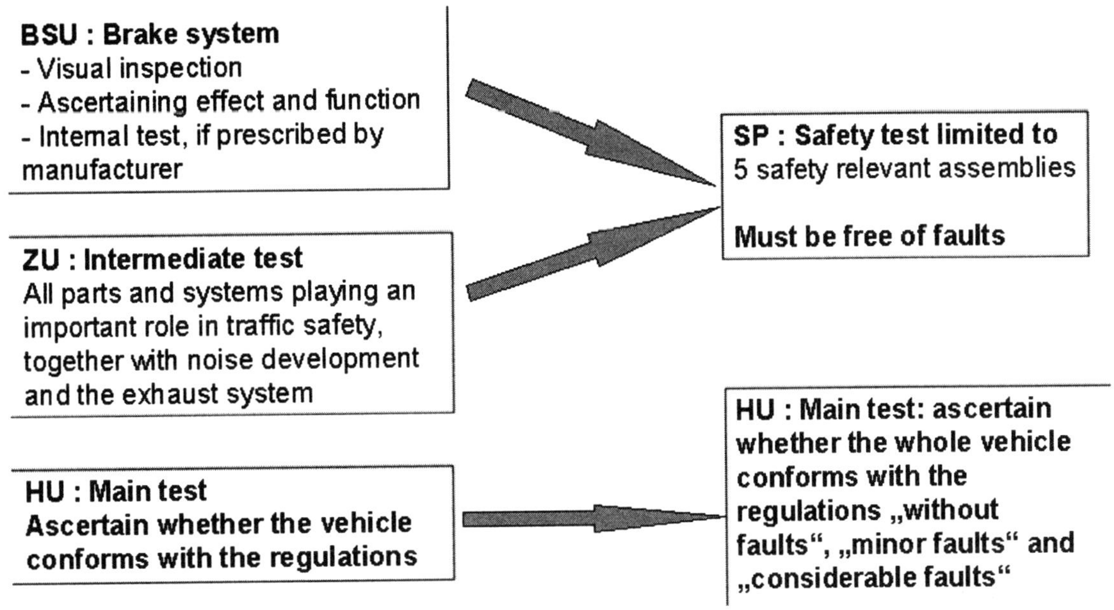

Fig. 15.1: The new safety test

For this (new) test, the vehicles are not examined in their entirety as in the HU, but only important vehicle components which are subject to wear and liable to repairs are checked, and various functions and effectiveness tests performed. The vehicle systems concerned here are the chassis, running gear, connection systems, steering, tires, wheels, brakes and exhaust system. There are concrete instructions as to which parts and systems are to be tested and how this should be assessed. The remaining vehicle parts or systems not included in this test scope are not tested. This means that the vehicle owner and driver now has a greater responsibility for checking and inspecting the vehicle. For example, this refers to the lighting system whose function can be easily checked by vehicle owner or driver. These vehicles are checked every twelve months in the main test anyway for functions and conformity with the regulations. To take the lighting system again, the main test will object to any headlamps, lights or other lighting equipment not mounted correctly, not certified for this vehicle or mounted without certification. By contrast, the SP ignores such aspects. In addition to streamlining the wide variety of tests (HU, AU, BSU, ZU, §57b) the whole procedure has been given a new dynamic aspect by altering the time periods involved (**Fig. 15.2**)

Commercial vehicles class N2 (> 7.5 t) and N3 (> 12 t)

before 30.11.1999

Type fo test	Years after first reg.
HU (incl. BOKraft)	1 / 2 / 3 / ...
BSU	1 / 2 / 3 / ...
ZU (> 9,0 t)	1,5 / 2 / 2,5 / 3 / ...
ZU (<= 9,0 t)	no test
AU	1 / 2 / 3 / ...

after 1.12.1999

Test type	Years after first reg.
HU (incl. BOKraft)	1 / 2 / 3 / ...
SP (N2 > 7,5 t)	3,5 / 4,5 / 5,5 / ...
SP (N3 > 12,0 t)	2,5 / 3,5 / 4,5 / ...
AU	1 / 2 / 3 / ...

Fig. 15.2: Comparison old/new regulations

These alterations to the time periods for the SP for new vehicles and then in regular intervals for older commercial vehicles were derived from the knowledge that the wear and repair vulnerability of the vehicles increases with increasing age and the usually associated higher mileage. This relieves the pressure on the owners of commercial vehicles by summarising the two ZU and BSU tests in the SP on the one hand, with the dynamic aspect of the SP intervals with regard to vehicle age on the other. Referred to today's stocks of commercial vehicles, around 1.4 million tests less per year will be required, which is sure to have an effect on certain commercial vehicle repair garages, as the safety test now stipulates a completely new quality and dimension for the new test.

Prescribed quality assurance systems, monitored and audited by concealed tests by the certification agencies, are to guarantee that the testing garages use the same rules and regulations and that there are no essential shifts on the market. By including the experts in the safety tests – this was precluded in the ZU and BSU tests, this has produced a healthy competition situation which can only promote the safety aspect, together with retraceable criteria thanks to the black/white appraisal method.

15.2 Test contents

In the HU, the vehicle is checked in its entirety for correct functions and conformity with the regulations, whereas the SP only deals with specifically designated components subject to extreme wear. Brake technology on a commercial vehicle plays a major role in this context. There are exact search routes and fault possibilities for every component (**Fig. 15.3**), so that in the end checklists are used and when there are no faults, the wanted test badge – here a year's arrow on the basic sticker with the numbers of the months – can be issued.

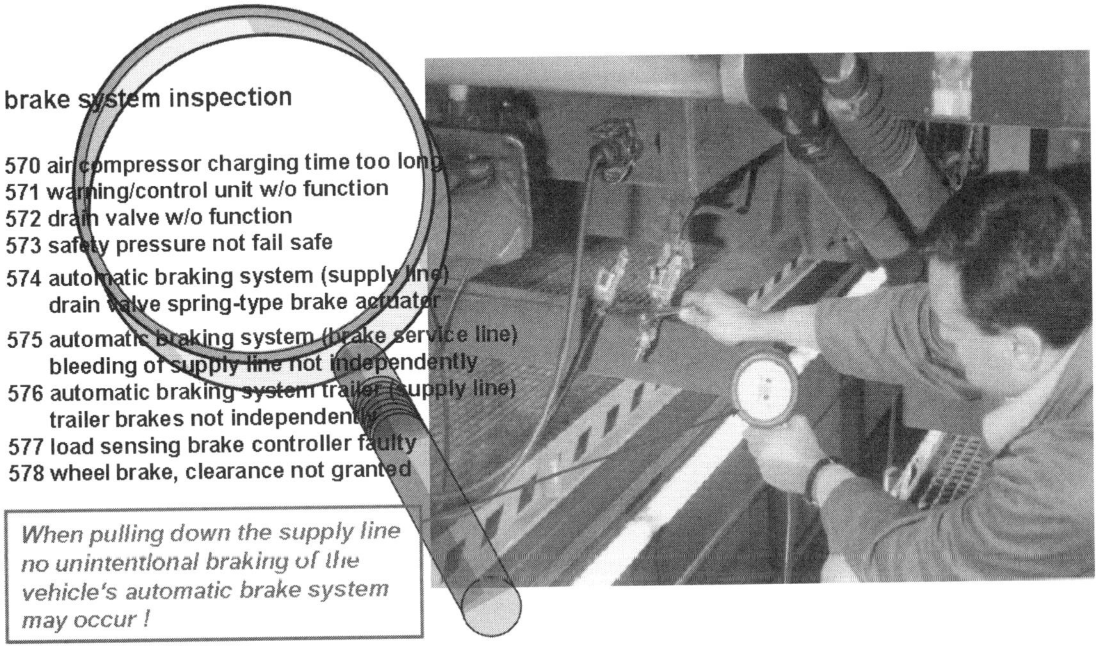

Fig. 15.3: Example: braking system

Documentation is also strictly reglemented, consisting of a safety test book, (**Fig. 15.4**), with continuous verification of all test dates and test reports for the whole service life of the vehicle.

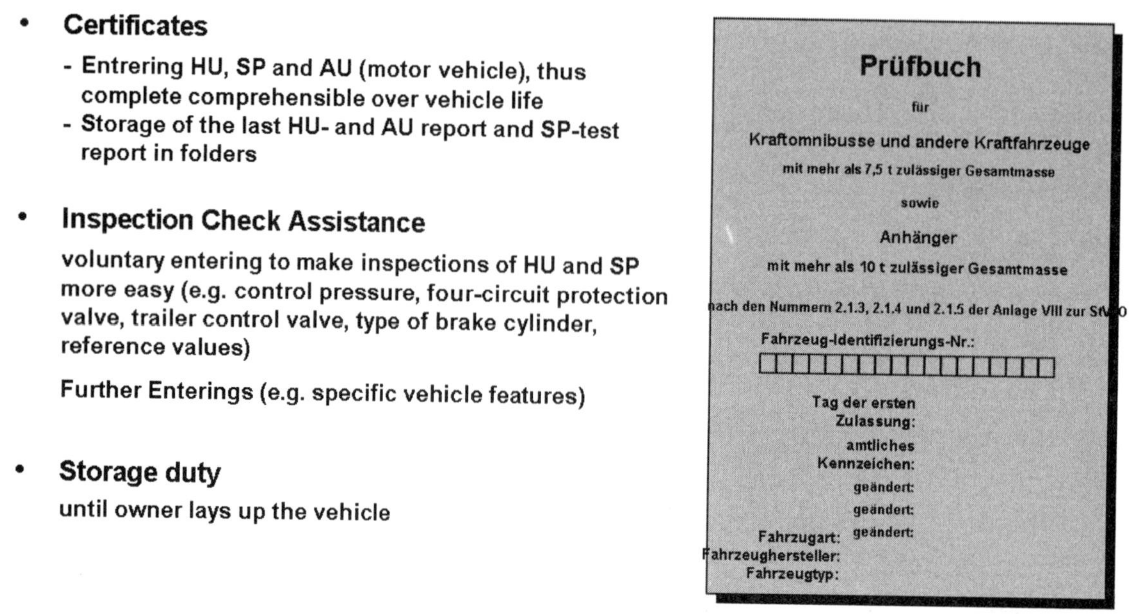

Fig. 15.4: SP safety test book

Any faults discovered in the vehicle (**Fig. 15.5**) have to be eliminated first be verified and monitored by external inspectors.

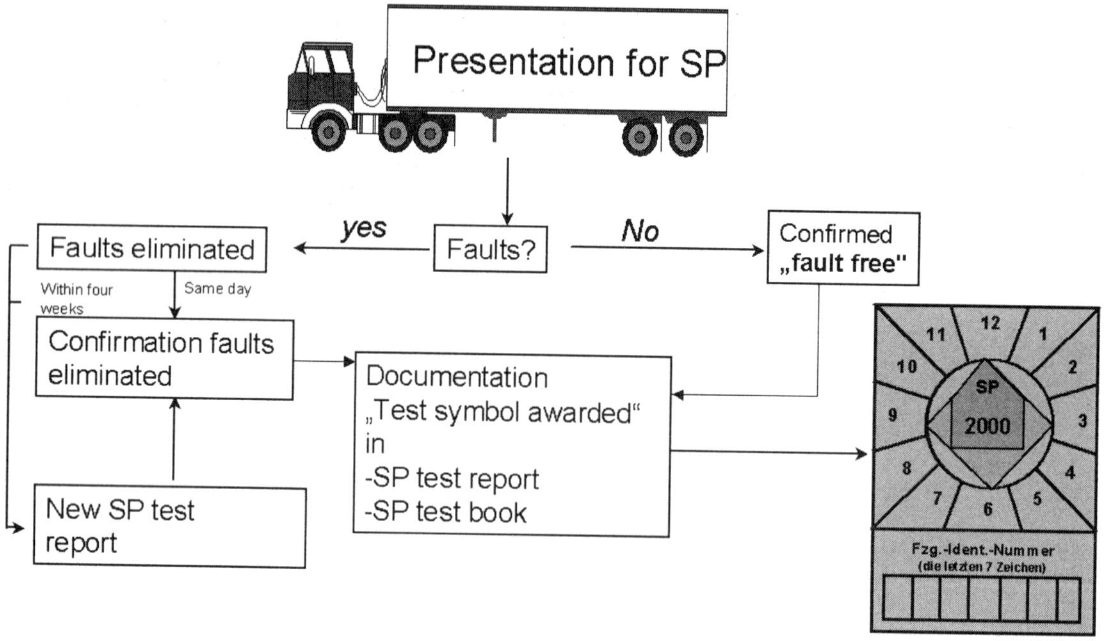

Fig. 15.5: Safety test procedure

A new SP test report certifies that the faults have been eliminated: once this is complete, the test badge can be awarded. Faults which are ascertained during the

SP but are not part of this particular test can be passed on to the vehicle owner as "information". They have no influence on being awarded the SP test badge. During the HU, there are practically no noticeable reductions in the test scope with reference to such a detailed test of the braking system as the SP, although here "minor faults" may be detected, which the vehicle owner then has to eliminate immediately even if he has received a test badge. It goes without saying that severe faults have to be eliminated before the badge can be awarded.

15.3 Test results focussing on the braking system

If all test results were summarised, this would result in a huge data file of around 1 million SP. Unfortunately, practical conditions are somewhat more tedious. The test monitoring organisations have managed to summarise their results by electronic means, but most of the safety tests are still stored on many different data carriers in the nearly 4,000 repair garages, right down to conventional handwritten duplicated sets. The ZDK has collected the individual "files" for 2000 through its guilds, but a central evaluation of the data is not yet available. We have therefore examined our own approx. 35,000 safety tests for 2000 and then in addition given particular attention to the braking systems.

The distribution of faults tends to fluctuate depending on vehicle type (**Fig. 15.6**).

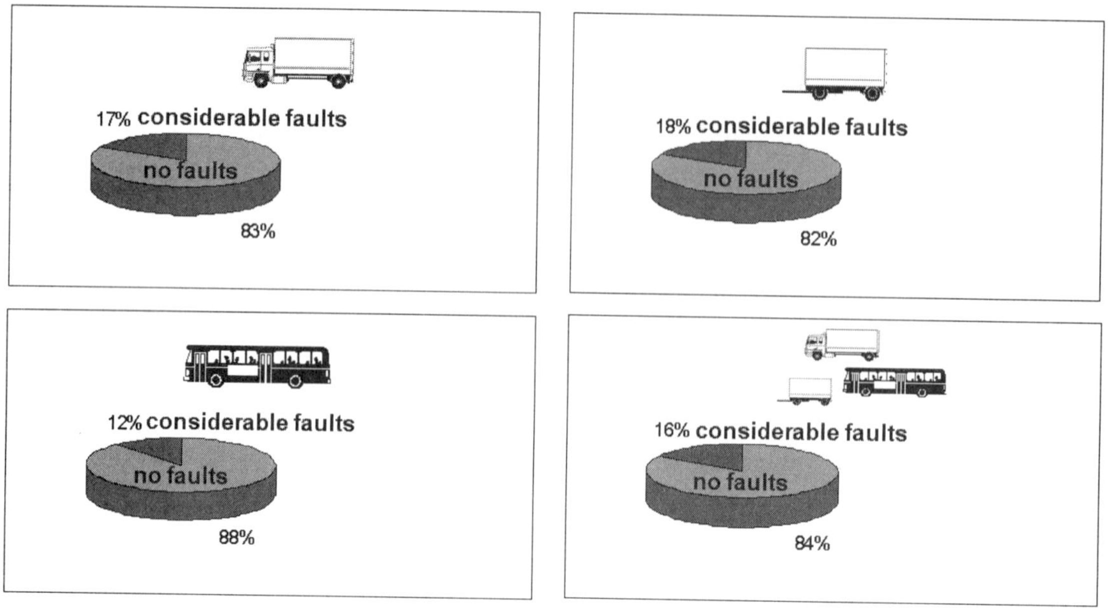

Fig. 15.6: Fault distribution SP (no faults / considerable faults)

For KOM, this is sure to be connected with the "more frequent" SP but also with the higher risk potential for passenger transport. An examination of the vehicles with faults then automatically produces the following figure (**Fig. 15.7**) with the vehicle faults doubling with increasing age.

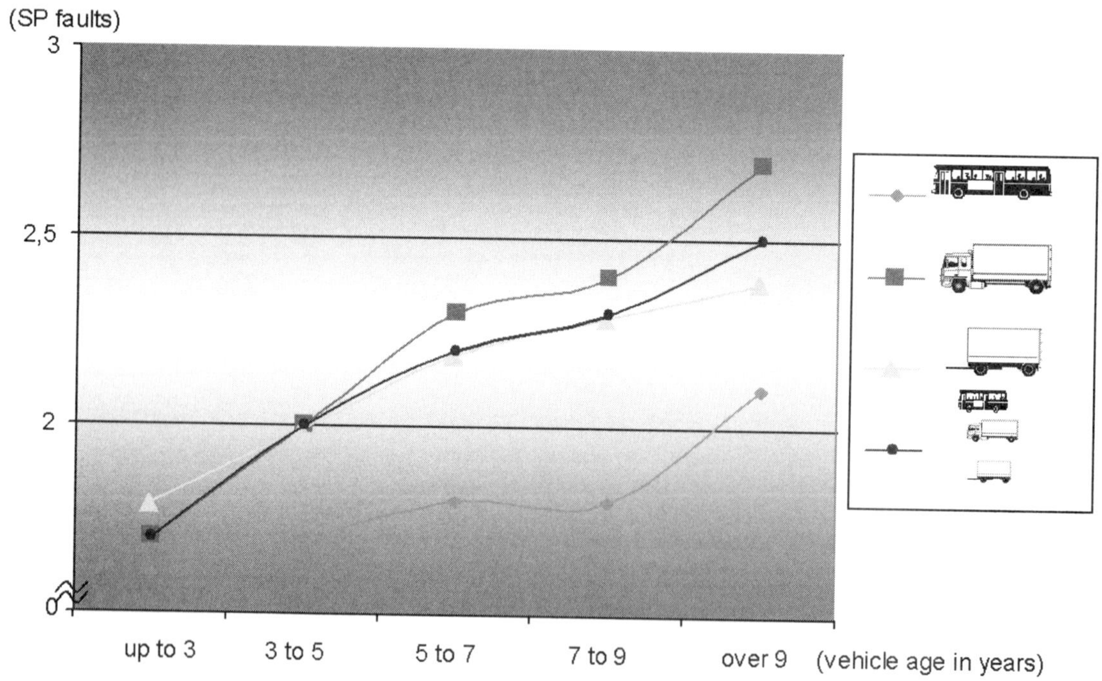

Fig. 15.7: SP faults depending on vehicle age

We have also checked a possible North/South gradient for the SPs of the ÜOs (**Fig. 15.8**).

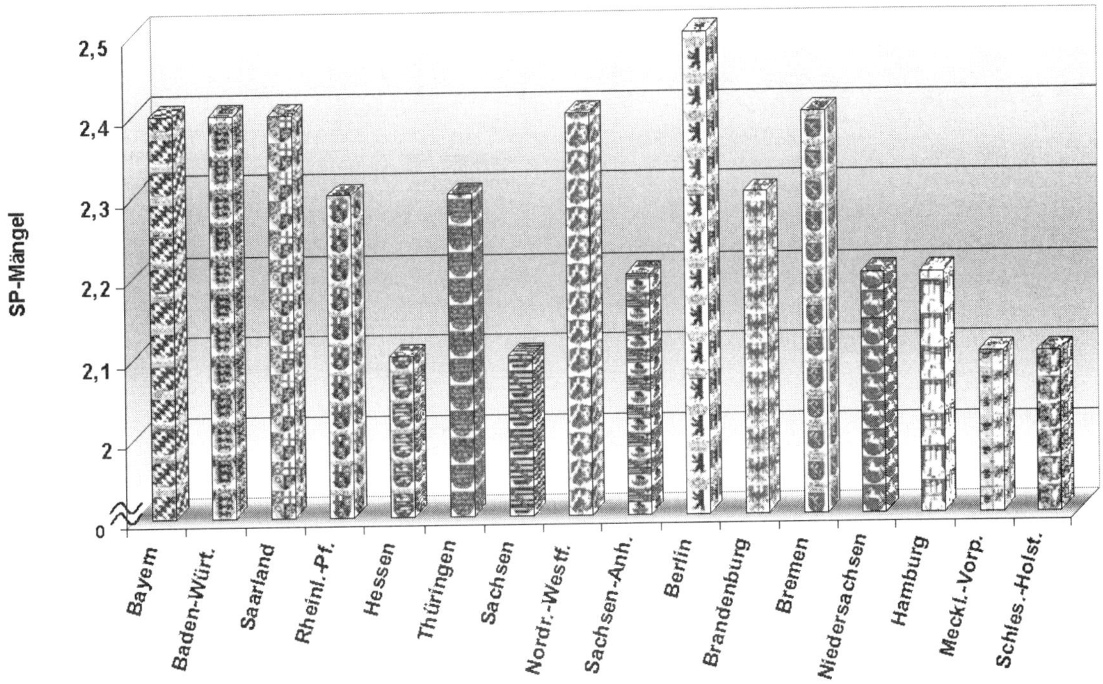

Fig. 15.8: SP faults per federal state

The fluctuation range is minimum and usually dependent on regional factors. This means that there are no safer regions or states, all transport companies and fleets boil the same water. So what are the individual faults in the individual tested brake components (**Fig. 15.9**)?

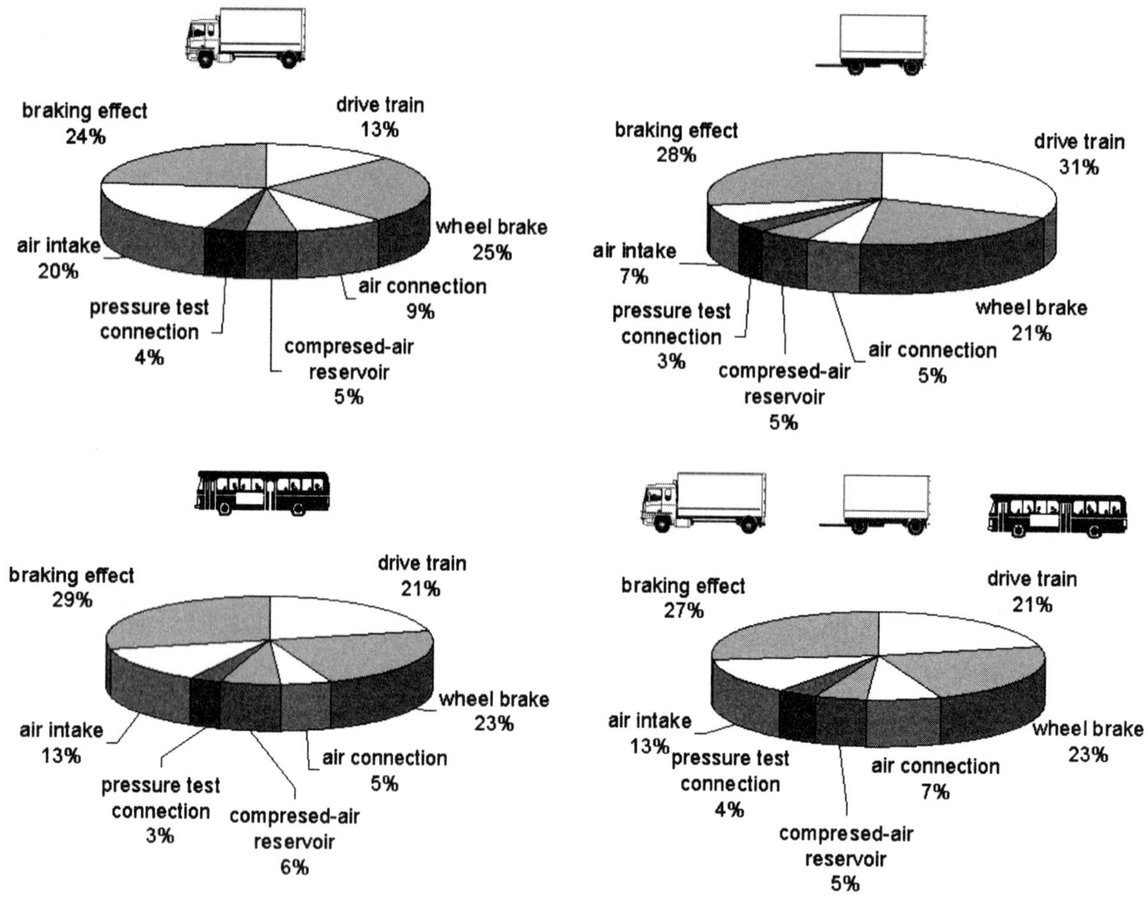

Fig. 15.9: SP fault distribution in braking systems

The vast majority was accounted for by all aspects of the braking effects, and most noticeably the purely mechanical parts involved in force transmission in the trailers (cables, rods, lines, hoses). Which is not surprising, as these vehicles are about three times as old as the trucks towing them. The wheel brake itself also accounts for a large share, this includes the brake cylinder, brake drum or disk and brake lining, as far as this can be seen without being dismantled.

15.4 Acceptance by vehicle owners

Compared to the BSU and ZU, the SP enjoys high acceptance among vehicle owners, as it not only contributes to traffic safety but also effectively to operating safety and thus to the stand-by readiness of a commercial vehicle. Many vehicle owners also use these tests as a means of quality control with regard to the external repair garage or even their own fleet garage, as the test contents of the HU and SP are at least similar with regard to high-wear parts, so that conclusions can be made regarding the test depth of the individual tests (**Fig. 15.10**)

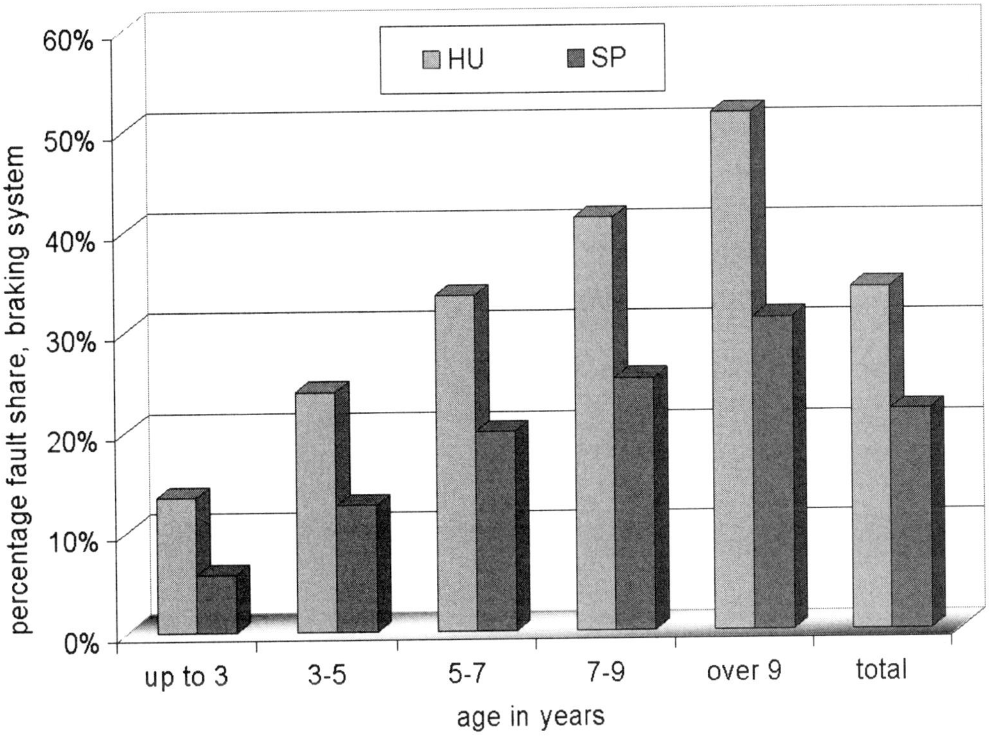

Fig. 15.10: Comparison HU-SP

The lower frequency of faults is easily explained, because these statistics are based on a far higher number of tests – 4 times as many as SP – and include vehicles which were not subject to an SP by the TÜV, in particular all local public transport vehicles and also a large number of older trailers.

But not every vehicle owner fulfils his test obligations voluntarily, particularly in view of the fact that failure to observe an SP date is a minor offence only incurring a fine of DM20.00 to DM80.00 without entailing any points in the Flensburg register. Vehicle owners and control authorities have got used to the HU dates, and not even the streamlined procedure for exceeding the date by backdating introduced from December 1, 1999 (**Fig. 15.11**) has changed this. A test badge i.e. the TÜV date is observed correctly by the vehicle owner, but in view of the fact that the SP date can be allowed to expire for 8 months without committing an offence and the vehicle is due for its next HU anyway 6 months after the SP the lawmakers have decided that in this case the SP is to be repeated, not at the full SP price but at only 60% of the SP price. Unfortunately, 28% of commercial vehicle owners are already utilising this possibility, and 12% of KOM vehicle owners.

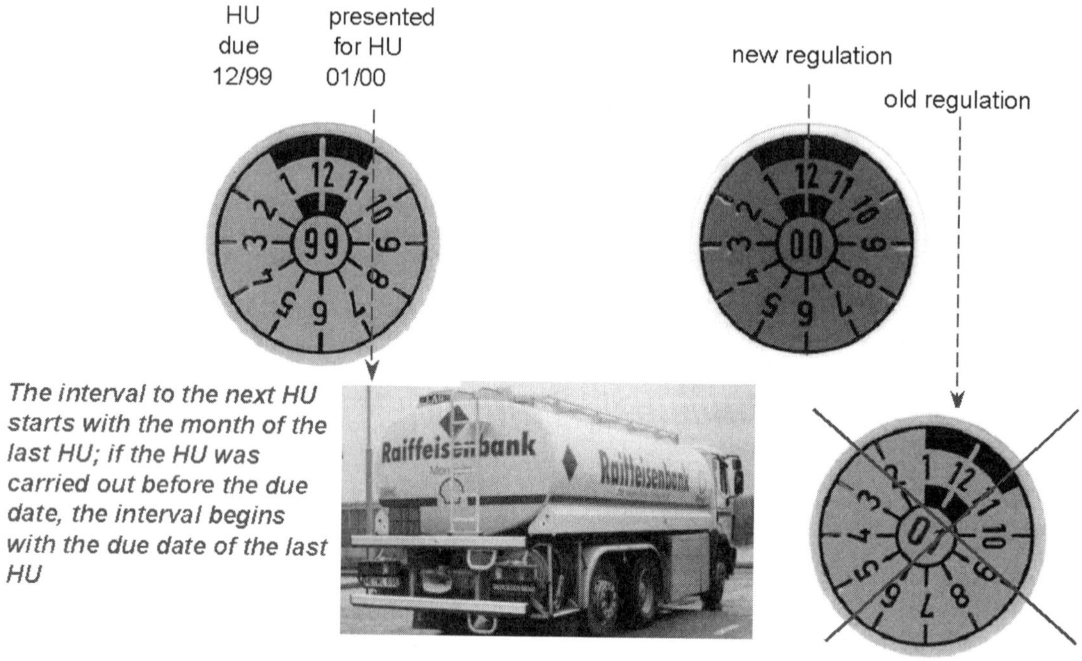

Fig. 15.11: Intervals between dates

Here there are relatively strong regional fluctuations as a result of more or less strict police controls. The fault distribution detected by us (**Fig. 15.12**) in the HU test referred to as "HU plus" when it includes parts of the omitted SP, has almost the same faults rate as for the SP when considered in isolation.

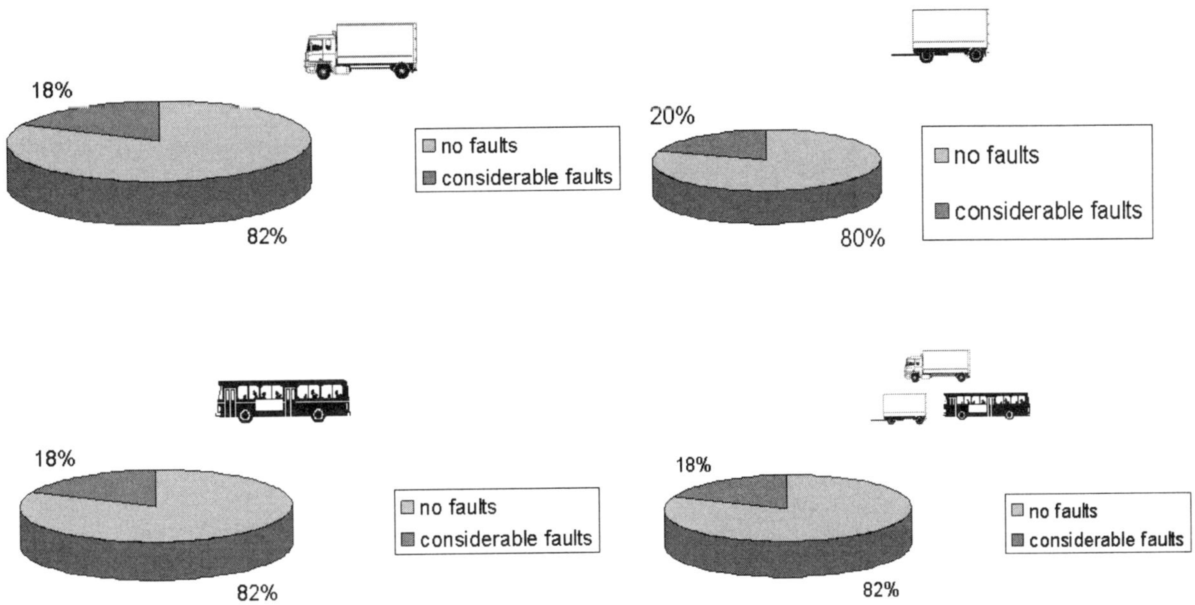

Fig. 15.12: Faults distribution HU plus

The trend to forget the SP is unfortunately on the increase, as the fleet owner saves a test date and 40% test costs, which he sets in relation to a small fine. But in doing so, he neglects the stand-by readiness of his vehicles.

15.5 International efforts

The next declared political objective is to make the SP accepted on an international scale or at least throughout the EU. But here the national data must be collected and evaluated first. There are already EU standards for the HU, but these constitute a minimum and can be tightened up by each country individually. In the interests of mutual quality assurance, the RiLi 2000/30/EU has been passed for the road-side control of commercial vehicles, which will have to be adopted by German law by August 10, 2002. The justification for this directive is that a one-off twelve-monthly technical test of commercial vehicles as prescribed by directive 96/96/EU (technical monitoring throughout the EU) is considered not to be sufficient for ensuring that commercial vehicles are in fact safe for road traffic throughout the whole year. This means that road-side controls are to be performed unannounced to check either just one or several of the following points (**Fig. 15.13**)

- **Test of control reports and other documents; examination of test results which are not older than three months**

- **Visual inspection of the maintenance condition of the commercial vehicle at a standstill**

- **Check for maintenance checks with particular attention to the braking systems and exhaust emissions**

Fig. 15.13: Test elements for road-side controls

The directive also introduces uniform regulations for:

- Sufficient road-side controls without differentiating according to nationality of the driver or country where the vehicle is registered,
- Introduction of a uniform checklist,
- Appropriate, effective sanctions for offences.

Compulsory reports are to be submitted to the commission every 2 years. We already know these international compulsory reports from the dangerous goods regulations, namely the dangerous goods control directive.

Along the lines of "trust is good, control is better", the directive 95/50/EU has been passed for "uniform procedures for controlling the transport of dangerous goods on the roads of member states" in cooperation with the ADR and with the objective of further improving safety in the transport of dangerous goods. In order to guarantee an adequate level for the corresponding tests, the directive includes a standardised check list and a standardised list of offences (**Fig. 15.14**).

1. Cargo not certified
2. Declaration of conformity missing
3. Leaks of dangerous substances
4. Certification document missing/not in order
5. No/incorrect orange warning signs
6. No accident instruction sheets
7. Vehicle/packaging unsuitable
8. Driver without training certificate
9. No fire extinguisher
10. No hazard notes
11. No transport documents
12. No bilateral/multi-lateral agreement
13. Tank over-full
14. Other infringements

Fig. 15.14: Offences

Every member state must submit a report to the commission about the controls performed in every calendar year. The suitability of the offences list is currently being discussed. At least the road-side controls introduced by most member states since 1997/1998 have proven to be very useful. There is no doubt that fear of the controls has forced most companies involved in road transport to conform with the regulations in advance. Unfortunately, these controls will have to be performed in an adequate frequency for them to act as deterrents.

Reports are available from practically all countries, with the frequency of controlled journeys varying between 0.025% and 0.5%. In Germany for example, about 1 journey in 500 was controlled, which is a frequency of about 0.2%. Depending on the country concerned, it transpired that between 10% and 80% of the controlled vehicles failed to conform with the statutory regulations, whereby the share of foreign vehicles controlled in transit countries amounted to up to 40% (**Fig. 15.15**).

The details presented here for the international transport of dangerous goods is reflected throughout the transport sector, regardless of whether passengers or goods are being transported.

In any case, weaknesses in implementing the rules can be detected here because they allow for a comparison between the individual EU member states with the possibility of introducing specific measures.

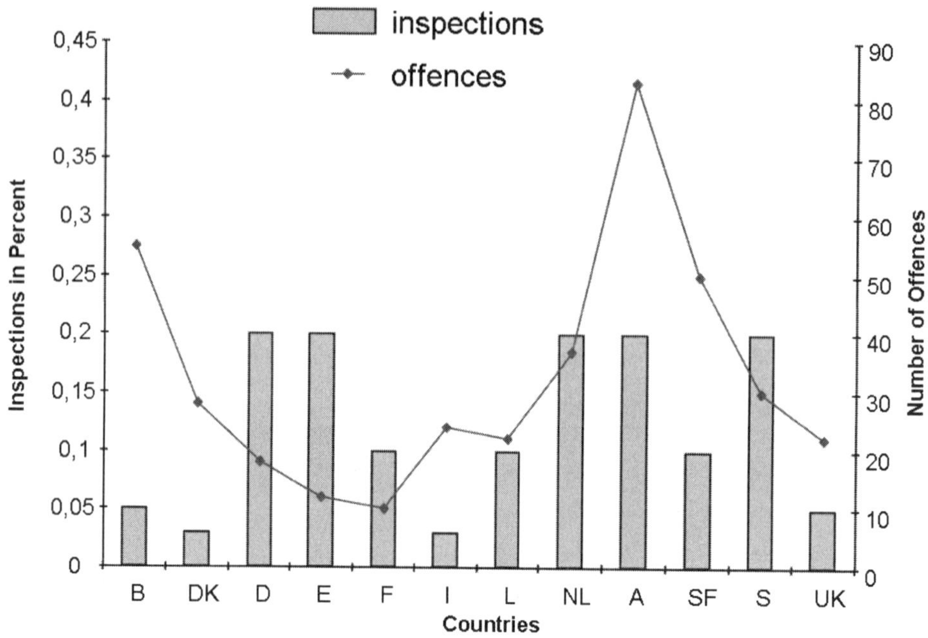

Fig. 15.15: Inspections / offences

15.6 Outlook

The SP and modernisation or adaptation of the HU for commercial vehicles to state-of-the-art engineering has proven successful and should certainly be pursued and updated on an international scale. Increased external controls are certainly also necessary. The EU road-side controls and a tightening of the offences situation (128th meeting of the on 26/27.06.01), whereby expired HU and SP dates of 4 to 8 months already result in fines of DM80.00 and 1 point, with DM150.00 and 2 points for vehicles with compulsory SP, are an initial approach to this problem, aiming at increasing the acceptance of these tests in view of the fact that these vehicles pose specifically higher dangers and risks. And the EOBD interface topic – currently only regulated for the exhaust emissions test, but no longer such a controversial topic when it comes to brakes and steering, will be able to build on the uniform QM system introduced for repair garages. Along these lines, I predict continued high safety and reliability when it comes to commercial vehicle braking systems, although in the heavy vehicle category with their high mileage and loads, it must be possible for this to be verified and monitored by external inspectors.

16 Brake Noise – A Comparison of Subjective and Objective Evaluations

Dr. W. Rumold, Dr. W. Keiper, Robert Bosch GmbH

Stuttgart, Germany; Drancy, France

16.1 Introduction

Driving Comfort and fun are attributes which came in the last decades more and more into the focus of drivers and the automotive industry. A crucial factor for the driving comfort is the noise behavior of the vehicle, which was optimized in recent time to a level that even sound events with comparatively low volume are noticed as being annoying. Therefore, the noise behavior became one of the most important criteria for the evaluation of braking systems. Brake systems consist of the foundation brakes, the actuation unit and, in most cases, the modulation system such as the Electronic Stability Program (ESP). As brake components initiate dynamic processes, noise is generated in any case. The task of the NVH (Noise, Vibration and Harshness) experts in braking system development is to minimize the radiated sound and/or to change the noise characteristics for a more convenient noise perception.

While the noise behavior of modulation and actuation components shows in general a high reproducibility and can be improved by optimizing the noise characteristics, one finds for the friction-induced brake noise a tonal sound characteristic arising from system instabilities at certain frequencies. Since the friction-induced instabilities depend on a large number of parameters which are often not known or measurable and therefore are difficult to influence, the reproducibility of the noise is in many cases very limited.

This affects first of all the development steps in improving the noise behavior – a noise which is difficult to reproduce is difficult to analyze and optimize. On the other hand, the low reproducibility makes as well the rating of the brake noise level problematic, which is the basis for an estimation of the situation in mass production and which is of major importance for the vehicle manufacturers. Therefore, the procedures for the evaluation of the noise behavior have come more and more into the focus of the vehicle manufacturers in the last years. The major criterion for noise evaluation procedures is that they give a realistic indication of the noise behavior in mass production, i.e. for a high volume of cars in completely different geographic and climatic conditions and a wide range of brake and friction material conditions.

The methods for the evaluation of brake noise can be divided into dyno testing and vehicle testing. Dyno testing is done on inertia dynos for corner modules or on roller dynos for the full vehicle. Usually matrix tests are run, see Blaschke und Rumold [16-1], which are designed to reflect the noise performance in a broad range of brake parameters. The benefit of dyno testing compared to vehicle testing is the higher reproducibility due to the matrix specification, the shorter testing time consumption and the lower cost. However, dyno test results do not always reflect vehicle test results, a question which is still under investigation.

The major criterion in customer sign-off procedures of the noise performance is up to now the result of vehicle testing, which is done in general with several cars over fairly large testing distances (from 5000 km). The test drivers give subjective noise ratings, which are summarized over the whole test period to an overall rating. The location and the test procedure is chosen to cover a large range of relevant conditions for the noise behavior, which is limited by the available parameter range e.g. with respect to the climatic conditions. According to the different environmental conditions and driving habits in the US and in Europe, different test procedures have been developed which are run in general in Los Angeles in the US and in Mojacar (Spain) in Europe.

In the last years the use of noise measurement systems has become popular in particular in the US. The initial intention was to generate a data basis for a subsequent noise analysis. On the other hand, the noise measurement systems enable a noise rating on the basis of measured 'objective' data, which can supplement the driver ratings influenced by a number of items such as psychological factors. However, an objective rating can only be accepted if it reflects an average human perception of the noise behavior as close as possible.

In this contribution, various approaches for the objective rating of brake noise are presented, the respective objective parameters are determined on the basis of measurements at the Mojacar noise endurance test and at matrix tests, and the results are compared to subjective driver ratings. Further, results from jury testing on the human perception of brake noise are presented and an outlook on future investigations for the development of an objective noise index is given.

16.2 Vehicle Noise Testing

16.2.1 Endurance Tests

The aim of an endurance noise test is the rating of the brake noise performance under realistic conditions. Due to the low reproducibility and the sometimes substantial differences of the noise performance of identical brake systems, for an estimation of the noise behavior a sufficient statistical basis has to be available. Therefore, noise endurance tests are usually carried out with several vehicles over large distances. With respect to the definition of the test track and the advised way of driving, a choice in between two extreme positions can be made: On the one hand, a test can be designed on the basis of an average driving style omitting extreme driving conditions or noise searches. On the other hand, the test can be designed especially for critical conditions with respect to brake noise such as intensive noise searches in cooling-down phases after high brake temperatures. Both directions have their pros and cons, for the testing reality a mixture can probably not be avoided. The European brake noise test in Mojacar (Spain), which is focused in the following, has a more 'worst-case' scenario tendency.

The area around Mojacar in South Spain is suitable for brake noise testing due to the climatic and geographical conditions, as it offers a hilly landscape as well as a coastal region and a fairly constant climate with little rain. The endurance test is run on public streets on a defined test track, a test round has a length of about 150-200 km. Depending on the companies involved, the details of the test route can vary.

Fig. 16.1 shows a typical test route of a Mojacar brake noise test. The route consists of a number of noise searches at different locations, which are marked in **Fig. 16.1**, and connecting tracks between the noise searches. In a noise search the driver tries to find brake noise, where in general low speeds and pressures are used. For each noise search a driver rating is given, the noise behavior during the connecting drives is usually watched less closely. The connecting drives are mainly used for the conditioning of the brakes, especially in the hilly landscape high brake temperatures are generated.

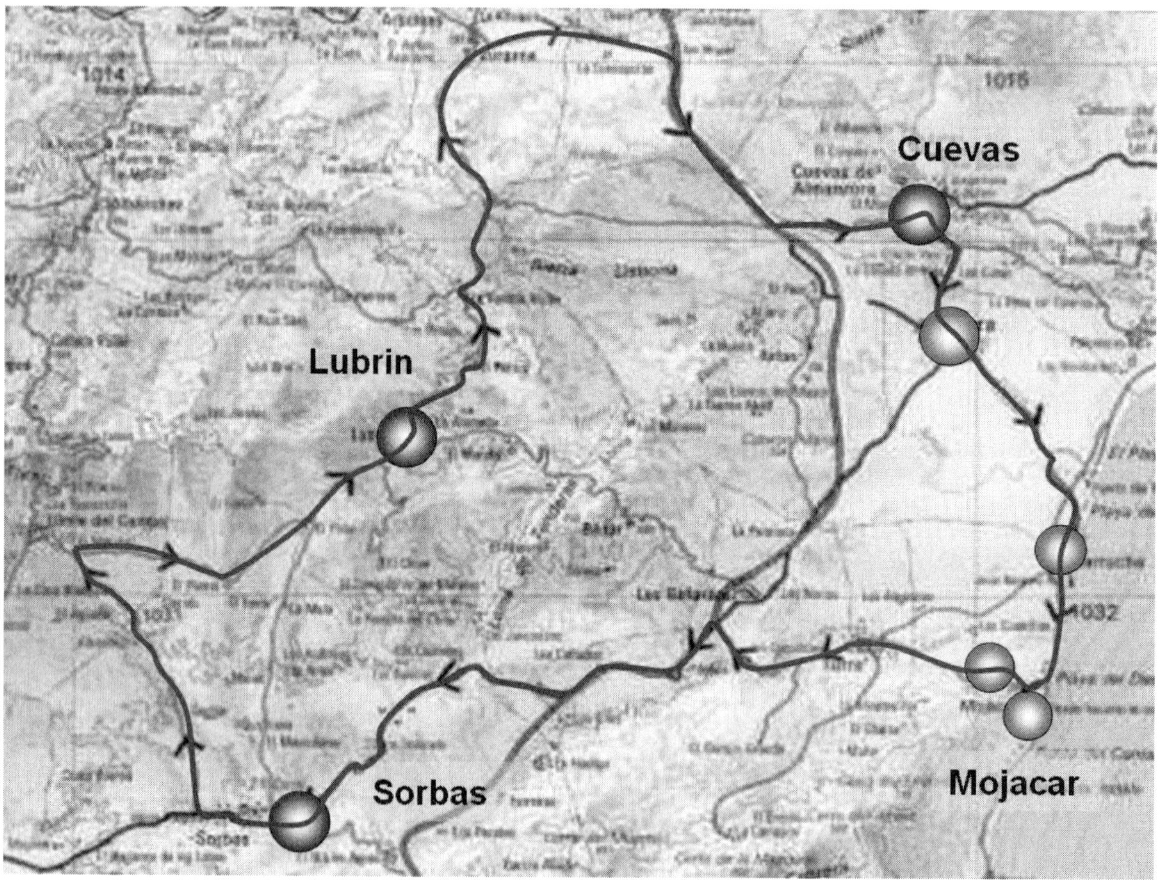

Fig. 16.1: Test route of a Mojacar noise endurance test. Noise searches are marked by circles

Fig. 16.2 shows a typical profile of brake temperature, vehicle speed and driving distance as a function of the brake events for one test round. The vehicle speed goes up to around 120 km/h, brake temperatures up to 260°C are measured here. From the vehicle speed profile the change between low-speed noise searches and high-speed connection drives can be detected.

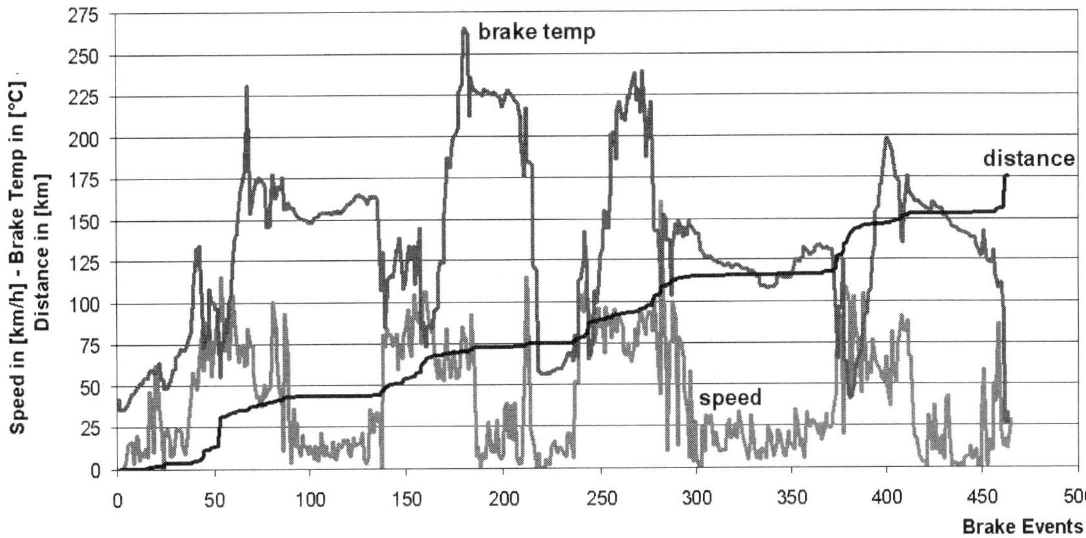

Fig. 16.2: Typical Mojacar noise endurance test profile of vehicle speed, brake temperature and distance

The rating of the brake noise by the test drivers is based on a scale called SNE (subjective noise evaluation) ranging from 1 to 10, see **Table 16.1**, where 10 means 'no noise' and 1 stands for extreme noise. Additionally to the SNE the drivers note down the source of the brake noise (front/rear axle), an item which is often difficult to determine. Further, the parameters vehicle speed, brake pressure and temperature for the noise events are requested as well as the kind of noise (high/low frequency squeal etc.). For a summary of the noise ratings, in general the average and the worst rating is given.

Table 16.1: Scale for noise rating.

	Subjective Noise Evaluation (SNE)							
	1-3	4	5	6	7	8	9	10
Perception	Extremely loud	very loud	loud	medium	quiet	very quiet	traces	no noise

16.2.2 Matrix tests

In contrast to noise endurance tests, in vehicle noise matrix tests each brake event is defined with respect to the brake parameters. The goal is to determine the noise performance in a wide parameter range. Contrary to the noise endurance tests, all brake events are rated separately with respect to noise, **Table 16.1** gives again the scale for the noise rating. Matrix tests are usually run on proving grounds. The benefits of matrix tests are the substantially lower timing (around 500 brake events), the higher reproducibility and the broader evaluation of the noise performance as a function of the brake parameters due to the systematic testing. However, compared to the noise endurance tests the relevance to the situation in mass production is in general lower due to the 'artificial' driving and the short duration of the test.

Fig. 16.3 shows a typical profile of the requested brake temperature. Each temperature sequence is run with another brake pressure, from a defined initial vehicle speed constant pressure stops are performed until full-stop. Before and in between the temperature sequences conditioning stops are done, which are omitted in **Fig. 16.3**.

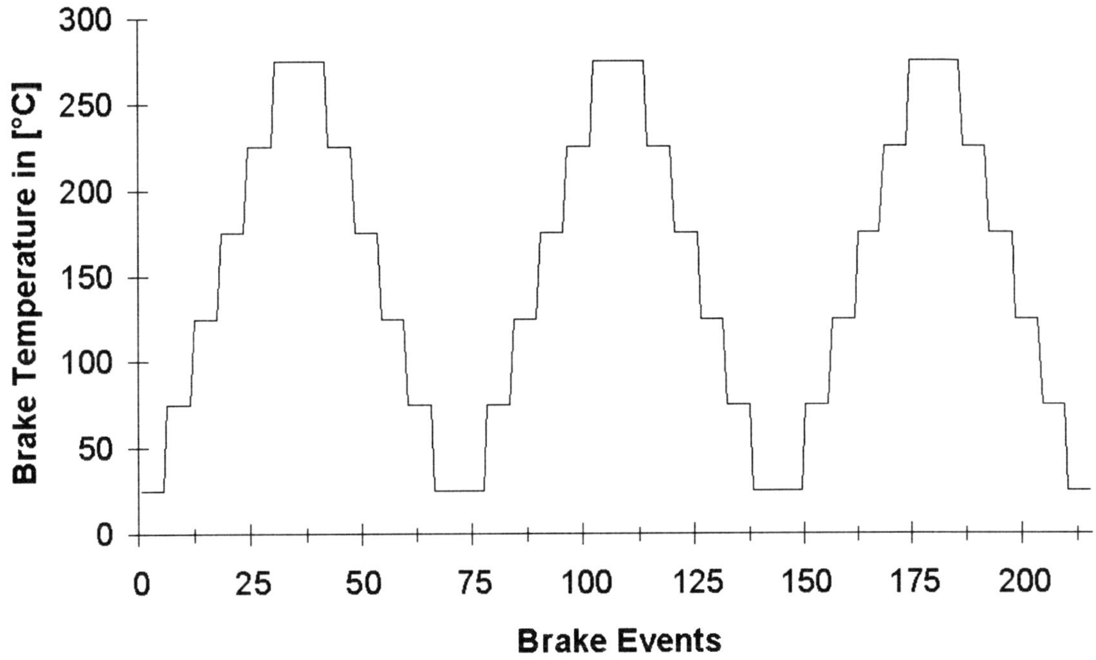

Fig. 16.3: Temperature profile of a matrix test

16.3 Comparison of subjective and objective evaluations

16.3.1 Vehicle noise measurement system

The progress of measurement technology for systems which are able to detect and localize brake noise is the key factor that enabled brake noise measurements during noise endurance testing in Mojacar. High robustness and straight-forward handling are major items for successful measurements in the fairly rough test conditions in Mojacar. Another decisive issue is the ability of the system to distinguish own brake noise from brake noise of other cars or further noise coming from the surroundings etc.

The actual recording of the brake noise is done by a microphone inside the drivers cabin, which can be supplemented by an exterior microphone. For the detection of the noise source one can use either further microphones close to the vehicle wheels, or accelerometers which are mounted on the brakes. The disadvantage of microphones for detecting the noise source is the cross modulation between the wheel sensors. The measurement system determines the noise source by a comparison of the measured spectra of the interior microphone and the accelerometers. The quality of the verification algorithm, which is not discussed in detail here, is the main item for the quality of the noise recognition, which is decisive for the usability of the measurement data.

During the measurement, the following data is recorded all through the testing:

- brake temperatures
- brake pressures
- vehicle speed and distance
- ambient temperature and humidity

Fig. 16.4 shows the measurement system [16-2] used to acquire the data which is discussed in the following. Brake noise is measured by a microphone in the driver cabin, the verification of the noise source is done by accelerometers at the brakes.

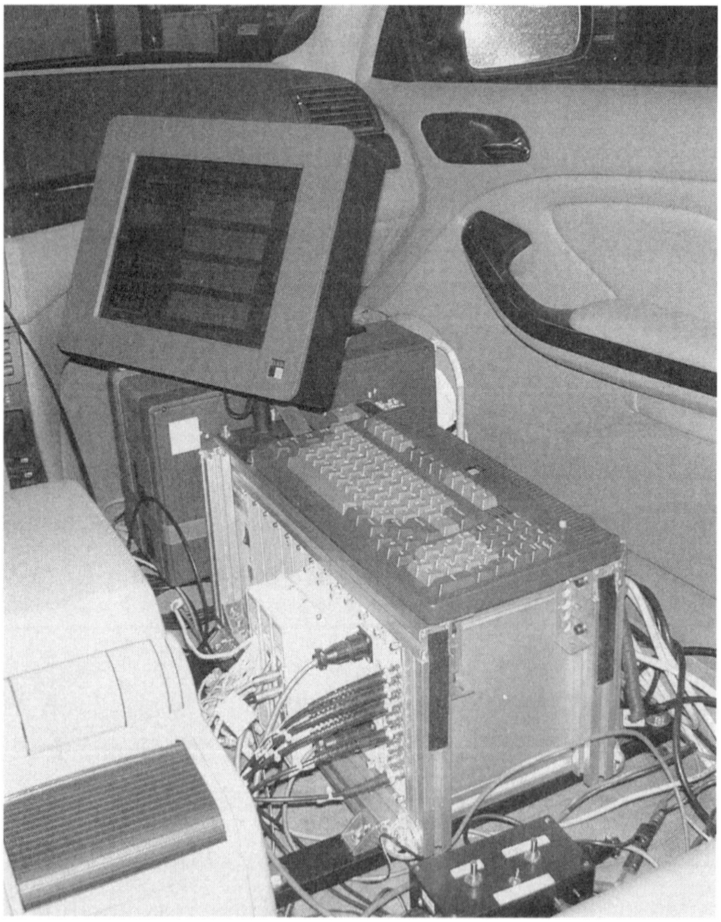

Fig. 16.4: Brake noise measurement system [16-2]

16.3.2 Brake noise measurements at the Mojacar noise endurance test

In the following, data from the Mojacar noise endurance test is analyzed and drivers noise ratings are compared to measurement data processed in various ways. The two Mojacar tests selected for the analysis are well suited for an investigation as the tests are done in an early project phase where considerable brake noise was yet detected. As it is common, the vehicles were equipped with brakes and lining materials from different manufacturers. A more detailed investigation of the data is given by Ansmann [16-3].

As mentioned above, the Mojacar noise ratings are concentrated on noise searches at specific locations. For each noise search with a number of brake events, one noise rating is given which comprises the brake noise performance for the whole noise search. One of the problems in the analysis is relating the measured data to the drivers ratings. In this investigation we used the fact that during the noise searches the vehicle speed is in general very low leading to a nearly horizontal line for the driving distance graph. The following analysis is carried out for the data measured in three noise searches, Sorbas, Lubrin and Cuevas hot, which are marked in **Fig. 16.5**. As there are two tests with ten days each and four rounds per day being analyzed, we are looking at 240 noise searches where noise searches without brake noise are omitted.

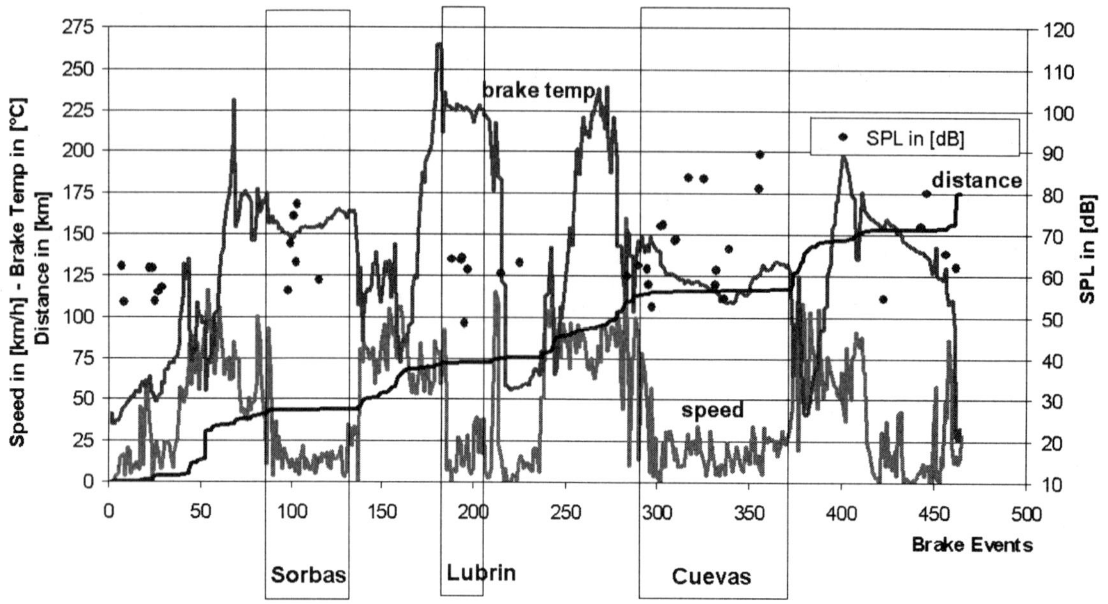

Fig. 16.5: Vehicle speed, brake temperature, driving distance and noise sound pressure levels (SPL) as a function of the brake events. The investigated noise searches are marked

Since we want to compare subjective driver ratings with objective data, the measurement data has to be condensed: for each driver rating a lot of data is available as all noisy stops are recorded. For the objective noise rating, a variety of possibilities are available from the measured noise data such as the noise sound pressure level (SPL), the noise frequency or the duration. The way of condensing the data such as averaging or the picking of maximum values adds another dimension to the possibilities in designing an objective noise rating. In the following, some of the ideas for determining an objective noise index are presented and compared to subjective ratings.

The most obvious approach for an objective noise rating is just to pick out the loudest noise in the noise search. **Fig. 16.6** left shows the comparison of the maximum sound pressure level and the driver ratings for each noise search. The subjective noise ratings analyzed in **Fig. 16.6** left range from 4 to 9, ratings with a SNE of 10 are not analyzed. The measured maximum sound pressure levels exceed 100 dB(A), the respective driver ratings for these high SPLs range from 4 to 7. From a quick look on the data no correlation of the subjective and objective data can be discovered. However, with a closer look one can at least get the impression of a tendency to lower SPLs with higher (better) driver ratings.

A better visualization of the data can be achieved by quartile representations, see **Fig. 16.6** right. For each subjective rating the lower and the upper bar show the minimum and the maximum value, the box indicates the 25% and the 75% level of the data and the bar inside the box shows the median. The median values in **Fig. 16.6** right show a tendency to lower SPLs for higher driver ratings, but one finds as well similar values for the driver ratings 4 to 6 and 7 to 8. The jump between the driver ratings 6 and 7 could arise from the information known to the drivers that the customer target was to have ratings better than 6.

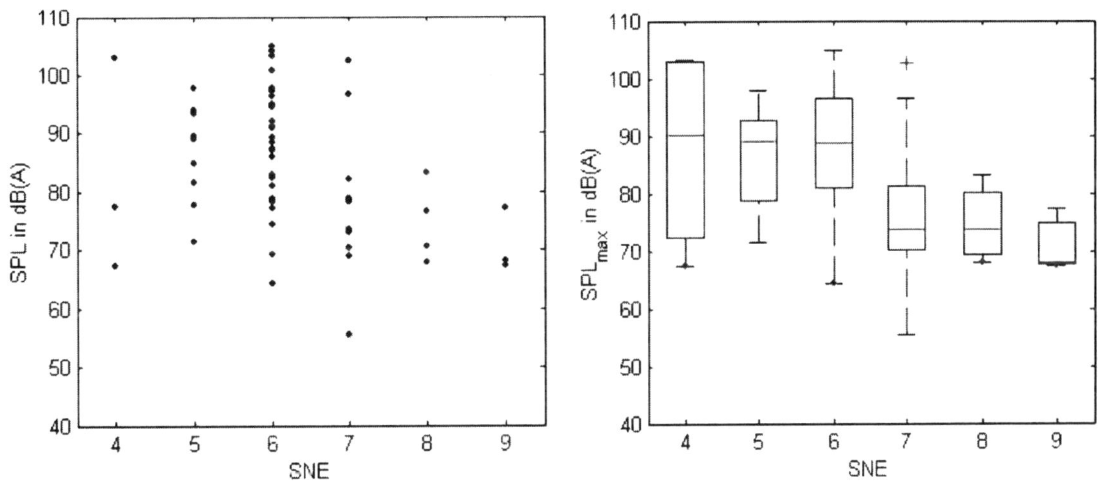

Fig. 16.6: Maximum sound pressure level (SPL) vs. subjective noise ratings (SNE). Left; scatter plot. Right: quartile representation

As the maximum SPL is picking out just one noise event of the whole noise search, it does not reflect the noise performance over the other brake events of the noise search, a fact which probably does not reflect human noise perception. The opposite approach is an evaluation on the basis of the number of noisy brake events for each noise search, where all brake events with an SPL over a certain threshold are considered equally. **Fig. 16.7** left shows the quartile representation of the percentage of noisy stops above 40 dB(A) in a noise search versus the subjective ratings. The investigated noise searches, which in general consist of 10 to 50 brake events, show up to 72% of noisy stops. We find again a tendency to lower medians of the percentage of noisy stops for higher (better) driver ratings. However, compared to the maximum SPL in **Fig. 16.6** one finds a lower level of correlation by visual inspection.

The results of the above evaluations encourage the assumption that the human perception of noise is a complex linkage of various influences, where both the volume and the occurrence or duration play an important role. A combination of the two parameters is the basis of a noise index called ONIF, which has been developed in joint investigations between Ford and Bosch on the basis of a dynamometer noise index. Here, noise events are weighted depending on the measured sound pressure level, see **Table 16.2**. The weights of the noise events in a noise search are summed up and divided by the number of all stops in the noise search, which introduces the occurrence of noises into the index. Ranges have been defined for the ONIF, where values up to 0.04 are considered to be in the 'green' range, the 'yellow' range goes from 0.04 to 1.2, and values above 1.2 are considered to be in the 'red' range.

Table 16.2: Weighting factors of the ONIF as a function of the sound pressure level (SPL).

SPL in dB(A)	Weight
p<50	0
50<p<60	1
60<p<70	3
70<p<80	10
80<p<90	30
p>90	100

Fig. 16.7 right shows the quartile representation of the ONIF versus the subjective driver ratings. The figure shows a steady decrease of the medians of the ONIF for better (higher) driver ratings. Another positive observation is the correspondence of the critical 'red' range of the ONIF higher than 1.2 to subjective ratings of 6 and lower based on the median values, which fits well into the general view of driver ratings below 7 as being critical. Aside from these positive observations we find as

well a fairly large spreading of the ONIF values and quite a number of extreme values (marked by the '+'-sign). A reason for the latter could be the exponential character of the weighting which emphasizes noise events with high sound pressure levels.

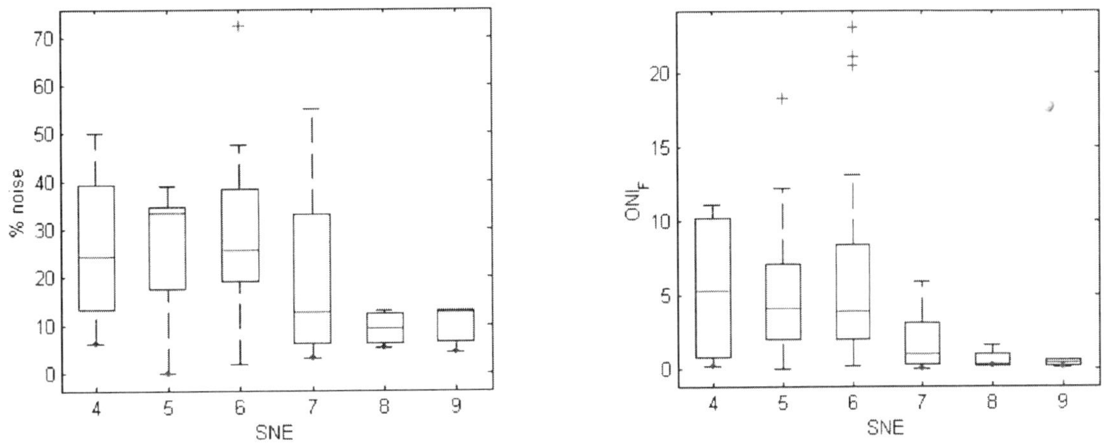

Fig. 16.7: Quartile representation of %noise (left) and the ONIF (right) versus driver ratings (SNE)

From the above analysis of the Mojacar endurance test data, we find limited correlation levels of the chosen, fairly straight-forward objective measures to the driver noise perception. Although the maximum SPL and the ONIF showed a basic correlation, probably more complex algorithms are required to model the human perception. On the other hand, the quality of the driver ratings is not clear as well, a question which could be investigated using e.g. jury testing see chapter 16.4. To reduce complexity, in the next step we will investigate matrix noise tests, where each driver rating corresponds to just one brake event.

16.3.3 Brake noise measurements in matrix tests

In section 16.2.2, the matrix test investigated here has already been described, which comprises around 400 brake events at different vehicle speed, brake pressures and temperatures. As mentioned before, in matrix testing each brake event is rated separately from the driver. Additionally, matrix tests are run by just one driver, variations of the subjective ratings due to different drivers are avoided.

The matrix test analyzed in the following was run with a vehicle having brake noise at the frequency 1.6 kHz and in the range from 10 to 14 kHz. **Fig. 16.8** shows the measured sound pressure level (SPL) versus the driver ratings, which lie in the range of 5 to 10, ratings of 10 are again skipped. The color bar gives the frequency of the noise. One finds noise events up to around 75 dB(A), higher SPL levels are measured especially for the 1.6 kHz noise.

The analysis of the data shows first of all that the driver rated the lower frequency noise as less unpleasant compared to higher frequency noise with similar sound pressure level. This again shows that the SPL can not be the single criterion for the human perception, other parameters must play an important role. However, there is a connection between the sound pressure level and the noise perception, which one finds in the tendency of higher SPLs for lower (worse) subjective ratings. When looking on the quartile representation of the SPLs versus the driver ratings, see **Fig. 16.9** left, this tendency is as well apparent: with the exception of the SNE of 9, the level of the medians is decreasing for higher (better) subjective ratings.

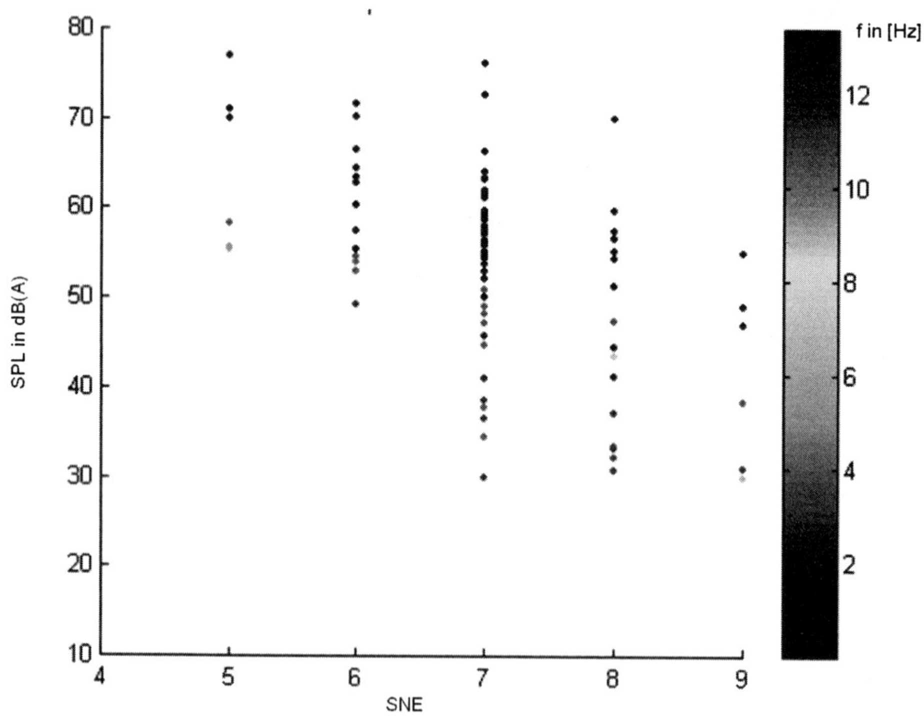

Fig. 16.8: Sound pressure levels (SPL) versus subjective ratings (SNE). The color indicates the noise frequency

For the Mojacar noise endurance tests we already added an approach on the basis of the noise occurrence to a pure SPL-based evaluation of the measurement data. Since we have here single-stop ratings of the noise, an evaluation of the noise occurrence is replaced by the noise duration during a single stop. **Fig. 16.9** center gives the quartile representation of the noise duration versus the subjective ratings. The duration of the measured noise goes up to 9 seconds. Again, the medians increase at least slightly in the direction of the worse (lower) subjective ratings, however, the spreading of the data and the high number of extreme values show the fairly strong limitations of the duration as an index for the subjective rating.

A linkage of the volume and the duration of the noise is achieved by a multiplication of the SPL and the noise duration, see **Fig. 16.9** right. One finds a nearly linear relation of the medians to the subjective rating, which indicates that the drivers perception could be governed by both the volume and the noise duration. Therefore, this approach could be the basis for a brake noise index. However, the spreading is still very high and strong extreme values appear. An increase in the data basis with substantially more measurement data included would be needed to show the potential of this approach.

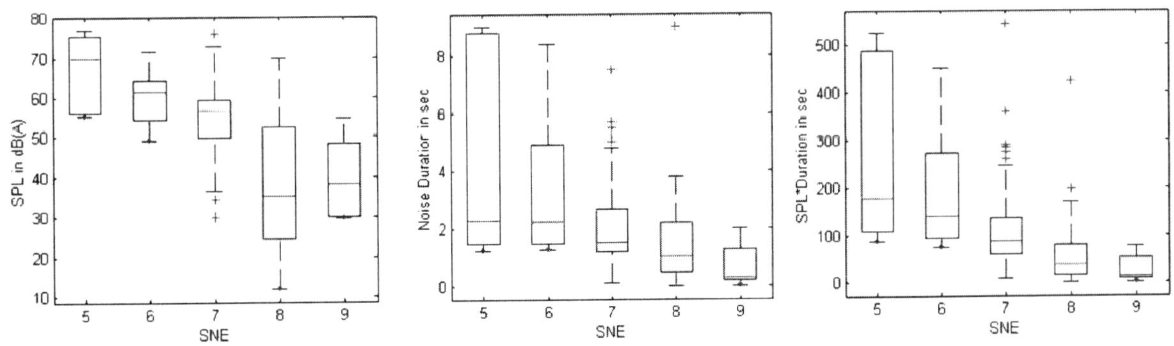

Fig. 16.9: Quartile representation of sound pressure level (SPL - left), noise duration (center) and SPL*noise duration (right) versus driver ratings (SNE)

So far the analysis of the objective noise data was based purely on parameters which can directly be measured like the SPL or the duration of the noise. An improvement of the correlation level to the subjective ratings could be achieved by using psycho-acoustic measures, which are designed to reflect the human perception of sound. In a pre-selection of a number of psycho-acoustic measures, the Prominence Ratio and the Tonality have shown a potential for indicating the perception of brake noise. For both parameters, the tonal components in the measured sound pressure level is determined, where the Tonality is determined for a frequency range and displayed as a function of time, whereas the Prominence Ratio is determined for a time range and displayed for a frequency range, see [16-3]. For both parameters, the respective maximum values are analyzed.

Fig. 16.10 shows the quartile representation of the Prominence Ratio and the Tonality versus the subjective ratings. Again we find decreasing median values for higher (better) driver ratings, a basic correlation to the subjective values is found for both psycho-acoustic measures. The median value at SNE of 5 for the Prominence Ratio does not fit to that general tendency, additionally the spreading of the data is fairly high for the Prominence Ratio. The Tonality shows a better reflection of the driver ratings, we find a constant increase of the median values for lower subjective ratings and a fairly small spreading of the data. In summary, the Tonality shows the highest potential of the investigated measures for an indication of the drivers perception. Whether it can be used as a noise index for brake noise has to be determined by further analyses of the correlation level to subjective ratings.

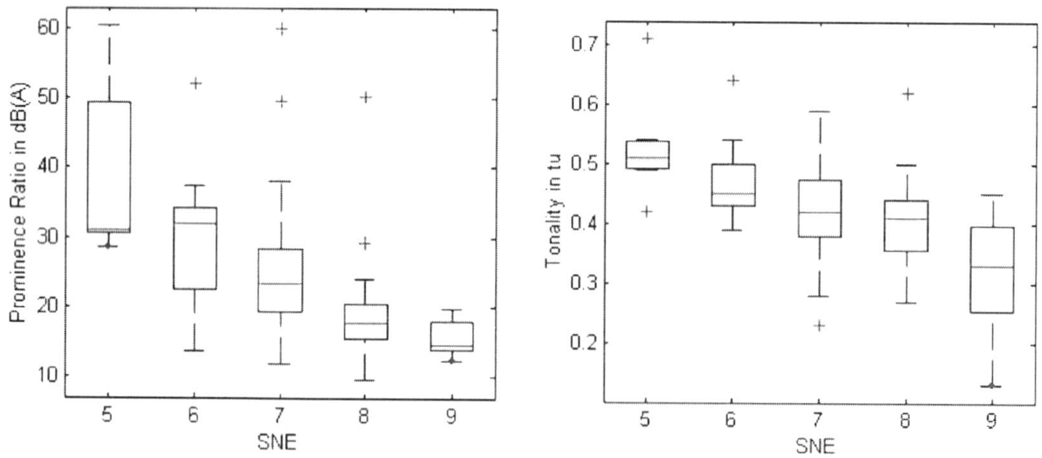

Fig. 16.10: Quartile representation of Prominence Ratio (left) and Tonality (right) versus subjective ratings

A summary of the ability to reflect the drivers noise perception for the investigated objective parameters is given in **Table 16.3**. The results presented here can only be a starting point for further, more detailed investigations. As a next step, further analyses of noise matrix tests would increase the data basis for the rating of the different approaches to evaluate an objective noise index. This could lead to further improvements or new ideas how to determine the objective index. In noise endurance testing, these suggestions could then be investigated with respect to their correlation quality to the driver ratings. On the other hand, it is necessary to investigate vehicle noise testing reality and the quality of the subjective noise ratings in the noise endurance tests. Only then a fair rating of a proposed objective noise index in the relevant Mojacar endurance test is possible.

Table 16.3: Ability to reflect drivers noise perception for the discussed approaches of measurement data processing.

Objective Approach	**Mojacar Noise Endurance Test**	**Matrix Test**
Max. SPL	–/o	–/o
Noise Duration / % Noise	–	–/o
ONI_F	o	
SPL * Noise Duration		o
Prominence Ratio		o
Tonality		o/+
Reflection of Drivers Rating: –: low o: medium +: high		

16.4 Driving collectives and objective goals

The preceding chapters showed the fundamental problem of the variance of subjective operator judgements concerning brake squeal, and the efforts towards objective interpretations of these results. A number of fundamental questions to this topic remain:

- are today's brake noise road tests relevant for, and can they reliably predict, the vehicle's brake noise during vehicle use by the end user / customer?
- are the subjective operator noise judgements during road tests reliable and reproducible ?
- which measurement-based criteria can be used for a clear, objective assessment of the noise behaviour ?

16.4.1 Today's road tests

A careful analysis and a comparative evaluation of the noise endurance tests in Los Angeles (LACT) and Mojacar was submitted recently by Mody, Rumold, Attia and Ansmann [16-4].

The Mojacar test employed by many (however not all) European car manufacturers is generally believed to represent a particularly severe condition for noise inclination (squeal in particular) of brakes. The round course is characterised by the hot, dry and dusty climate on the plateau of SORBAS, salt-loaded and rather moist air along the coast, as well as the acoustic conditions between the 'hard' house walls in the villages. The round course is rich on curves and on ascending and descending sections. The drivers usually drive fast. According to experience, under these conditions one finds brake squeal events which are not heard otherwise at the vehicle.

However a successful Mojacar brake noise test gives strictly taken only the guarantee that the checked vehicle brake is quiet and acceptable in hot climate and within that driving collective. Upon more detailed analysis, it turns out that the final evaluation of the Mojacar test is often dominated almost exclusively by the subjective evaluations of the so-called ' noise search ' (e.g. in Lubrin) – thus very special conditions.

The Los Angeles City Traffic test LACT contains extensive town traffic portions with 'stop and go'. According to the opinion of the authors of the SAE paper [16-4], the LACT test is more representative of a typical collective of American vehicle use by the final customer.

LACT and Mojacar test have the advantage that they can be driven almost all year round under similar conditions. Both tests do not cover 'cold and wet' weather conditions. For these conditions the brake and vehicle manufacturers run their own specific additional tests.

16.4.2 How reliable are the drivers' judgements?

As is the case for all subjective judgements, the assignment of evaluation notes for noise events is a highly complex process in the human mind. The vehicle operators (they represent a test subject) connect and combine subconsciously and automatically a multiplicity of sensory input data, psychoacoustic evaluations, as well as other cognitive information (also fore-knowledge and expectations). It lies in the nature of the thing that such subjective judgements are partly determined by factors that are difficulty to define, including daily form and general condition. The judgements of individuals are usually subject to strong fluctuations.

Therefore jury tests are essential as starting point for the search for an objective noise index for brake squeal. Jury tests are now common for many subjective evaluations, also for noise quality. Jury tests are evaluation experiments with several test subjects, who assess, compare and judge noise events under guidance of the test manager and under controlled, reproducible test conditions. Examples: evaluations on selected absolute rating scales, or better/worse evaluations in the pair comparison. [16-6] The noises which are to be evaluated are reproduced in constant quality. By means of suitable test sequence, it is possible to evaluate among other things the reliability/repeatability of the test persons' judgements, and their ability to discern different noises.

In an extensive study at Bosch (Bosch research, Schillerhoehe), about 150 brake squeal events were evaluated during noise tests on the internal test route by the drivers, and recorded at the same time in high signal fidelity. These data were made available for jury tests. The jury tests were done in the Bosch audio laboratory with more than 20 test persons of different origin. As evaluation scale the usual SNE absolute rating scale with marks from 1-10 (Subjective Noise evaluation, 10 = best) was used. **Fig. 16.11** shows a view of a group of six jury test persons.

Fig. 16.11: Jury Testing of brake noises

Fig. 16.12 shows the result of a careful statistic analysis: The comparison of the evaluations of the road test drivers during road test with the jury test results. The jury test shows a somewhat broader distribution of the judgements. During the quietness and concentration of the jury test in the audio laboratory the test listeners differentiate a little more finely than the operators with the double load of vehicle driving and noise evaluation. However, the jury test participants evaluate the brake noises generally in the same way as the test operators on the road.

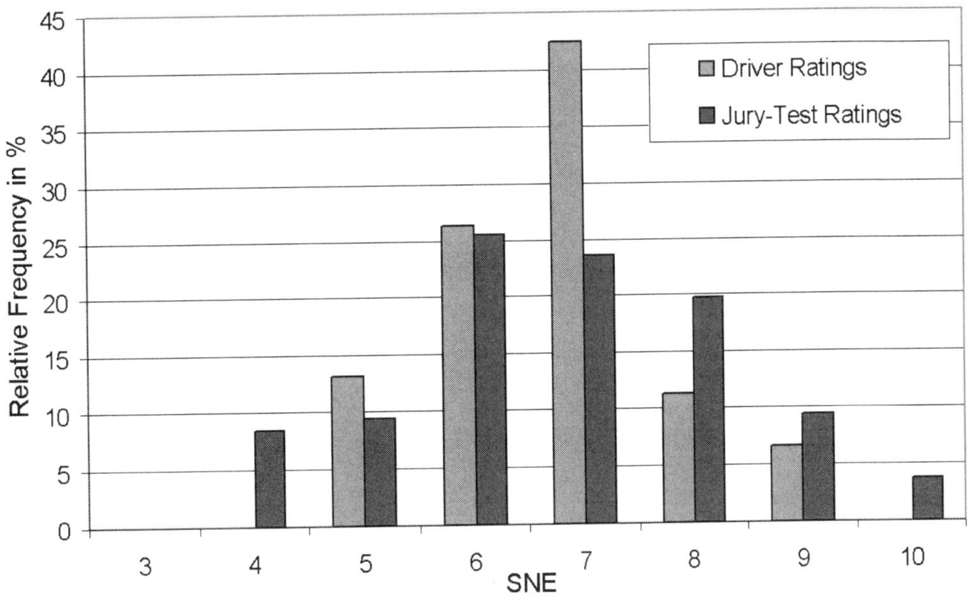

Fig. 16.12: Comparison of the subjective evaluations (SNE) of drivers on the road and of the jury

The data now obtained by high-fidelity signal recordings and subjective evaluations [16-5], validated by means of jury tests, are the necessary basis for the work on a new objective brake noise index.

16.4.3 Objective criteria

In the work running at present – building on the results already presented – further physical and psycho-acoustic parameters, and combinations thereof, are checked for their suitability as objective index for brake squeal. The database must be still extended, and environmental data and braking conditions (e.g. brake pressure, deceleration etc..) must find consideration. An objective index should be applicable equally for all brake applies, for small and large vehicles, for different vehicle types and for all user target groups. The goal is it to make available a measuring method for characterising brake noise that is well correlated with customer satisfaction. As intermediate objective an index is to be developed, which represents well the current vehicle test driver judgements. Some progress was achieved, here; correlation coefficients in the order of 0.8, as they are achieved for instance for engine or drive train noises, can be found for objective brake noise evaluations, too.

- There are indisputable benefits of a generally recognised brake noise measuring method that reflects the subjective evaluation of the passenger. (Brake Objective Noise index). Some thoughts on the application of the technique:
- definition and specification of noise goals of brakes by means of characteristic target values or limiting values of the BONI,
- application during the development phase of brake systems as evaluation/optimisation criterion,
- determination of the characteristic noise values on the inertia dynamometer, and even more so on the air-conditioned roller bench (as weatherproof addition of road testing)
- partial reduction of noise endurance testing on the road,
- homogeneous evaluation also of different vehicle noise tests and procedures,
- development of a new test procedure (definition of driving collective, climatic/environment conditions, data acquisition, evaluation technique...)
- validation and noise acceptance checks / noise sign-off for brake systems on an objective basis.

16.5 Summary

Noise endurance test, in particular the Mojacar test often driven in Europe, serve the evaluation of the noise behaviour of drum and disc brakes under hot-weather 'worst case' conditions. For the profile of speeds, brake temperatures etc pursued thereby, the test drivers pass a judgement on the brake noise on the SNE scale (Subjective Noise index) with values of 1 (extremely loud) to 10 (noise-free). Another form of the subjective brake noise evaluation in the vehicle test is the evaluation of single brake applies on test track or road. Like all subjective judgements this noise evaluation is subject to individual fluctuations. A reliable objective brake noise index on basis of measuring data is not available at present yet.

In several series of tests with Bosch the brake noises and the accompanying data were recorded and compared with the subjective SNE evaluations. Sound pressure levels, for instance, do not correlate sufficiently with the operator evaluation, hence they are not suitable as objective measurement for brake noises. Combinations of physical measured variables, like for example the product of sound pressure and noise duration, show at least the correct trends of the SNE evaluation. The correlation of selected psycho-acoustic parameters, for example the tonality, with the SNE is promising. A unique link between calculated parameter value and subjective judgement has not yet been identified.

The subjective operator judgements were refined and further specified with jury testing methods. They are now available for the search for a generally applicable BONI (Brake Objective Noise Index). Such an index as core of a brake noise assessment method can be also used for the definition of brake noise targets and their validation in the vehicle, as well as in the development process of quiet brakes.

16.6 References

[16-1] Blaschke, P.; Rumold, W.: Global NVH Matrix for Brake Noise – A Bosch Proposal. SAE 1999-01-3405. Proc. 17th Annual Brake Colloquium & Engineering Display, 1999.

[16-2] Roush Anatrol BrakeDAQ Manual. Roush Industries, 2001.

[16-3] Ansmann, S.: Untersuchung und Vergleich objektiver und subjektiver Daten aus Geräuschmessungen. Diplomarbeit, Fachbereich Mechatronik und Mikrosystemtechnik, Fachhochschule Heilbronn, 2001.

[16-4] Mody, P.; Rumold, W.; Attia, F.; Ansmann, S.: Mojacar and Los Angeles City Traffic Vehicle Testing: A Comparison & Analysis of Subjective Ratings and Objective Measurements. SAE 02Brake-45. Proc. 20th Annual Brake Colloquium & Exhibition, 2002.

[16-5] Tippkemper, H.: Contribution to the Development of a Brake Noise Evaluation Procedure of On- and Off-Road Brake Tests. Diplomarbeit, Fachhochschule Münster, 2002.

[16-6] Otto, N., Amann, S., Eaton, C., Lake, S.: Guidelines for Jury Evaluations of Automotive Sounds. SAE-paper 1999-01-1822 pp 1315, 1999

17 Porsche Brake Development „From the Race Track to the Road“

Dipl.-Ing. D. Neudeck, Dipl.-Ing. R. Martin, Dipl.-Ing. N. Renzow, Dr.-Ing. h.c. F. Porsche AG

Stuttgart, Germany

Abstract

Companies engaged in motorsport often justify their commitment by arguing - both publicly and internally - that the results of their race car developments can also be put to good use in regular road cars. Rare are the cases, however, where this strategy has been as consistently followed as in Porsche's brake development.

For Porsche, this uncompromising approach is not just a sales gimmick but rather an indispensible condition which makes our cars suited for both, every-day driving on regular roads and racing on closed circuits.

Therefore, the demands on brake engineering must be defined in a way so as to meet today's increased demands on comfort and, at the same time, take into consideration that our cars can be run on racing circuits.

The Porsche „brake family tree“ shows the succession of the various brake developments over the last decades and their interrelationships in a clearly understandable way.

17.1 Introduction

Dr. Ing. h.c. F. Porsche AG Stuttgart is one of the few automotive manufacturers worldwide, who decided, decades ago, to turn this vital aspect of active safety into one of their core competences. The consistent, future-oriented idea of fitting Porsche production cars with the best and most powerful brakes available, was born in 1977 already. At that time, the 911 Turbo was the first series car to come with pure-bred race brakes carried over from the Porsche 917. This brake system consisted of a four-piston aluminum fixed caliper combined with perforated, internally ventilated brake discs which provided the 300 HP 3.3 liter engine with the powerful deceleration it needed. Porsche brake engineering has found its temporary climax in the PCCB Porsche Ceramic-Composite-Brake, which was presented to the public for the first time in 2000 and had its production debut in the 911 GT2, in 2001. Also in the future, we shall allow our production cars to benefit from progress in motorsport in order to further optimize our products and pave the way for continued innovation. Taking the example of the „brake family tree“ (**Fig. 17.1**) the following paper explains the relationship between the racing brake technology and the production brake development derived therefrom.

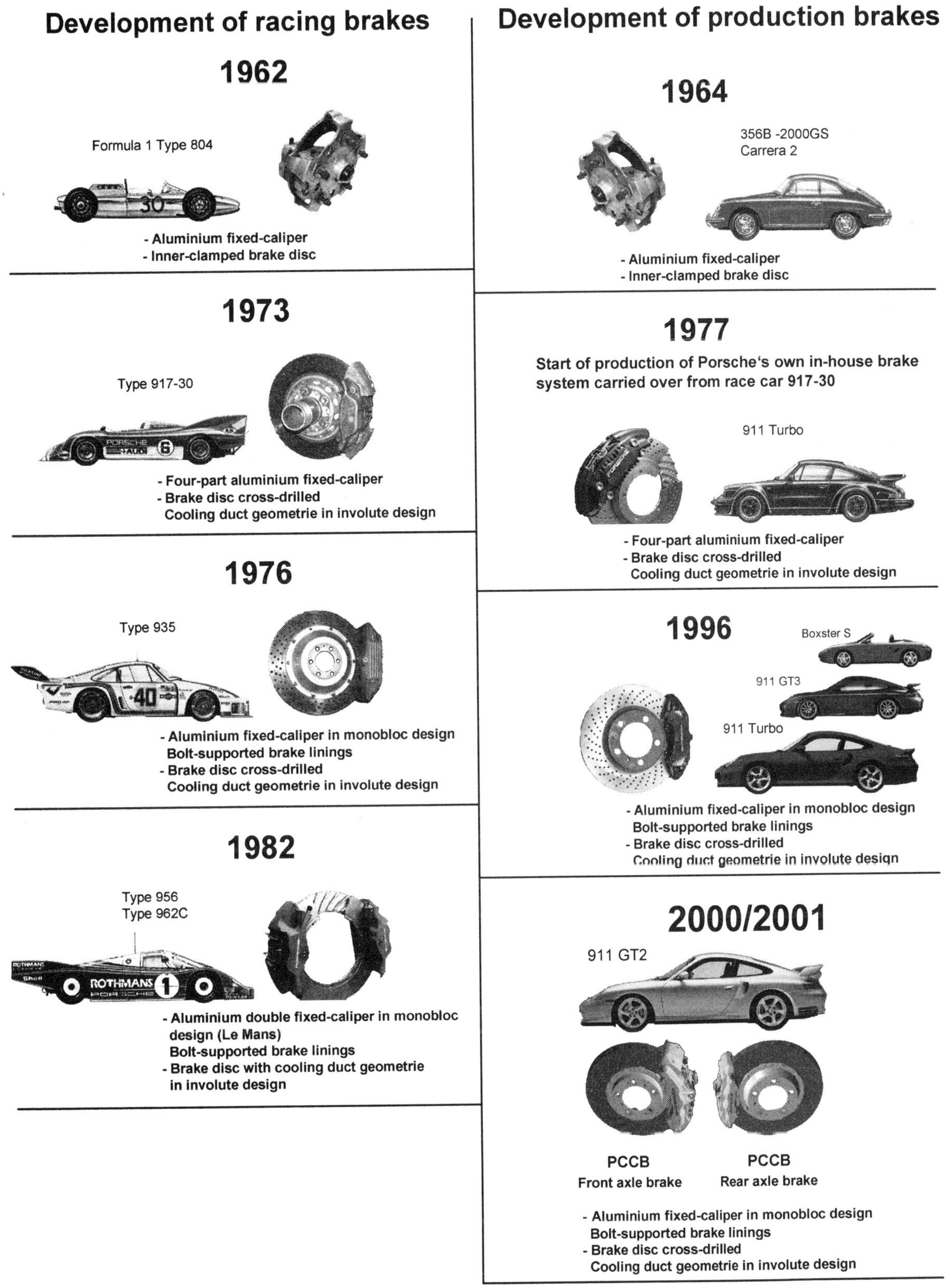

Fig. 17.1: Porsche brake family tree

17.2 1962: Brake system with internally clamped brake disc

The first offspring of the family tree was the brake system with an internally clamped brake disc (**Fig. 17.2**) which Porsche developed for the Formula 1 Porsche 804 F1 (**Fig. 17.3**) and which was later used in the production 356 B-2000 GS. The main benefit of this system over a conventional disc brake was the better utilization of the available packaging space. It allowed the outer disc diameter and the lining surface to be increased by about 20 percent without changing the rim size.

Fig. 17.2: Brake system with internally clamped brake disc

Fig. 17.3: Formula 1 Porsche 804

17.3 1976: Aluminium monobloc fixed-caliper brake

The Porsche 935 (**Fig. 17.4**) was the first race car to come with aluminum monobloc fixed-caliper brakes with bolt-mounted brake linings (**Fig. 17.5**). This newly developed system excelled by its high component stiffness and operational reliability. And it was the first one to feature external connecting lines for better brake fluid cooling.

Fig. 17.4: Porsche 935

Fig. 17.5: Aluminium monobloc fixed-caliper brake

17.4 1977: The first production car to come with a racing brake system

Porsche decided not to fit the 3.3 liter 911 Turbo (**Fig. 17.6**) with a mass-produced brake system but to combine the car's powerful 300 HP engine with a brake system carried over from motorsport instead.

Fig. 17.6: Porsche 911 Turbo

This decision marks the birth of Porsche's own production brake engineering. The brake system of the 911 Turbo, developed at that time, had been derived from the race-proven Porsche 917. It featured aluminum fixed-calipers and brake discs with perforated friction surfaces and involute-shaped cooling ducts (**Fig. 17.7**).

Fig. 17.7: Porsche 917 brake system

The aluminum fixed-caliper concept, which Porsche had specifically developed for motorsport, offers the following advantages over the floating caliper designs which are currently used for high-performance cars:

- The component weight relative to the braking power is lower (pressure/volume displacement-to-brake force ratio).
- The structural stiffness is higher which results in a lower volume displacement.
- The brake lining pressure is more homogeneous with lower taper wear.
- Heat dissipation is more efficient thanks to the use of high-temperature-resistant aluminum alloys for the brake housing and pistons and to exposing the brake fluid line to the cooling air flow.
- The brake linings are adjusted more evenly. This reduces the time the brake disc needs to cool down and prevents thickness variations of the friction surfaces
- Corrodibility is lower thanks to the pin-supported brake linings

In the late sixties - when doing brake disc investigations for motorsport - Porsche engineers discovered another decisive method of realizing outstanding decelerations. They found out that a particular involute-shaped cooling-duct geometry could reduce the brake disk temperatures during fading tests by about 40°C when compared with conventional internally ventilated brake discs with radial cooling ducts (**Fig. 17.8**). A further temperature reduction by another 60°C was achieved by perforating the brake discs. Apart from their better cooling characateristics, perforated brake discs have another decisive advantage: they are less heavy and respond better on wet road surfaces. In 1967, Porsche obtained a patent on this internally ventilated perforated brake disc with involute-shaped cooling ducts.

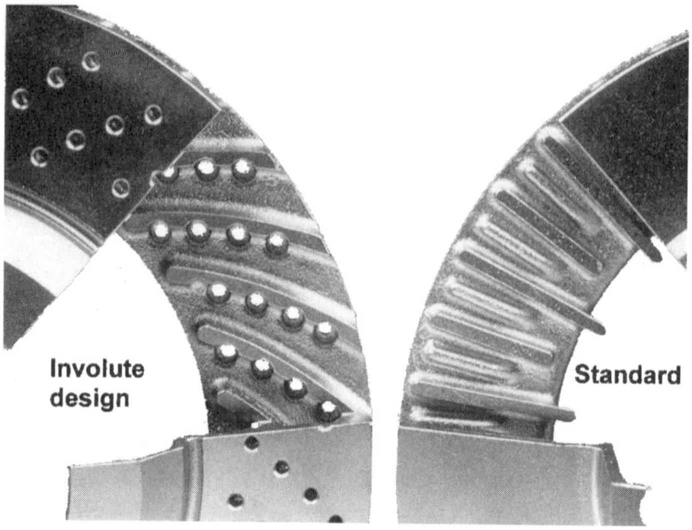

Fig. 17.8: Brake disc cooling duct geometry

17.5 1982: Aluminum double fixed-caliper brake system

After having been fitted with the aluminum double fixed-caliper monobloc brake system (**Fig. 17.9**), which had been specifically developed for high-speed racing, the 956/962 race cars (**Fig. 17.10**) succeeded in winning five LeMans victories in succession. The aluminum double fixed-caliper brake, which had twice the regular lining volume, allowed intervals between lining replacements to be considerably extended.

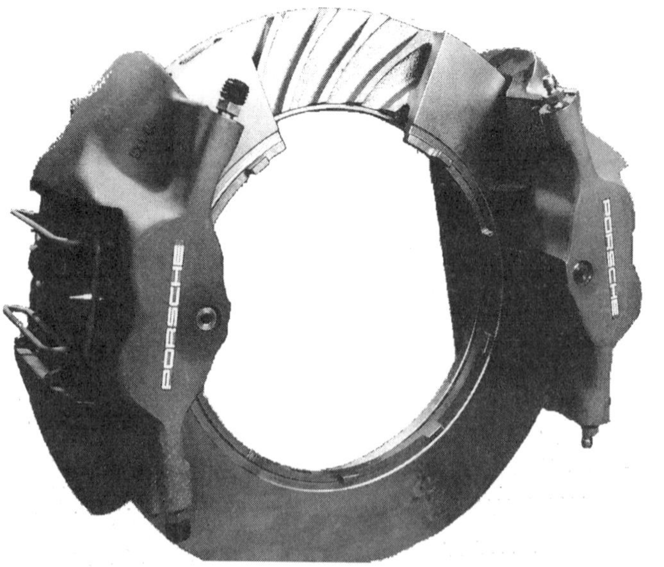

Fig. 17.9: Aluminum – double fixed-caliper monobloc brake

Fig. 17.10: Porsche 956

It is from this monobloc fixed-caliper housing and pin-mounted brake linings, that the current generation of monobloc fixed-caliper brakes for the Boxster and Carrera platforms have been derived.

17.6 1996: Brake concept for the new generation of Porsche cars

This wheel brake concept combines and optimizes the experience gathered with the aluminum monoblock fixed-caliper brakes in the 935 and 956/962 race cars. Special attention was focussed on (**Fig. 17.11**):

- the brake pistons with their different diameters to compensate for the diagonal wear of the linings,
- the high component stiffness which results from the monobloc design,
- the low brake-fluid temperatures resulting from the exposure of the connecting lines to the cooling air flow,
- the pin-mounted brake linings (three-point mounting) and
- the low weight of the aluminum components which makes caliper connecting bolts unnecessary.

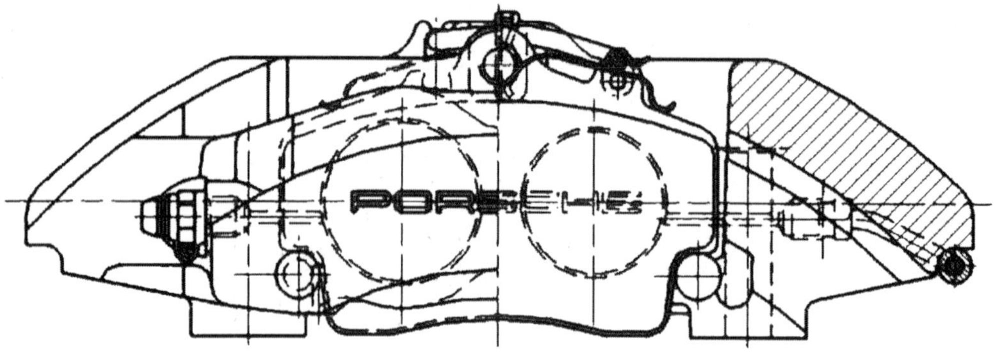

Fig. 17.11: Porsche aluminum – monobloc – fixed caliper

17.7 2000: Porsche Ceramic Composite Brake

The temporary climax of Porsche's brake engineering program is the Porsche Ceramic Composite Brake (**Fig. 17.12**).

Fig. 17.12: Porsche aluminium monobloc fixed caliper

This concept, too, has been carried over from motorsport:

Its carbon discs – which are still used in motor racing today - show clear advantages over their conventional grey-cast-iron counterparts, which are particularly important in racing, that is:

- extremely low weight
- stable friction values even under extreme thermal loads
- good response
- As far as everyday suitability is concerned, however, there are some serious drawbacks:
- inferior low-temperature friction coefficient,
- poor response on wet road surfaces,
- strong wear under high temperature loads

When developing the PCCB brake system, these drawbacks had to be eliminated without giving up the inherent advantages of the carbon materials.

The outcome of this project has been a totaly new material and new production process. In combination with a special brake and pad material this system is known by the designation „PCCB“.

Its specific qualities perfectly complement the aforementioned advantages of the carbon disc and are of vital importance for the safe operation of road vehicles. The carbon disc stands out for

- high heat dissipation
- good response on wet road surfaces,
- outstanding hardness an wear resistance
- very high dimensional stability also at high temperatures.
- Friction level nearly not influenced by temperature and high speed
- high vibation damping
- no stick – slip effect (no squeak noise)

17.8 Reasons for the application of racing technologies in Porsche production cars

Quite frequently, the question arises, whether it makes sense to use race car technologies in regular road vehicles.

For Porsche, the answer is yes.

According to Porsche's sports car philosophy, the target is to provide utmost driving fun while guaranteeing maximum active safety - and this on both, regular roads and race circuits.

Therefore, the demands on brake engineering must be defined in a way so as to meet today's increased demands on comfort and, at the same time, take into consideration that our cars can be run on racing circuits.

17.9 The Porsche fading test

Fig. 17.13 shows a typical disc temperature curve of a Porsche fading test.

Among brake specialists, the Porsche fading test is known and dreaded for his brake-killing qualities.

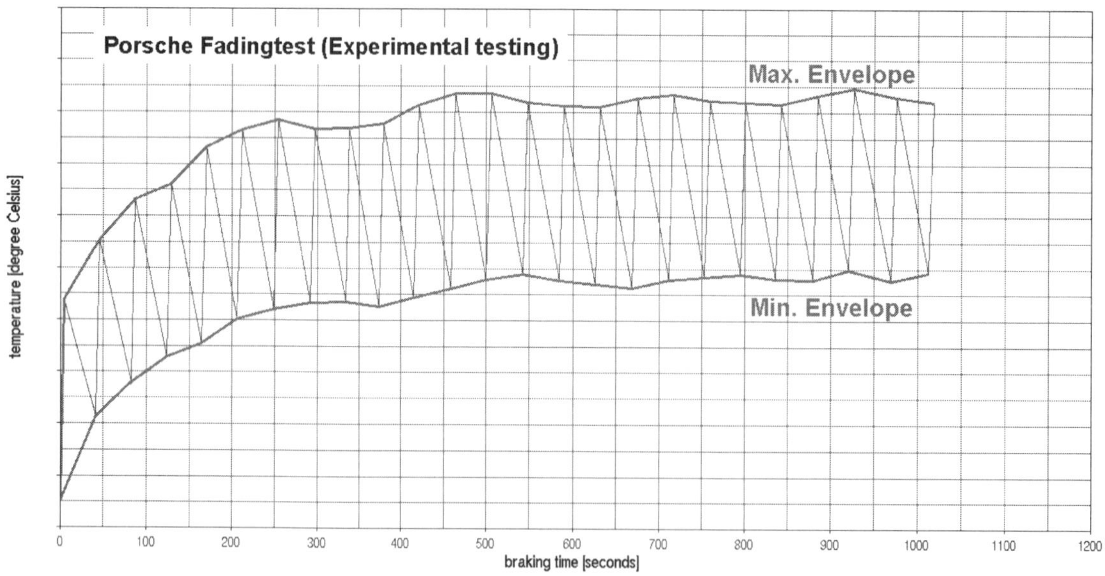

Fig. 17.13: Brake disc temperature response during a Porsche fading test

To us, it is nothing more but a method to make sure that our brake systems are suited for high performance conditions. By defining this method, we wanted to create a reproducible test with stresses, which are equivalent to those experienced on a racing circuit.

How is that put into practice?

The task is, for example, to develop and launch a more powerful version of an existing model.

The basic data for the evaluation of modifications are provided by a real fading test carried out with a current production car (**Fig. 17.13**) . The validated simulation model (**Fig. 17.14**) is fed with the modified parameters, such as acceleration performance, weight etc. Then, the results of the fading simulation are compared with those obtained with the production car and evaluated (**Fig. 17.15**).

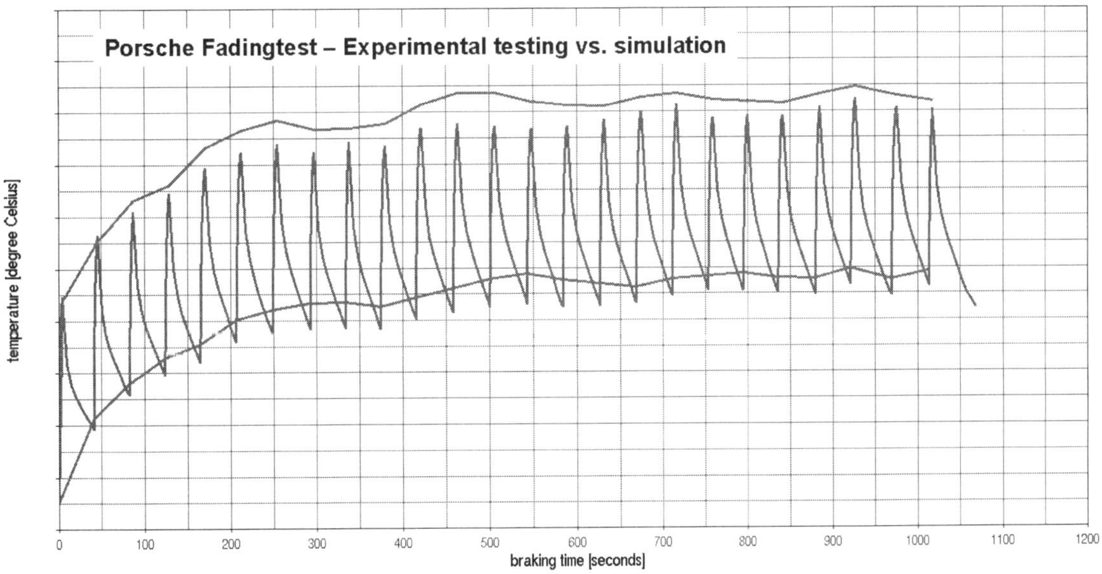

Fig. 17.14: Porsche Fadingtest – experimental testing vs. simulation

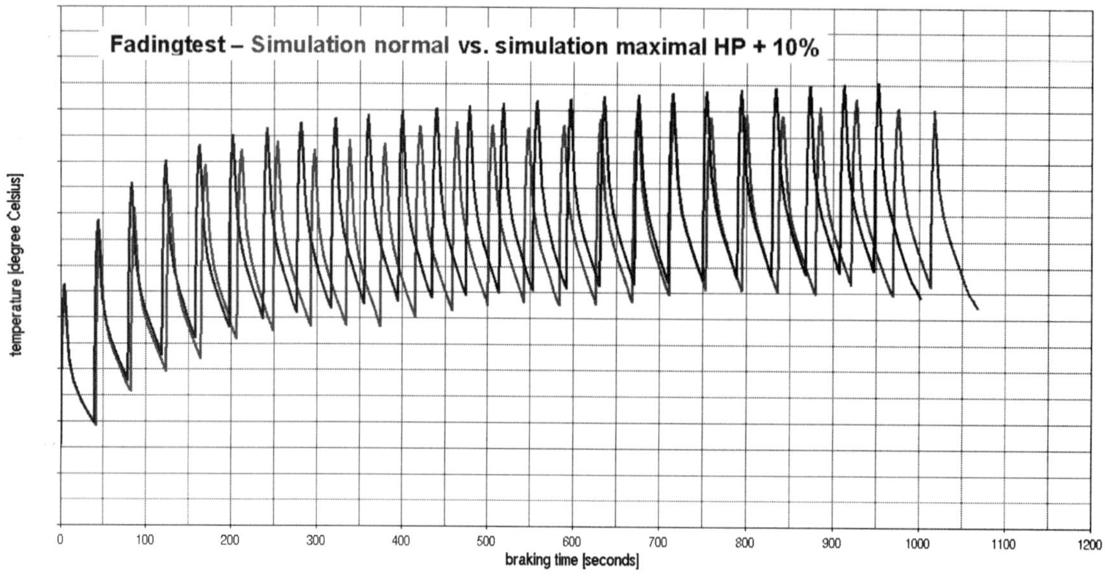

Fig. 17.15: Fadingtest – simulation normal vs. simulation maximal HP + 10%

Further, the different boundary conditions on the various circuits are checked for their influence. This is done again with the help of simulation. Validated simulation models of three different racing circuits are used, that is: the Porsche test track, the short Hockenheim circuit and the Northern loop of the Nürburgring which help to evaluate such variables as brake disc temperature responses, deceleration and acceleration times and so on (**Fig. 17.16**).

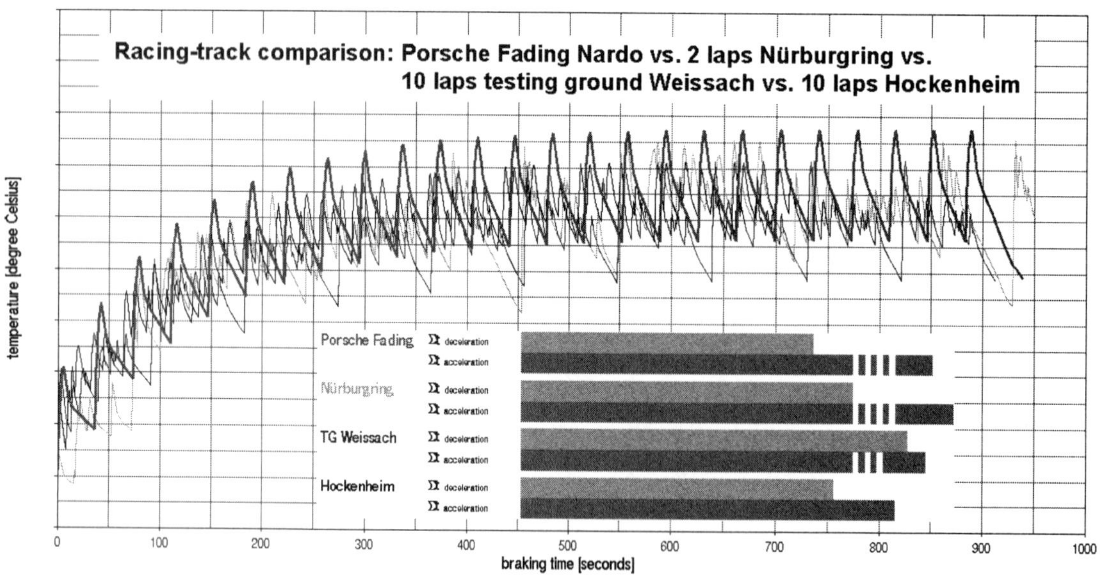

Fig. 17.16: Race-track comparison

The results of such simulations are analysed and then used to stipulate the measures needed to reach the planned targets. It is thanks to these development loops that reliable information can be rapidly obtained about the effects of a modification on the performance of the brake system.

17.10 Summary

So, we have come full circle:

Race car technologiy is carried over into road vehicles which, in turn, must be suitable for everyday use and - at least as far as Porsche cars are concerned - also be fit for circuit racing.

In Porsche Brake Engineering, this goal is reached through real fading tests and runs on the racing circuit, which provide, amoung others, the basis for simulation calculations.

That Porsche Brake Engineering is on the right way shows the success of the PCCB Brake . An important contribution to increasing active safety.

Part IV

Future Aspects and Outlook

18 Research on Tribological Characteristics between Al–MMC Brake Disc and Friction Material

K. Shibata, K. Kawasaki, Honda

Tochigi, Japan

Summary

Recently weight reduction of automobiles has become more important because of higher demands for reducing CO_2 in the atmosphere. Especially weight reduction in body and chassis are considered because of their effect. In this situation alternating the conventional cast iron brake disc to AL-MMC has been considered by many brake and automobile manufacturers as a way of saving weight. In the last five years much research about the making process or thermal property of AL-MMC brake disc has been presented. However one of the most important problems in applying AL-MMC brake disc to automobiles is the difference in frictional characteristics from cast iron disks. So, we have researched the tribological characteristics between AL-MMC brake disc and friction material.

From this research the technique of friction material that is suitable for AL-MMC brake disc has become clear. In this presentation I will introduce the key points of this technology.

18.1 Introduction

The need for environmental protection has increased throughout the world in recent years, and tightening of environmental controls is progressing at a rapid pace. In the automobile industry as well, in addition to enhanced levels of performance as in the past, increasing importance is being placed on the development of technology that is able to accommodate environmental issues. Examples of technical issues in the automobile environment are as indicated below:

- lower levels of harmful substances in exhaust gas with respect to air pollution and acid rain
- improved ease of recycling with respect to protection of resources and waste treatment
- improved fuel consumption and changeover to alternative energy sources with respect to global warming and energy conservation needs.

In particular, reduction of CO_2 levels through further improvements in fuel consumption is an important issue for automobiles. **Fig. 18.1** shows the proportions of energy consumed by automobiles. As shown in this figure, 50% of the energy consumed by automobiles is lost to the engine drive train, 20% to running resistance and 30% to acceleration resistance. In this figure, 90% of the acceleration resistance is attributable to inertial weight. Thus, efforts to reduce vehicle weight by reducing inertial weight are the most important in terms of improving fuel consumption.

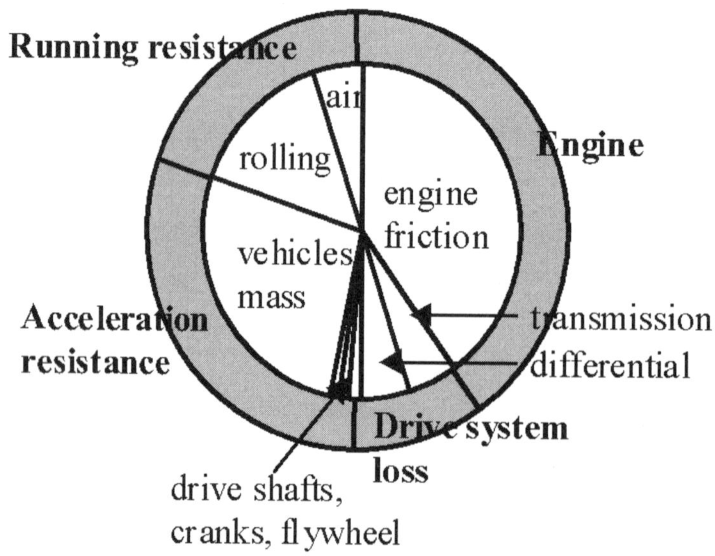

Fig. 18.1: Details of automobile energy consumption

Fig. 18.2 shows means of accommodating improvement of fuel consumption. In recent years, there has been a growing demand for the use of aluminum, magnesium and plastic in the automobile body or frame for the purpose of reducing vehicle weight. The use of aluminum in brake parts is also proceeding rapidly. In particular, reducing the inertial mass of the chassis as well as improving acceleration by reducing the rotational mass of the suspension and wheels or improving kinetic performance by reducing the under-spring weight are expected to be effective in reducing the weight of a rotating body in the form of the brake disc, and attention has been focused on aluminum MMC in particular for achieving this objective.

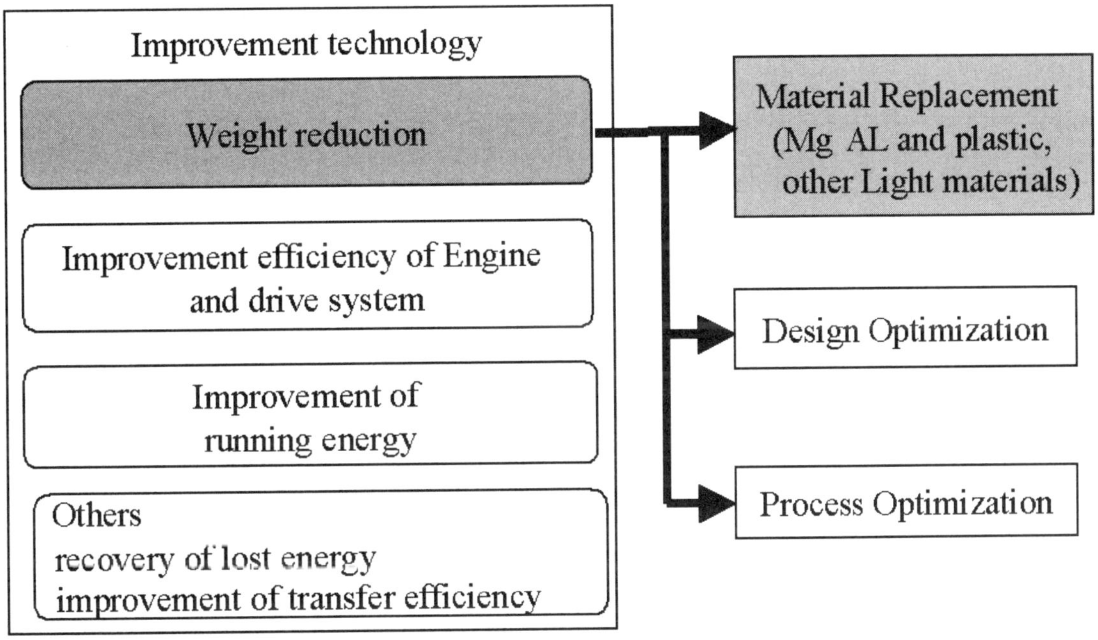

Fig. 18.2: Improvement methods of fuel consumption

However, there are many aspects of the thermal limitations of aluminum MMC, and friction material that can be applied as brake materials, that remain unknown. Examples of which include its heat resistance and friction mechanism. These factors are making it difficult to apply these materials to brakes. Thus far, we have proceeded with the development of an aluminum MMC brake disc and friction material while focusing on the characteristic friction properties of MMC in both microscopic and macroscopic observations.

This paper therefore describes friction material technology established by clarifying the differences with conventional systems and identifying ensuing problems, and by focusing on the effects of key friction material components, based on analytical results of the tribological characteristics of MMC.

18.2 Overview of Aluminum MMC

Aluminum MMC is comprised of dispersing and compounding a reinforcing material such as ceramics (particles and fibers of SiC, Al_2O_3 and so on) in an aluminum matrix. The compounding of this material improves the mechanical strength, wear resistance and thermal expansion of the aluminum matrix while maintaining its thermal conductivity.

In this research, aluminum MMC was used in which the disc material contained 22% alumina for the reinforcing material and aluminum (containing less than 1% of magnesium) for the matrix, and which satisfied the mechanical strength required by brake materials produced by the infiltration method. The microstructure of this aluminum MMC is shown in **Fig. 18.3**, while its typical properties are shown in **Table 18.1**. Aluminum MMC makes it possible to obtain excellent tribological characteristics as a result of inhibiting shear fracture of the aluminum matrix with hard alumina particles, thereby making it possible to use aluminum as a brake material. However, on the friction surface, as a friction material will be in contact with the aluminum matrix in addition to alumina, it is necessary to give ample consideration to the development of the corresponding friction material.

Table 18.1: Mechanical properties of brake disc materials

		Cast Iron	AL-MMC
Density(g/cm3)		7.00	3.00
Specific heat(J/kg/K)		0.502	0.981
Thermalconductivity (W/m/K)		52	114
Thermal expansion ($\times 10^{-6}$/°C)		11	17.7
Most operating temperature(°C)		1200	650
Young's modulus (Gpa)	R.T	108	98.1
	200°C	98.9	90.2
Elongation(%)		0.5	1.9
Tensile strength (Mpa)	R.T	208	200
	200°C	199	162
Fatigue strength(Mpa)		73	123
Sharpy impact Value(J/cm2)		2.64	5.00

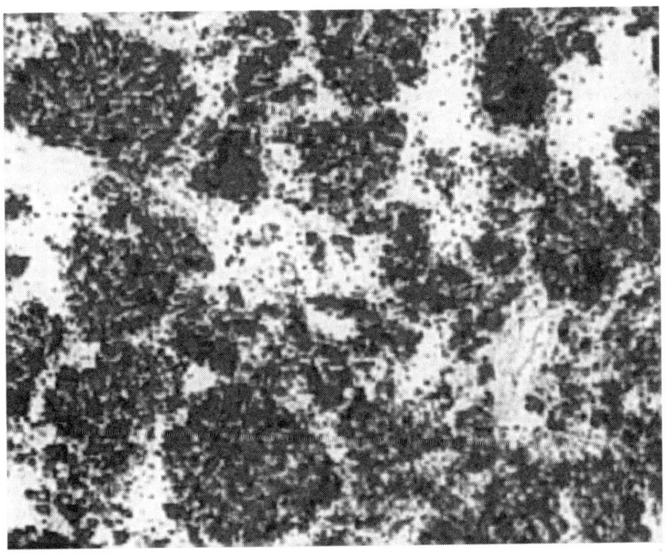

Fig. 18.3: Micro structure of Al-MMC

18.3 Types of Brake Friction Materials

Conventional brake friction materials are classified into non-steel organic materials (NSO materials), low-steel organic materials (LSO materials), semi-metallic materials and sintered metallic materials. Among these, NSO materials have come to be frequently used in four-wheel vehicles in recent years, and are formed into aramid fibers, ceramic fibers, non-ferrous metallic fibers as well as various lubricants, ceramic powders, metal powders and so on, using phenolic resins as a binder. Since these NSO materials do not contain steel fibers, they have a low degree of disc attack and have favorable brake noise characteristics. Conversely, they are inferior to LSO materials in terms of μ level and fade resistance, and for this reason, LSO materials are widely used in Europe where vehicles are subjected to heavy loads and high speeds. Since sintered materials do not use a resin binder, they have even higher heat resistance. However, due to their inferior brake noise characteristics, they are limited to use in two-wheel vehicles, racing and railways that require high μ levels and heat resistance.

18.4 Analysis of Friction Characteristics and Related Issues

In the development of a friction material for aluminum MMC, sliding tests were conducted on various friction materials to analyze the friction characteristics of aluminum MMC. An inertia-type friction tester was used for the testing device. After initial sliding together 200 times using a disc having a diameter of 215 mm and thickness of 10 mm, friction materials measuring 30 × 30 × 10 mm, inertia of 0.5 kgm^2 and effective sliding radius of 88 mm. This was followed by measurement of μ at test conditions combined with disc temperatures of 100, 150, 200 and 300°C, rotating speeds of 423, 845 and 1,014 rpm, and torque ranging from 6 to 12 kgm (at 2 kgm intervals). In order to confirm fading, the interval was adjusted at a torque of 10 kgm and the change in μ was measured resulting from repeatedly braking 20 times with the aim of achieving a maximum disc temperature of 500°C. Moreover, an additional test was performed in which μ was again measured at 100°C following the fade test to confirm the changes in μ following thermal hysteresis.

18.4.1 Study of Various Brake Friction Materials

Tablc 18.2 summarizes the results for NSO, LSO and sintered materials with respect to friction materials for typical cast iron disks. For all of these friction materials, μ tended to decrease in the case of the MMC disc as compared with using a cast iron disc. Although there were no large differences between the friction materials with respect to μ when using an MMC disc, the NSO material was relatively favorable in terms of the surface properties following friction. With the LSO material, there was an extremely high degree of transference, and with the sintered material there was remarkable damage to the disc resulting in high degree of surface roughness. Although both the LSO and sintered materials are considered to be worth examining as candidates for use as high μ friction materials, in this research, the NSO-based material was selected as the base friction material for the aluminum MMC disc because it is widely used in four-wheel vehicles and has relatively favorable tribological characteristics and surface state.

Table 18.2: Comparison for friction properties of various friction materials to each disc

material	NSO-based material	LSO-based material	Sintered
Friction coefficient to Cast iron disk	0.40	0.45	0.60
Friction coefficient to AL-MMC disk	0.29	0.25	0.30
Visual characteristics	relatively good surface 	extensive adhesion 	roughening surface 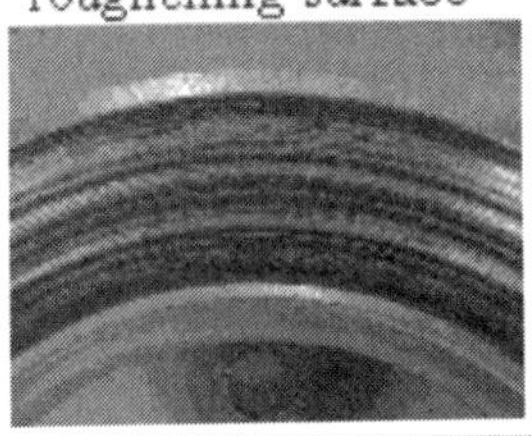

18.4.2 Friction Wear Characteristics of MMC

18.4.2.1 Wear and Transference

This section describes an examination of the tribological characteristics of the NSO-based friction material. As can be seen in **Fig. 18.4**, compared with the cast iron material, the MMC disc has a transfer film formed uniformly over the entire friction surface, which gives its characteristic appearance. Although transference also typically occurs with mass-produced cast iron disks, remarkable transference is hardly ever observed.

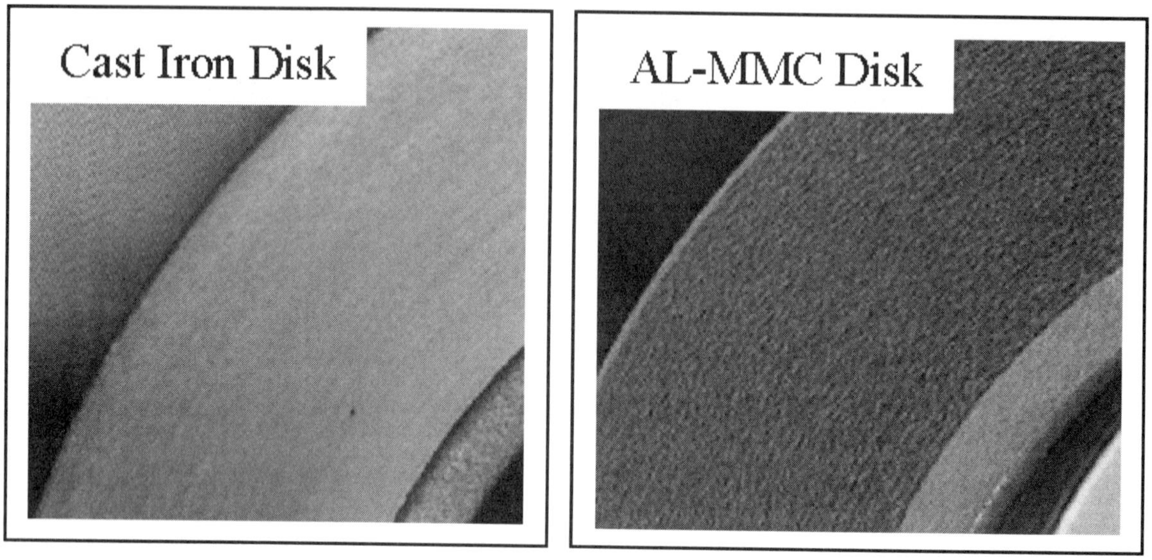

Fig. 18.4: Typical friction surface of disc

Fig. 18.5 and **Fig. 18.6** show the wear changes caused by temperature per 1000 sliding movements of the friction material and disc, respectively. The amount of wear of the friction material tends to increase with temperature in both cases, and there are no large differences observed. However, disc wear differs considerably between cast iron and MMC. In the case of a cast iron disc, wear increases with temperature in the same manner as the friction material, demonstrating wear of 57 μm at 300°C. In contrast, in the case of the MMC disc, there was either no wear or disc thickness increases by about 20 μm. This is believed to be due to the fact that, in the case of MMC, in addition to wear resistance of the MMC base material itself being improved by the ceramic powder, the stable formation of a thin transfer film on the sliding surface protects the disc surface while simultaneously increasing the thickness of the transfer film, thereby resulting in reduced disc wear. The mechanism by which a stable transfer layer is formed on the MMC disc is thought to involve the formation of a uniform and stable transfer film due to the higher degree of bonding at the interface of the disc base material and transfer layer.

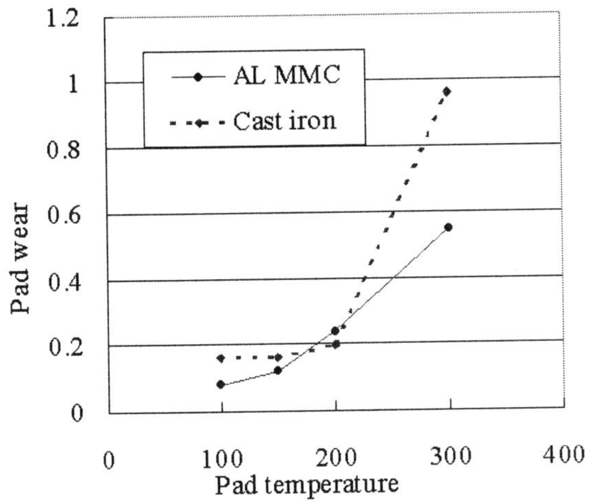

Fig. 18.5: Pad wear vs. pad temperature

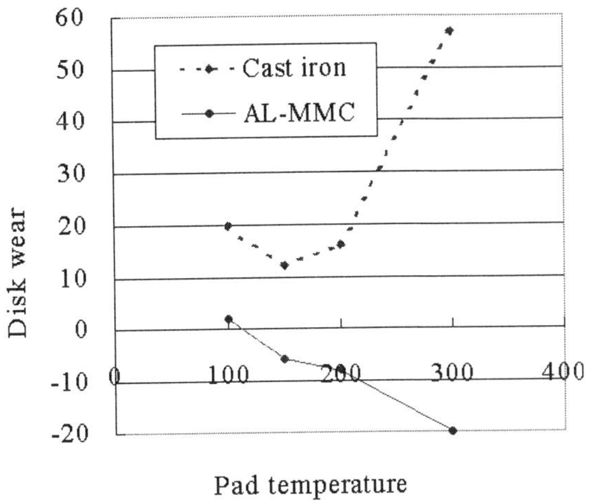

Fig. 18.6: Disc wear vs. pad temperature

In the case of cast iron disks, on the other hand, due to the low degree of bonding with the transfer layer, the difficulty in forming a uniform film, and the cast iron disc itself being relatively susceptible to abrasion, a transfer film is not formed in the manner of MMC, thereby resulting in greater wear.

In any case, the formation of this transfer film is believed to have a significant effect on the sliding properties of MMC, and control of that transference is thought to be one of the keys to developing a friction material for MMC.

18.4.2.2 Comparison of μ for MMC and FC Materials

The following describes a comparison of the characteristics of μ of various types of NSO-based friction materials. **Fig. 18.7** shows a comparison of mean μ at a disc temperature of 100°C for various types of NSO-based friction materials when using a cast iron disc and MMC disc for the corresponding material. Although μ values ranged from 0.32 to around 0.48 when a cast iron disc was used for the corresponding material, μ decreased for all friction materials when MMC was used for the corresponding material. Moreover, when a cast iron disc was used for the corresponding material, even for friction materials having large differences in their friction coefficients, the use of MMC for the corresponding material was observed to cause the friction coefficient to stabilize around 0.3. Similar trends were confirmed at disc temperatures of 150 and 200°C as well. In addition, the transfer film described in the previous section was uniformly formed in a stable manner on the friction surface of the MMC for all of the friction materials.

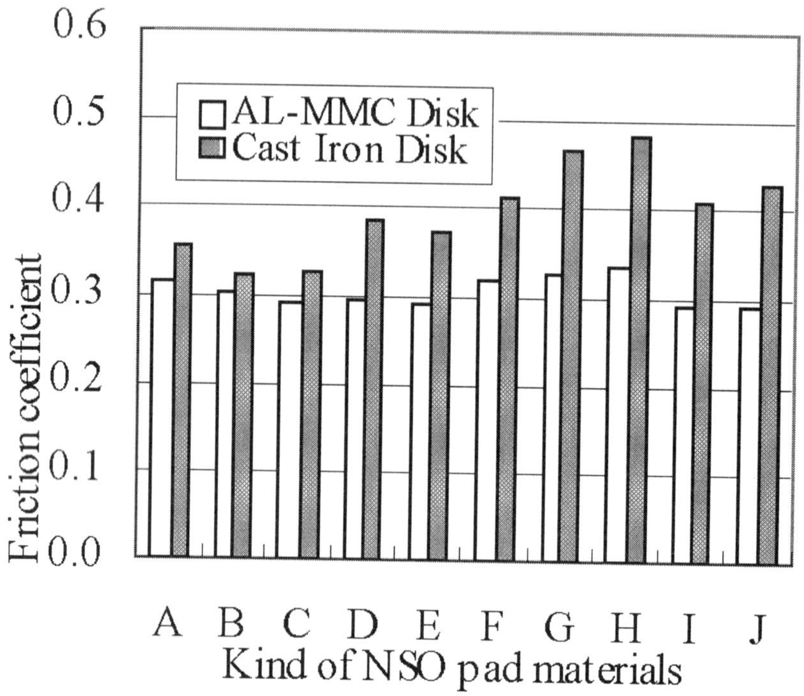

Fig. 18.7: Comparison for friction coefficient at 100°C of several NSO pad materials to each disks

Here, frictional force F, which is generated due to transfer and shearing at the friction interfaced, can be expressed as $F = A \times S$, where A is the contact surface area and S is the shearing strength. Since numerous metals become deformed if shearing is small, contact surface area ends up becoming large. On the other hand, where hard metals are used to decrease contact surface area, shearing strength becomes large. However, Bowden et al. explained that [18-1], when a hard metal is covered with a soft thin film, since the hard metal supports the load and shearing occurs within the soft film, both contact surface area and shearing strength are reduced. As a result, frictional force becomes smaller. At this time, the shearing strength of the friction interface can be approximated by S Sf, and in the case the base material also demonstrates plastic deformation, the friction coefficient μ can be expressed as $\mu = F/W = S \times A/W = Sf/Hs$. Here, W is the weight, Sf is the shearing strength of the thin film, and Hs is the hardness of the base material. Accordingly, this shows that the harder the base material, the smaller the shearing strength of the thin film, and the lower the friction coefficient.

On the basis of these findings, in an MMC disc, sliding occurs that is mediated by a transfer film having relatively low shearing strength that covers the hard MMC. It is assumed that, in addition to abrasive friction produced by direct contact with the friction material and the MMC base material itself, friction between sheared transfer film is also applied, thereby bringing about a reduction in μ. [18-2,18-3]

18.4.2.3 Change in μ Following Thermal Degradation

As the temperature of the disc rises, the thickness of the transfer film increases which tends to increase the plate thickness. Therefore, we next focused on what types of changes are brought about in μ by increasing the transfer film following thermal hysteresis.

Fig. 18.8 shows the changes in μ before and after thermal hysteresis for both cast iron and MMC disks. Although μ decreases for the same reasons as previously explained in the case of using the same friction material, after thermal hysteresis, the difference was observed to become even larger.

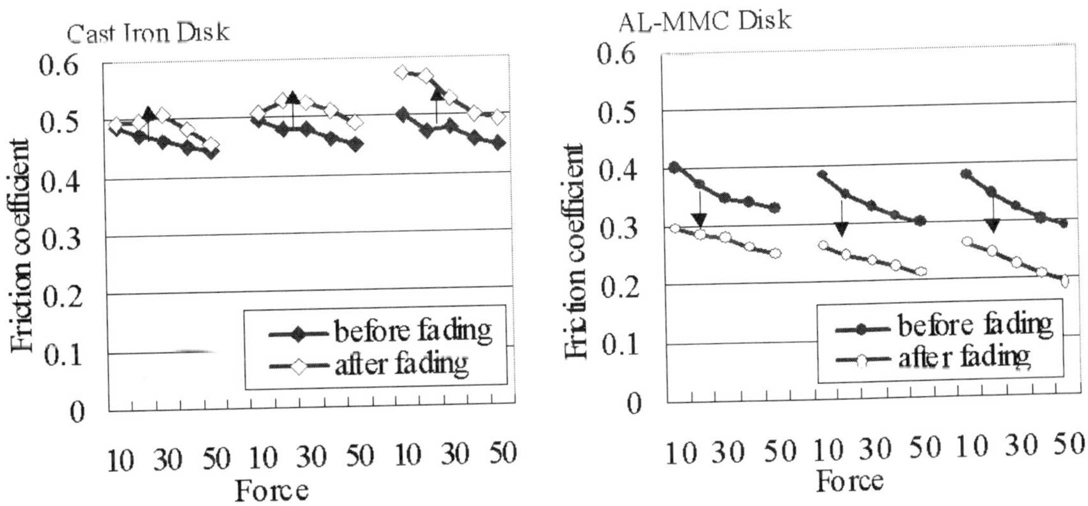

Fig. 18.8: Comparison of friction coefficient after fading

This is assumed to be due to increases in temperature promoting transference to the disc thereby resulting in an increase in film thickness. Consequently, the effect of abrasive friction primarily with the reinforcing material of the MMC, is relatively small, while nearly the entire effect is the result of friction produced by shearing with the transfer film. Therefore, the friction coefficient tends to become even lower following thermal hysteresis. [18-2, 18-3]

18.4.3 Analysis of Friction Surfaces

18.4.3.1 Analysis of Disc Surface

It has been previously described that the transfer film formed on friction surfaces has an effect on the tribological characteristics of MMC. Therefore, we next attempted to conduct a microscopic analysis of the friction surface. This analysis was conducted on an MMC disc heated to temperatures of 200 and 500° by friction.

Fig. 18.9 and **Fig. 18.10** show the results of SEM and EDX analyses of the surface. Although partial separation of the surface is observed, due to the formation of a transfer film, a uniform and smooth surface is presented regardless of temperature hysteresis. In addition, although peaks corresponding to Cu, Ba, Ca, S, Si, C, O and other elements are observed from the surface according to the results of EDX, since these components are completely absent in the MMC base material, they originated primarily in the metals and ceramic powder contained in the friction materials and are thought to indicate the components of the transfer film.

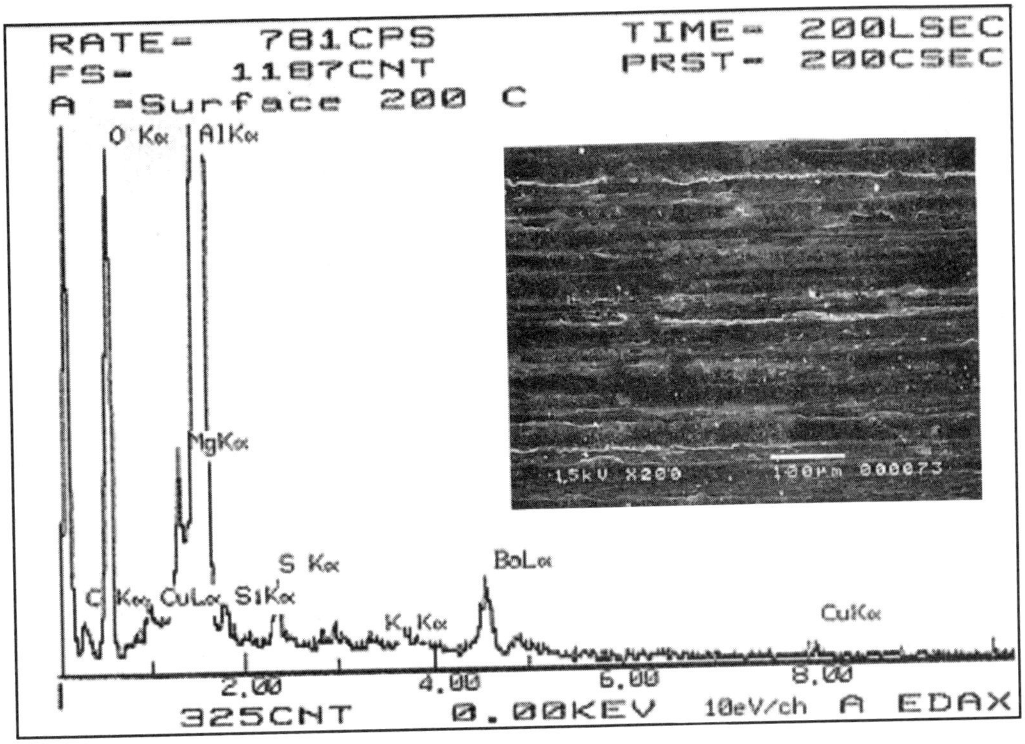

Fig. 18.9: EDX chart and SEM photograph of friction surface tested at 200°C

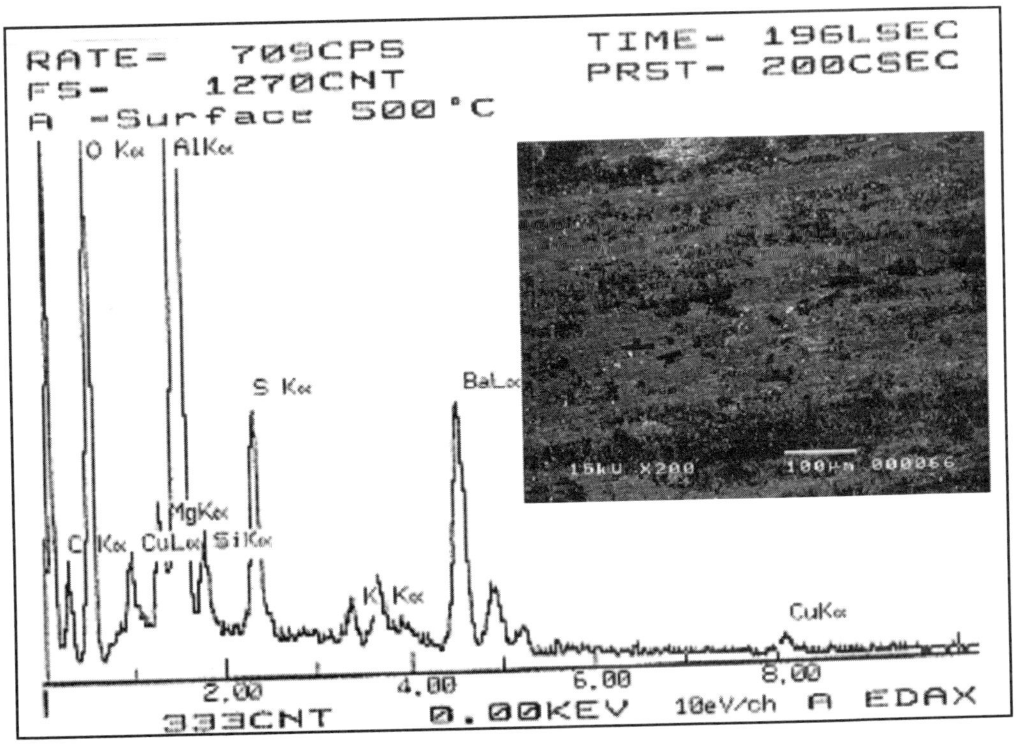

Fig. 18.10: EDX chart and SEM photograph of friction surface after thermal hysteresis at 500°C

In addition, when temperature hysteresis is increased from 200°C to 500°C, the peak intensities of the friction material components tended to increase. This indicates that the thickness of the transfer film that contains the friction material components increases, and closely agrees with the wear trends according to temperature.

Moreover, the surface was also analyzed by FT-IR. The analysis chart is shown in **Fig. 18.11**. Since peaks originating in phenolic OH groups, sulfates and aromatic rings were detected, this indicates that not only components of metal and ceramic powders, but also components of the phenol resin used as binder are also transferred.

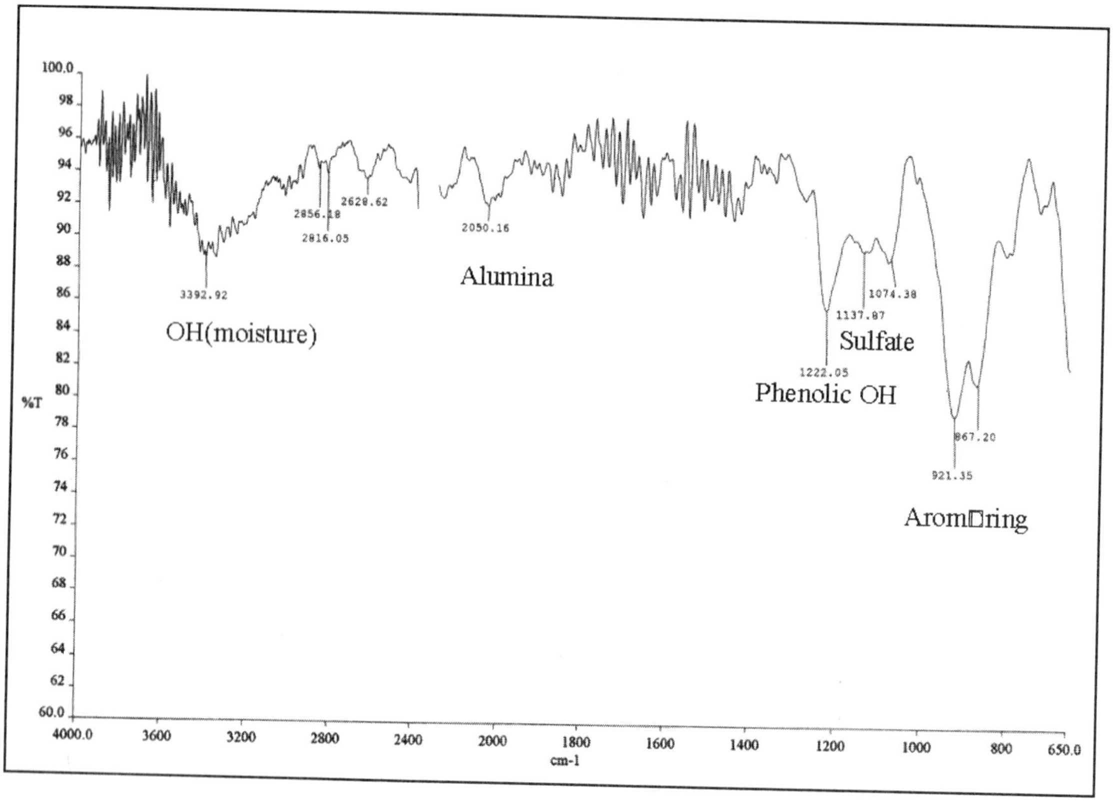

Fig. 18.11: FT-IR chart of friction surface after thermal hysteresis at 500°C

Although the surface apparently maintains a smooth surface in conjunction with the MMC having a high shearing strength, there was also the occurrence of grooves and scratches as seen in **Fig. 18.12** at some locations on the surface. Although wear resistance is improved by the ceramic powder and the friction surface is protected by the formation of a transfer film, there is the potential for the friction material to constantly be in contact with the soft, highly reactive aluminum matrix. Thus, these grooves and scratches are thought to occur as a result of the matrix itself being removed from the disc base material by scuffing and cutting, and continuously grinding away the disc. Consequently, when blending friction materials, they must be added while taking into account such factors as compatibility and reactivity with the matrix.

18.4.3.2 Analysis of Cross-Sections

Fig. 18.12: Scratch and transfer particle on friction surface

Next, we attempted to conduct an analysis of the transfer film based on its cross-section. Cross-sectional SEMs of a sliding portion are shown in **Fig. 18.13** and **Fig. 18.14**, respectively. In a disc that was heated to a temperature of 200°C, there was relatively little damage and a transfer layer was unable to be confirmed in this cross-section. In addition, damage was particularly prominent at locations where alumina had aggregated. In a disc that was heated to a temperature of 500°C, there was even more separation damage of the surface, and what was thought to be a transfer layer was observed in the surface layer. The alumina reinforcing material can also be observed in some parts of the transfer layer. In addition, partial separation was also observed, suggesting that transfer wear consisting of repeated transference and separation was occurring preferentially. As was previously mentioned, in the use of MMC, not only is friction produced by shearing with the transfer layer, but also the occurrence of abrasive friction in the form of contact with the disc reinforcing material. It was surmised from this SEM observation that, although there is little transfer film, and friction with the disc base material is expected at a disc temperature of about 200°C, due to the increase in the transfer film in disks subjected to temperature hysteresis at 500°C, contact with the disc base material subsides and μ ends up decreasing.

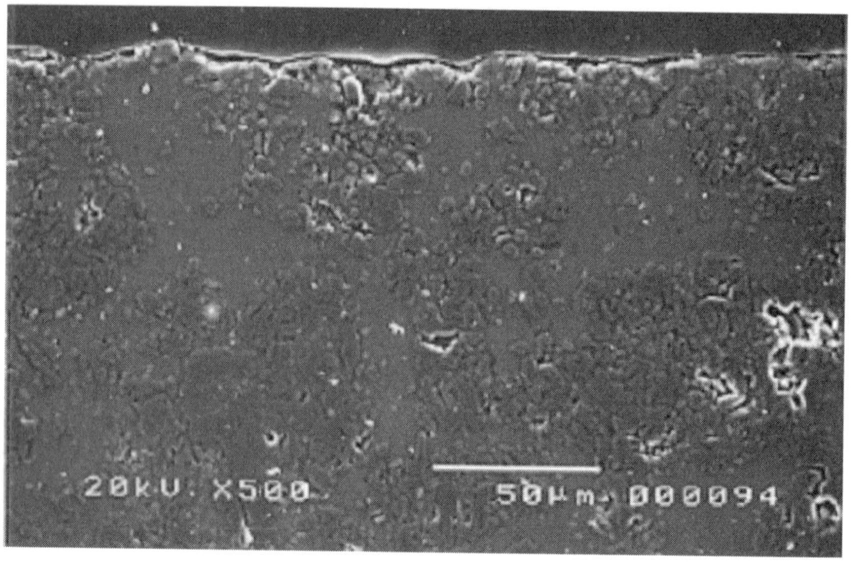

Fig. 18.13: Cross section of disc surface tested at 200°C (SEM)

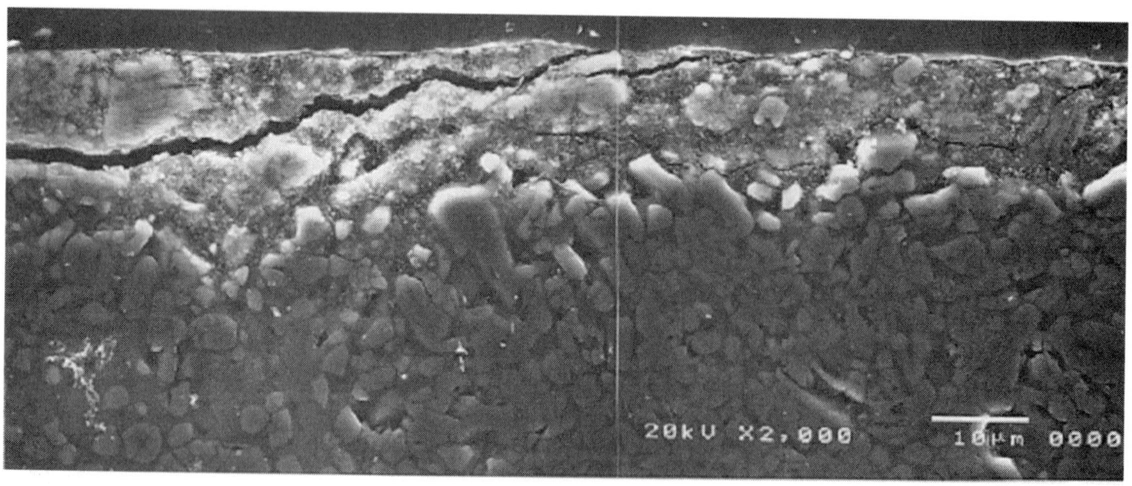

Fig. 18.14: Cross section of disc surface after thermal hysteresis at 500°C (SEM)

Next, the results of analyzing the transfer layer by EDX are shown in **Fig. 18.15**. With the exception of Al, peaks were observed inside the layer that are thought to be friction material components in the same manner as the surface, examples of which include Cu, Fe, Ba, Ca, S, Si and C. Since the alumina reinforcing material and so on were also observed within the transfer layer, it was suggested that friction material and disc material components become mixed in the transfer layer, and become powerfully bound due to friction, resulting in the formation of a stable transfer layer. Although transfer films were unable to be confirmed in cross-sections when disc temperature was raised to 200°C, judging from the results of surface analysis, it is clear that a transfer layer exists, and is believed to be firmly transferred at the interface with the disc in the same manner as transfer films present in disks subjected to thermal hysteresis at 500°C.

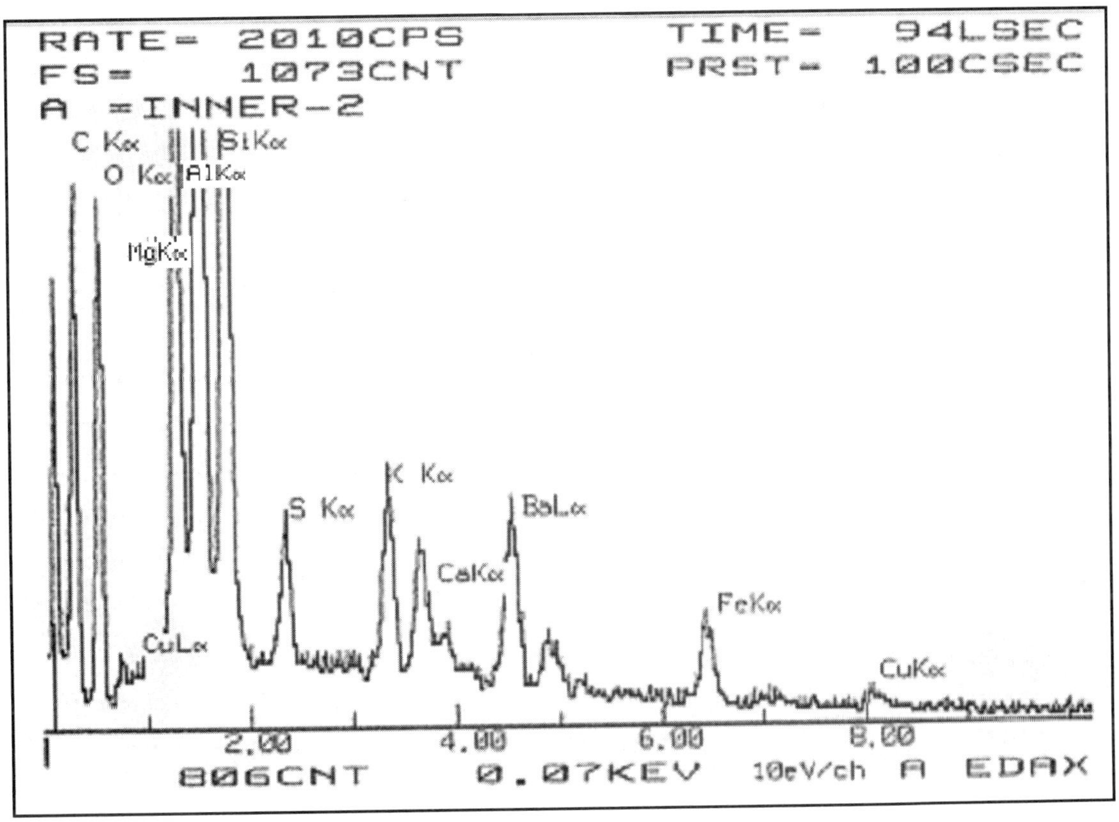

Fig. 18.15: EDX chart of transfer film of cross section after thermal hysteresis at 500°C

18.4.4 Friction Mechanism and Important Factors

The tribological characteristics of the MMC disc and their important factors were summarized on the basis of the analysis of results accumulated thus far. First, an active, fresh aluminum surface is exposed as a result of sliding then friction material and abrasive powder make contact with that aluminum surface causing it to be crushed and resulting in firm transference. Initially, since there is little transference, and due to the effect of abrasive friction of the disc base material containing hard particles in the case of a thin film, the decrease in μ is inhibited to a certain extent. However, since the transfer film becomes strongly bonded and the disc material itself is resistant to grinding, the transferred amount increases when heat is generated. Since this leads to friction between the friction material and transfer film having relatively small shearing strength, the decrease in μ is thought to become more prominent. Important factors relating to MMC brake disks can be summarized in the manner shown below.

- μ decreases easily due to the formation of a transfer film.
- μ decreases further due to increase in transfer film thickness in the presence of heat.
- There is the possibility of damage due to the aluminum matrix itself being attacked and reacting.

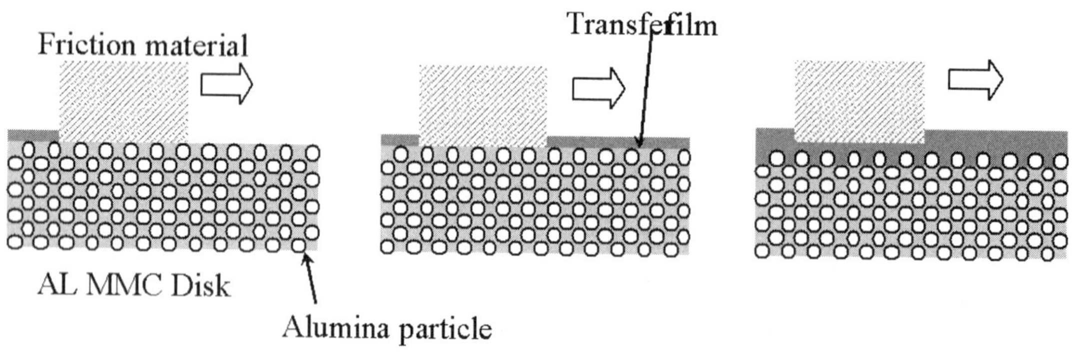

Fig. 18.16: Schematic diagram of friction mechanizm of Al-MMC

It is therefore essential to determine the blending ratio of friction materials by carefully considering these factors. However, although the formation of a transfer film is considered to be unavoidable to a certain extent (from the viewpoint of bonding with the aluminum and protection of the friction surface), an attempt to improve μ must be made by inhibiting the increase in film thickness by grinding as much as possible. In addition, since there is simultaneously always the possibility of contact with the soft, highly reactive aluminum matrix, in order to maintain surface properties, considering the compatibility with the aluminum matrix is an important factor in terms of developing friction materials for MMC. Moreover, although perhaps obvious, it is necessary to adjust the composition of friction materials to maintain the proper balance of the brake material by taking into consideration various properties such as brake noise and judder.

18.5 Effects of Various Friction Components and Development of Friction Materials

We examined the effects of various friction material components based on important factors of tribological characteristics. Although only a small number of examples are indicated here, those major components that will serve as the key to MMC friction material technology are described.

To begin with, the amount of alumina particles (mean particle size: 0.3 μm) added in the friction material was changed over the range of 4.5...10 wt%, followed by investigation of the effects on friction coefficient and lowering of μ following thermal degradation. Particularly fine alumina was added in consideration of disc damage. As shown in **Fig. 18.17**, μ improved as the amount of alumina particles included increased, and was found to be the most effective for improving μ. In addition, although there is concern over brake noise and judder when the amount of hard particles is increased particularly in cast iron disks, in the case of MMC, there were no problems with performance caused by the addition of fine alumina particles.

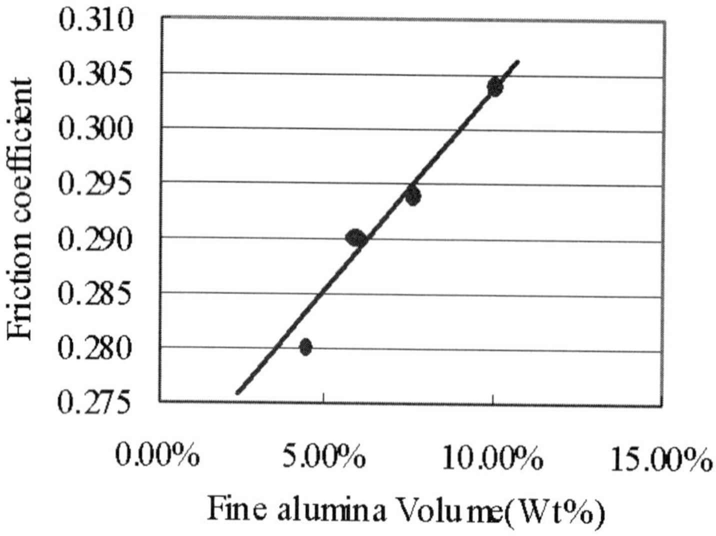

Fig. 18.17: μ vs. fine alumina volume

Moreover, the decrease in μ following thermal degradation (**Fig. 18.18**) was also observed to improve due to addition of fine alumina particles. It is believed that the addition of these fine alumina particles is effective in improving μ due to cutting of the transfer film and disc base material. In addition, since excessive transference is prevented, decreases in μ due to heating was able to be inhibited without impairing other performance.

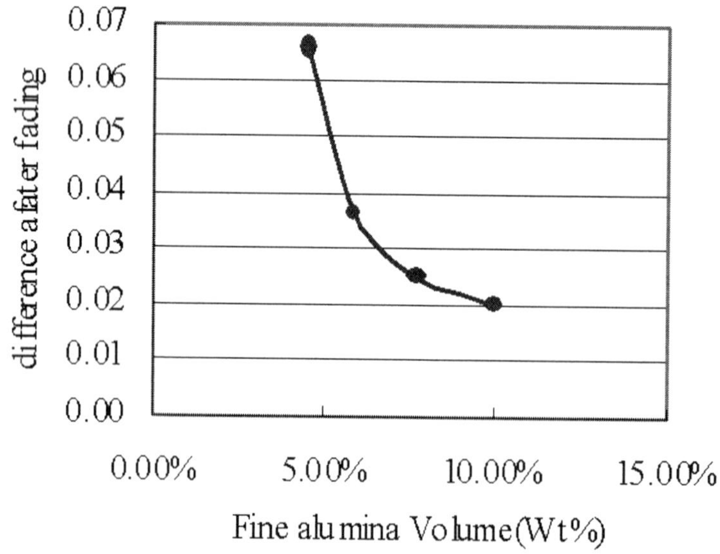

Fig. 18.18: μ difference after fading vs. alumina volume

We next conducted an examination of the metal components. In general, adhesion with metal is dependent on the ease of forming a solid solution, and in particular, the same types of metal are known to adhere extremely easily. [18-4,18-5,18-6] With respect to aluminum, although aluminum forms a solid solution extremely easily, metals such as Cu or Fe form solid solutions with difficulty.

Here, we examined the effect on lowering μ by adhesion with respect to Cu fiber in particular, as an example of a metal that is considered to demonstrate little transference, from the viewpoint of sliding. **Fig. 18.19** shows the results of examining thermal degradation of μ by adding Cu fiber over the range of 0...15%. Although the initial μ tended to decrease when the amount added was increased, there was no decrease in μ observed following heating, and conversely, a slight increasing trend was observed. This indicates that the addition of Cu fiber, because it adheres with greater difficulty, is effective in preventing decreases in μ caused by transference. However, since the initial μ decreases, the amount of Cu fiber must be adjusted in consideration of the balance with μ following thermal hysteresis.

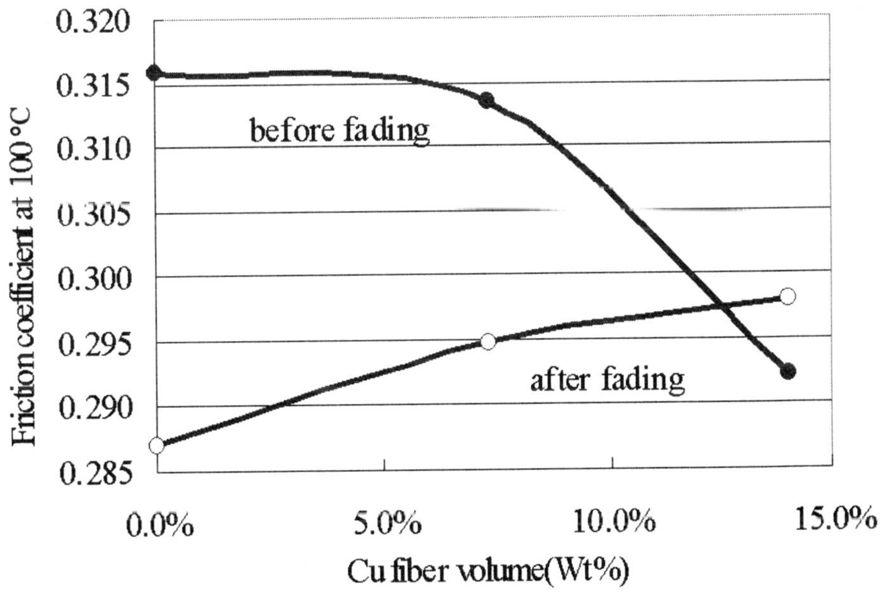

Fig. 18.19: Relationship between Cu fiber volume and μ

Next, we examined the lubricating component (ZnS). When a lubricant having for its main component ZnS was added, the surface condition of the disc was clearly demonstrated to deteriorate as can be seen in **Fig. 18.20**. It is surmised that the ZnS in the lubricant reacts with the aluminum matrix causing the disc to be worn away.

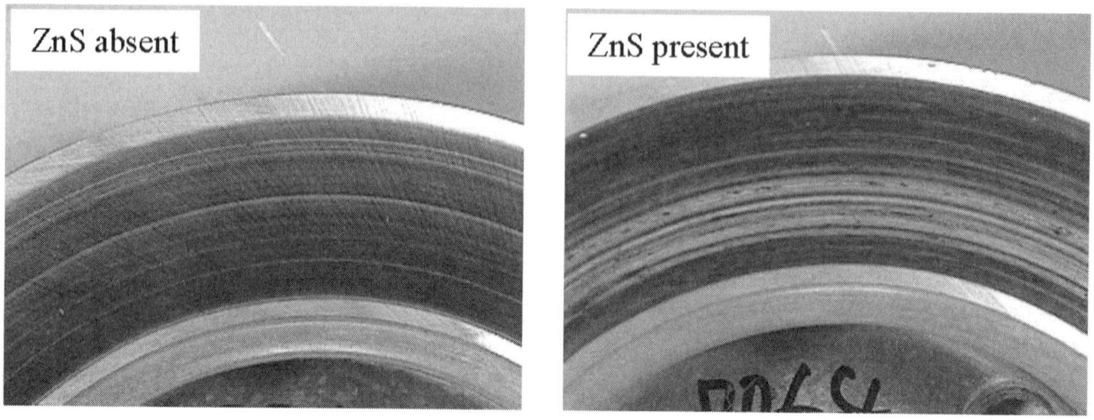

Fig. 18.20: Comparison of friction surface with various pad materials

Similarly, roughening of the sliding surface occurred even with the use of alumina silica fibers. Since surface roughening did not occur in the case of alumina particles or aluminum borate, it is believed that this had a similar effect on surface condition as ZnS since the silica was deposited causing the aluminum matrix to be worn away. In addition, since silica also causes μ to decrease at high temperatures, silica was not considered preferable as a friction material component for MMC.

Studies were also conducted on the effects on tribological characteristics when using various combinations of other components in addition to those mentioned above. These studies made it possible to establish a friction material technology of MMC for practical application in brakes.

18.6 MMC Brake Performance

A friction material for MMC brakes was developed that comprehensively satisfies various aspects of performance, including brake noise and judder, based on the results of examining the effects on MMC tribological characteristics for various friction material components. The basic μ performance of this MMC brake is shown in **Fig. 18.21**. The mean μ is equal to or better than current cast iron brakes, and fluctuations in μ due to deceleration and speed are relatively small indicating satisfactory stability. Decreases in μ following thermal degradation, considered to be a potential problem, were also inhibited. In addition, the friction surface demonstrated a satisfactory surface.

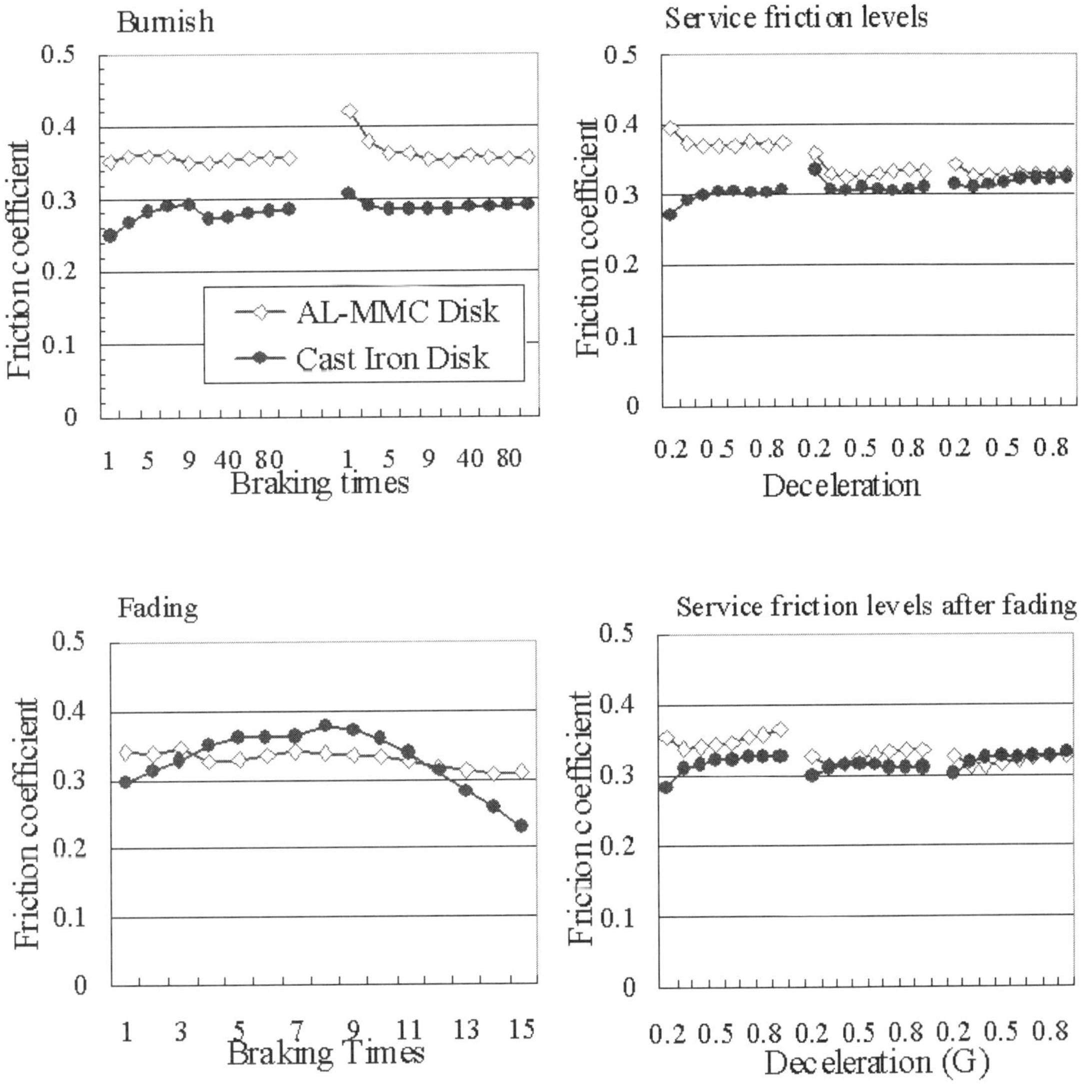

Fig. 18.21: Friction properties of newly developed Al-MMC brake

18.7 Conclusion

With respect to the development of a friction material, considered to be an important factor in the practical application of aluminum MMC brake disks, control of the transfer film was clearly shown to be a key technology as determined by conducting analyses of its tribological characteristics and the friction interface, the results of which yielded the findings indicated below.

1. Friction material components are strongly transferred to the friction surface of MMC, and a transfer film is formed having a thickness of several μm to several tens of μm. In addition to disc wear decreasing, this also tended to facilitate lowering of μ. Moreover, as the temperature rises, friction becomes even lower due to the increase in thickness of the transfer film.
2. Grinding by fine hard particles of alumina and so forth along with metals that have difficulty in forming a solid solution with respect to aluminum were found to be effective for controlling the transfer film, and the addition of Cu fiber in particular improved μ characteristics following thermal hysteresis.
3. In addition, since the aluminum matrix itself fuses (resulting in the occurrence of cutting that enables the formation of scratches and grooves), components such as ZnS or silica that fuse easily are not considered preferable.

However, since there are still numerous aspects of the tribological characteristics of MMC that remain unknown, further analysis of friction phenomena will be absolutely essential. We intend to establish material technology that surpasses the current level of brake technology.

18.8 References

[18-1] F.P.Bowden and D.Tabor : The Friction and Lubrication of Solids, Clarendon press, Oxford(1986)112.

[18-2] R.D.Arnell and F.A.Soliman: Thin Solid Films, 53(1978)333.

[18-3] J.Halling:Tribology International,12(1979)203.

[18-4] E.Rabinowicz:Wear Control Handbook(M.B.Peterson and W.O.Winer eds.),ASME(1980)491.

[18-5] Bond:J.Opt.Soc.Ame.,44(1954)429.

[18-6] A.E.Roach,C.L.Goodzeit:Trans.ASME,78,11(1956)1659.

19 Thoughts on the chassis of the future

Prof. Dr. U. Seiffert, Braunschweig University of Technology

Braunschweig, Germany

Abstract

All aspects of how the chassis of the future should look cannot be dealt with at first glance. The application of systems accepted by the driver such as ABS, ASR, ESP and also the strengthened application of electric-electronic systems will continue. In this case, the way of a higher integration density and network of singular components is indicated.

However, there are still many questions to answer concerning the driver assistance systems in relation to technology (reliability), legality and the man/machine interface. The task that lies before us is shaped out of the necessity to reduce accidents on the street to the factor 5...10. Driver assistance systems in combination with communication systems are an important building block for these tasks. A precondition for success is the mastery of the man/machine interface. **Fig. 19.1** shows the prognosis of how large a proportion could be for the systems coming on the market in 2010.

- ABS/ASR/ESP/ASC → 100%
- ESB → 100%
- brake assist → 20%
- electronic servo-assisted steering/ electro-mechanical servo-assisted steering → 100%
- electro-hydraulic brake → 30%
- electro-mechanical brake → ?
- four-wheel drive for road vehicles → 3%
- pneumatic suspension with adaptive damping → 5%
- distronic/ACC → 7%
- four-wheel drive steering → <0.5%
- intervention in steering and braking by multi-eye or external communication systems → ?

Fig. 19.1: Share of suspension systems coming onto the market in 2010 in Europe

19.1 Introduction

The requirements for automobiles have become considerably more stringent over the last few decades. This means that successful offers as the total of the characteristics are positive for the customer, if we are talking about volume models. **Fig. 19.2** provides a general overview of the individual areas.

For vehicles which are sold in small quantities, the purpose for use is often tailored to a specific application profile, e.g. soft top, sports, luxury limousine, along with the requirements of high-level safety and low exhaust emissions, as well as maximum quality and reliability.

The prognosis for the future chassis is not altogether simple. However, certain trends are evident.

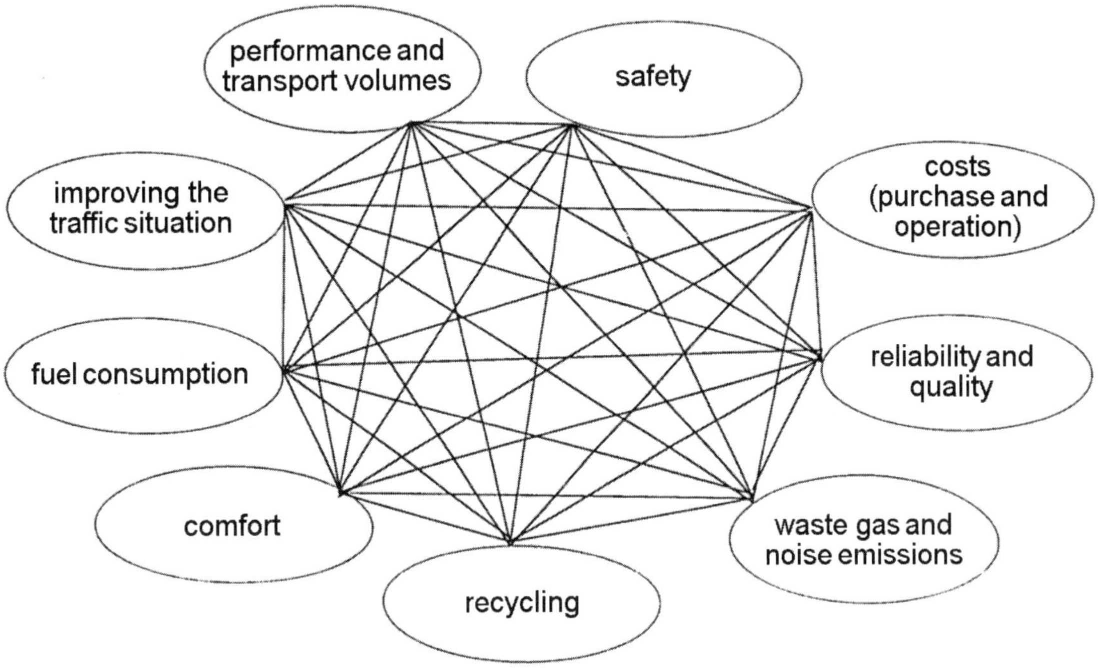

Fig. 19.2: General requirements for automobiles

The last 10 years especially have been characterised by the amazing speed at which the chassis has been changed. Many developments can be described as "fine tuning". It is therefore perfectly appropriate, from many other solutions, as a successful example of the chassis development for the most recent past, to indicate solutions such as the twist-beam rear axle, **Fig. 19.3**, as the result of a continuous and cost-optimising development of rear axles.

The most recent changes support the fact that work on the chassis area has never stopped and the entire chassis will be significantly developed over the next decade. It is fairly safe to say that the wheels will be a fundamental component of the chassis for a long time to come.

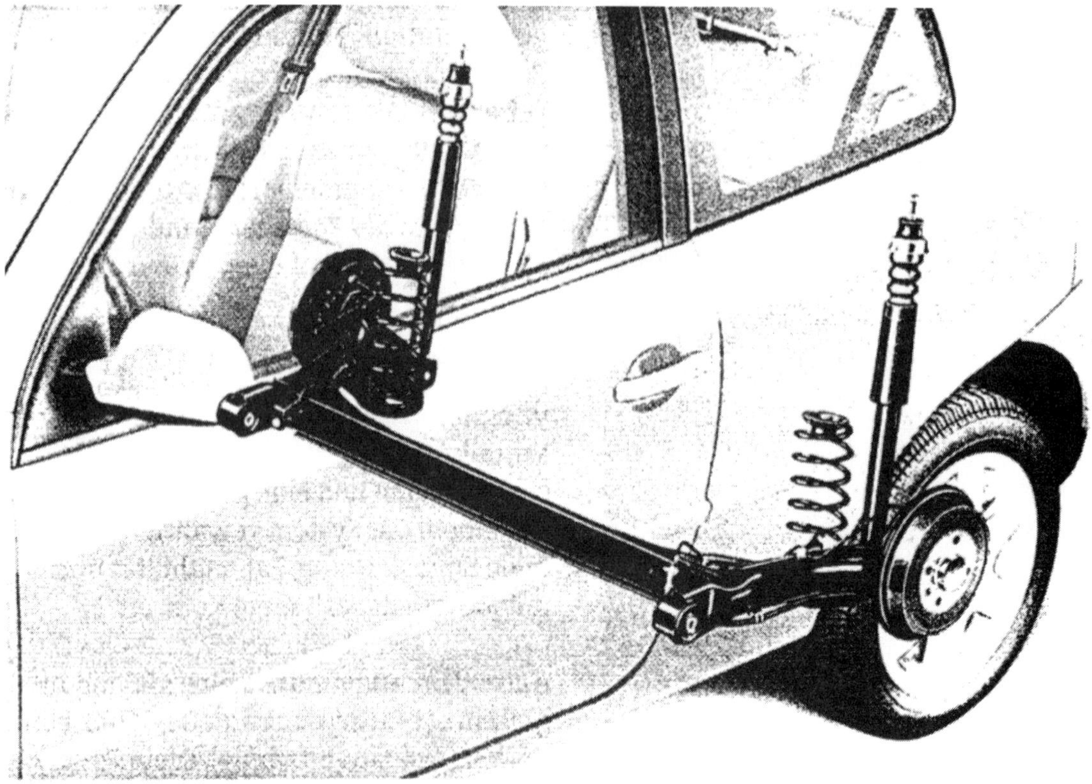

Fig. 19.3: Twist-beam rear axle with disintegrated spring-and-shock absorber unit [19-1]

19.2 Influences

Customers' wishes and other requirements for the chassis set by the legislators for example have increased consistently. Some of these requirements are shown in **Fig. 19.4**.

Particularly the requirement for a higher level of vehicle safety and the progress made in the area of electrics/electronics, including the sensor/actuator technology require and facilitate new technical solutions. In contrast to reducing the outcome of an accident, where approx. 90% of the initiatives to be implemented for the passenger have been realised, in the area of accident prevention, the percentage implementation of potential initiatives has reached approx. 50%. **Fig. 19.5** shows the development stages in the area of accident prevention.

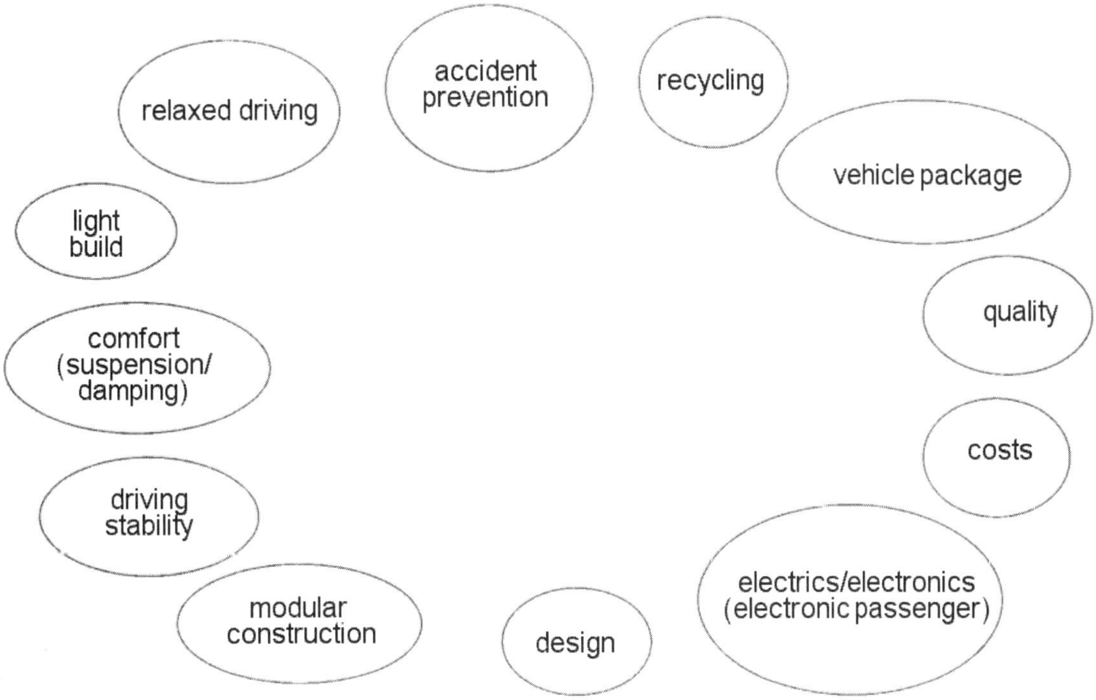

Fig. 19.4: Influences on the suspension

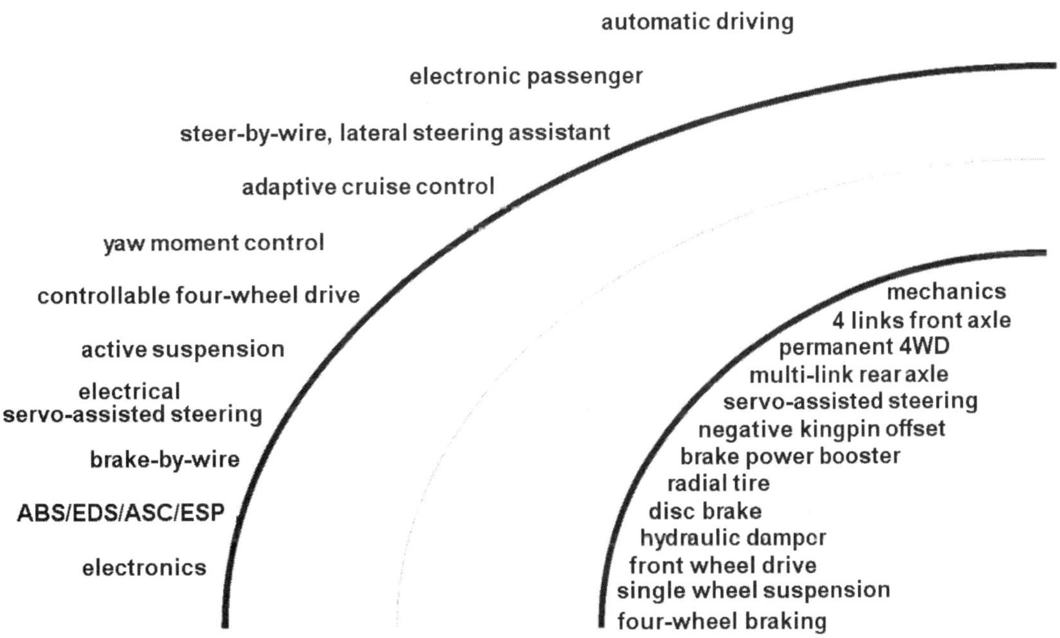

Fig. 19.5: Innovations for accident prevention

As is the case with the chassis and many other areas of the automobile, the pure mechanics have been neglected and the electrics/electronics have taken on numerous functions. This is demonstrated by an analysis of accident causes, which reveal that over 80% of accidents are caused by human error. This error frequency can be reduced by an appropriate assistance system. A typical example is ESP = Electronic Stability Programme, the effects of which can significantly reduce serious side impact. Since many assistance systems affect the driver either directly or indirectly, interference with the driver's function must take place by careful agreement with the human/machine interface. Moreover, the new vehicle models demonstrate that even in the case of smaller, lighter vehicles, the highest requirements of comfort and safety are satisfied, some of which have already been realised.

19.3 Tires

As already stated, tires are becoming increasingly important. **Fig. 19.6** shows an overview of road, tire and vehicle in relation to quality.

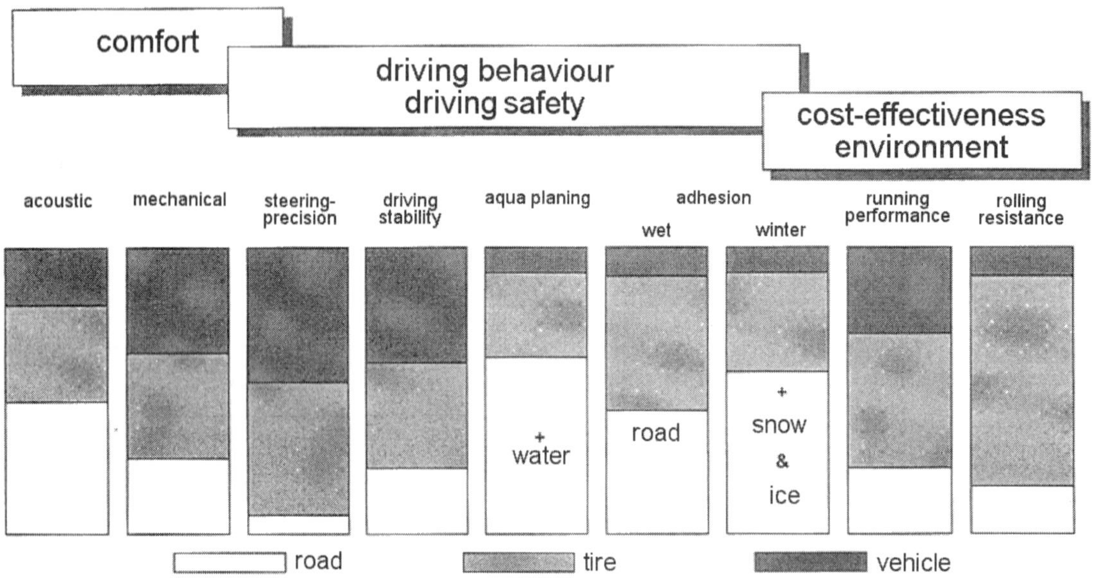

Fig. 19.6: Influence of road, tire, and vehicle on the quality [19-8]

In addition to these factors, costs, recycling and particularly tire/road noise should be taken into consideration. Even today, tire/road noise makes up 50% of the entire external noise. In respect of car tires, the customer has the following requirements:

Table 19.1: Customer requirements for car tires.

Fuel consumption	80 %
Ecological acceptance	80 %
Safety in the wet	70 %
Wear-rate	50 %
Price	50 %
Driving enjoyment	40 %
Aesthetic	30 %

(More than one choice is possible)

19.4 Material/Weight

In general, vehicles have become significantly heavier over the last twenty years. With comparable interior ranges, the weight sometimes increases by more than 25%. There are several reasons for this, e.g. high levels of comfort and safety, improved emission treatment, low external noise and wider interior ranges. As a result of the clear correlation between weight and fuel consumption, a reduction in the total weight of the vehicle and therefore also the chassis is vital, since the weight of the chassis, excluding the fuel tank can be more than 20% of the total vehicle weight. If the individual components and the materials used are analysed, aluminium comes out best. **Fig. 19.7** shows an overview of wheel rim types as an example.

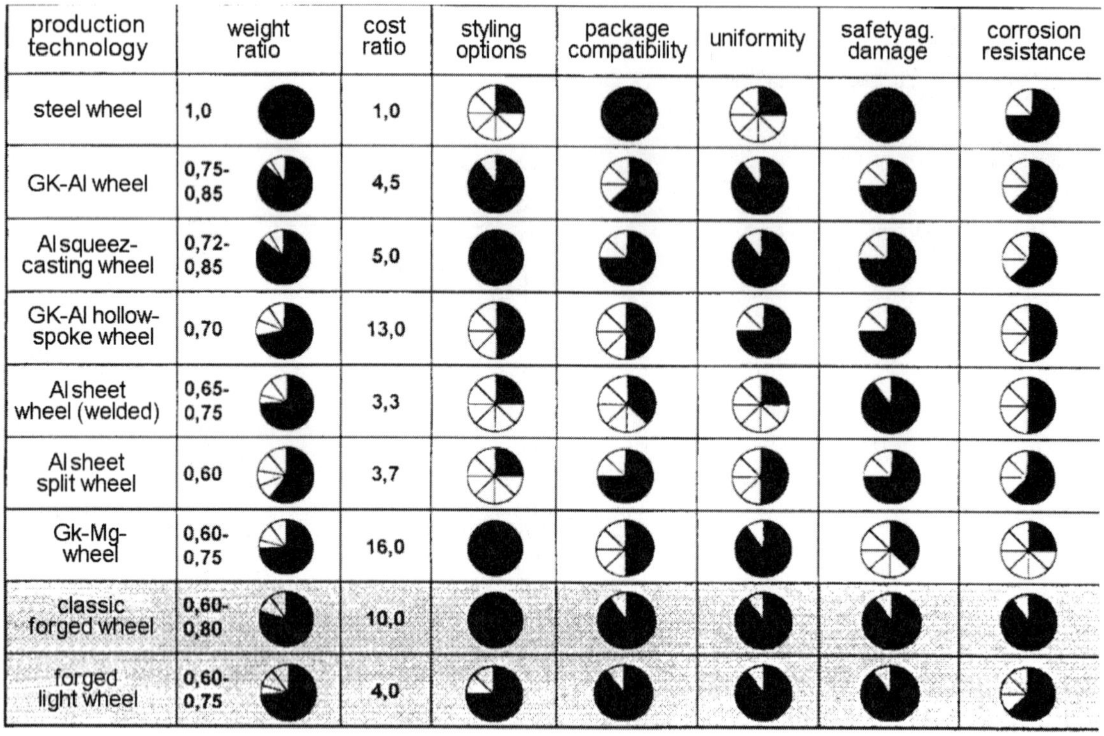

Fig. 19.7: Comparison of various wheel rim types (size 6Jx15, inset 45mm, wheel load 550 kg)[19-2]

It is clear that after the welded steel rim with a 25 to 35 % weight saving, in terms of costs, the next best version is the welded aluminium rim. A particularly good example of aluminium's weight saving potential is shown by the data on the BMW 5 series rear axle. In this case, the aluminium construction **Fig. 19.8** reduces the weight from 74.5 kg in the steel version, to 51.9 kg, i.e. a reduction of 30.3 % [19-3].

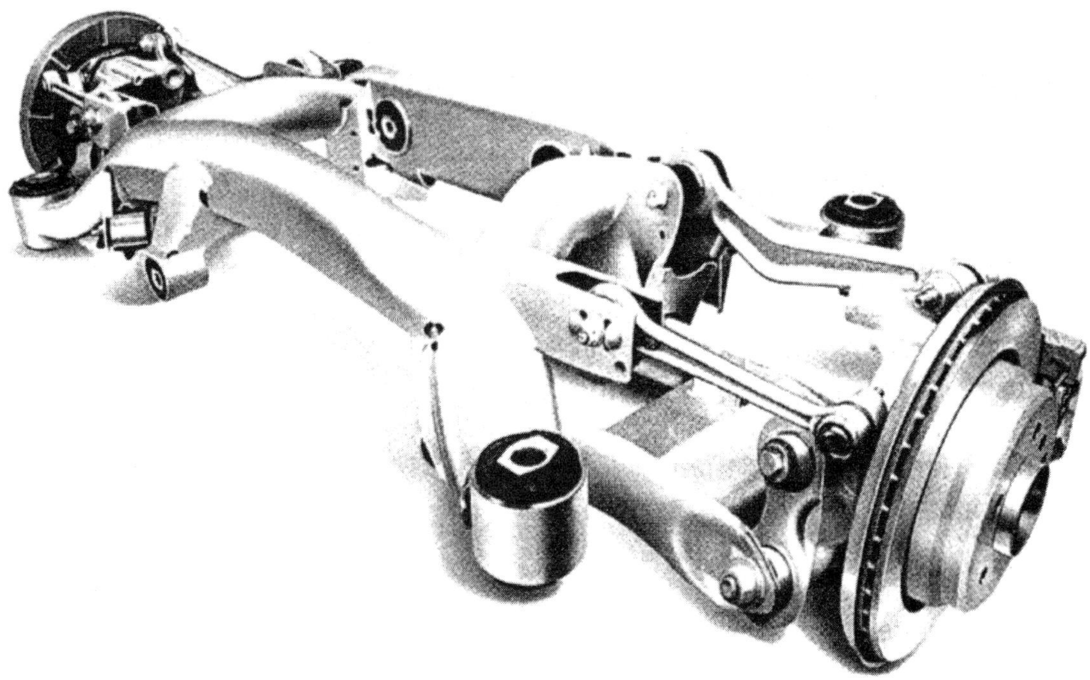

Fig. 19.8: Rear axle of the BMW 5-series [19-3]

Due to the low elongation at break, the use of an increased amount of magnesium in chassis materials has been a long time in coming. One exception is the steering wheel frame, which as a result of using magnesium, the weight can be reduced by approx. 50% in comparison to the conventional component.

In conclusion, it can be assumed that for some time after the turn of the century, aluminium will be the dominant material used in chassis building.

19.5 Chassis developments in consideration of the electrics / electronics

Customers are becoming increasingly discerning in terms of what they expect from a chassis. Over recent years, the goalposts have moved so far that for A Class and Ao size vehicles, increasingly expensive solutions are being used. One example of this trend is the rear axle of used in the Ford Focus [19-4], **Fig. 19.9**.

Fig. 19.9: Rear axle of the Ford Focus [19-4]

In this case, there is an attempt to uncouple the functions, paying particular attention to the driving comfort and impact absorption.

Once ABS systems were used or offered almost universally for every class of vehicle, an expansion of the ABS systems to cover "driver support" was very quickly implemented. Although some other company-specific descriptions have been used, the systems can be split into the following categories:

- ABS: Anti-locking brake system
- ASR/FDR/ASC: Traction and stability control
- MSR: Engine torque control

Public discussions on the "Elch test" have made the ESP system very popular. It has already been integrated into the A Class. ESP = Electronic Stability Programme, which is also known for example as CBC = Cornering Brake Control [19-5], stabilises and/or prevents extreme yaw of the vehicle, **Fig. 19.10**.

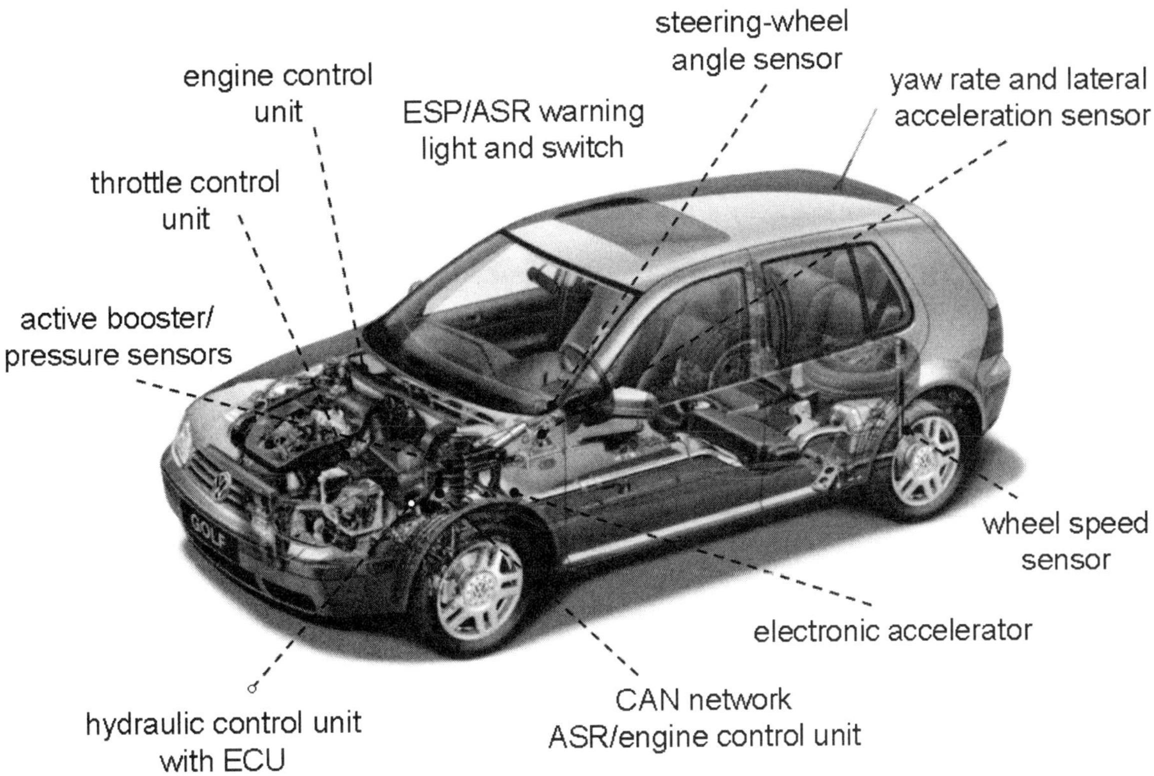

Fig. 19.10: Electronic Stability Program (ESP) [19-6]

It is interesting to see how the various vehicle manufacturers handle intervention of the system. Some intentionally allow more severe vehicle movements and trigger the counter-control relatively late.

Even for this one aspect, a prognosis can be ventured on the level of use. It is safe to predict that by 2010, the above systems will be used in 100% of new vehicles.

This does not necessarily apply to the braking assistant, which depending on the emergency braking situation, is able to build up the necessary, maximum braking force support more quickly. This technical solution is supported by the fact that numerous drivers step onto the brake pedal without applying the required even high pedal force.

Further support for the driver is provided by the electronically actuated parking brake, which once activated, allows a slight uphill creep. The last two categories should be placed in the "nice to have" category. It is anticipated that their general use by 2010 will be less than 20%.

The all-wheel steering system will presumably be retained by vehicles with a very long wheelbase and should essentially improve the parking function. Accordingly, their level of distribution is likely to be very low.

As the examples above show, more and more hydraulic functions are being replaced by electrical functions or a combination of the two. Three are of particular interest:

- electrical servo steering
- electrical-hydraulic brake
- electro-mechanical brake

The pure electric servo steering system and the electric/hydraulic steering system came about as a result of the need to save energy and the requirement for future vehicle drive technology, e.g. fuel cells. These systems are in general use.

A change also seems to be looming for the braking system. The electro-hydraulic brake is far more technically advanced than the electro-mechanical brake. With the electro-hydraulic brake, **Fig. 19.11**, the hydraulic energy supply ensures that the individual wheel brakes can be supplied under pressure through a modulator.

Should the external energy supply fail, a stand-by main cylinder facilitates direct impact of the wheel brake at a pressure proportionate to the pedal force, i.e. in principle as is currently the case. The advantage of this solution is the improved efficiency of the entire system with a higher level of safety and a more simple integration of ABS, ASR, FDR and ESP.

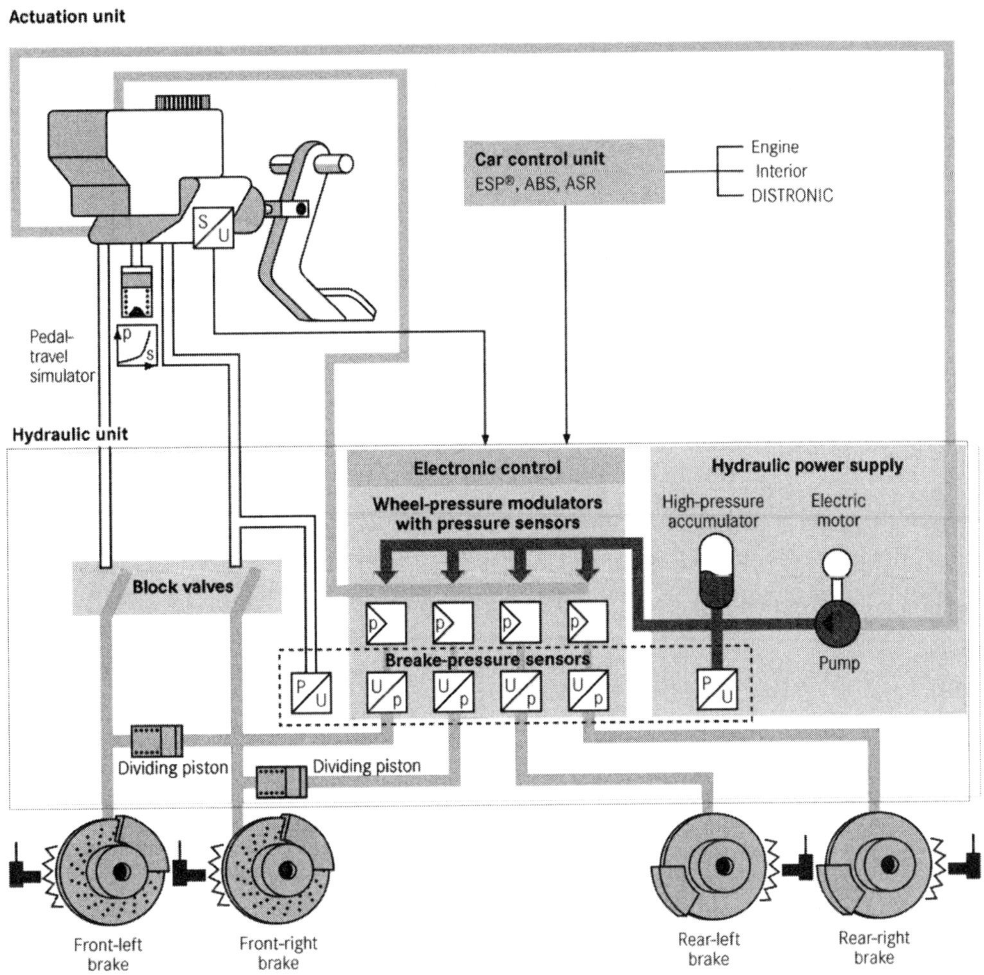

Fig. 19.11: EHB – Electro-hydraulic braking system

The electro-mechanical braking system, **Fig. 19.12**, requires a new dimension in respect of the safety and performance requirements of the electro-mechanical brake actuators and also the vehicle electrical system.

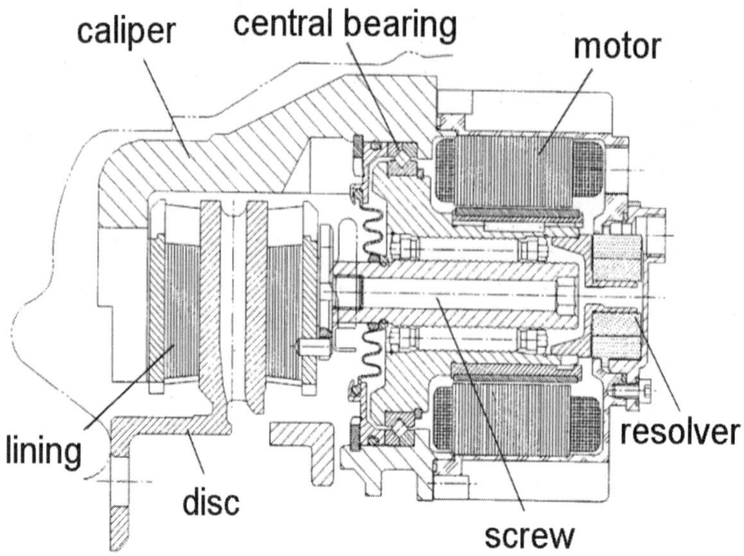

Fig. 19.12: Electro-mechanical wheel brake [19-7]

This system [19-7] described at the Electronics Conference in Baden-Baden in 1998 shows the complexity of such a solution. The authors have published the illustration in **Fig. 19.13** relating to the development of the braking systems.

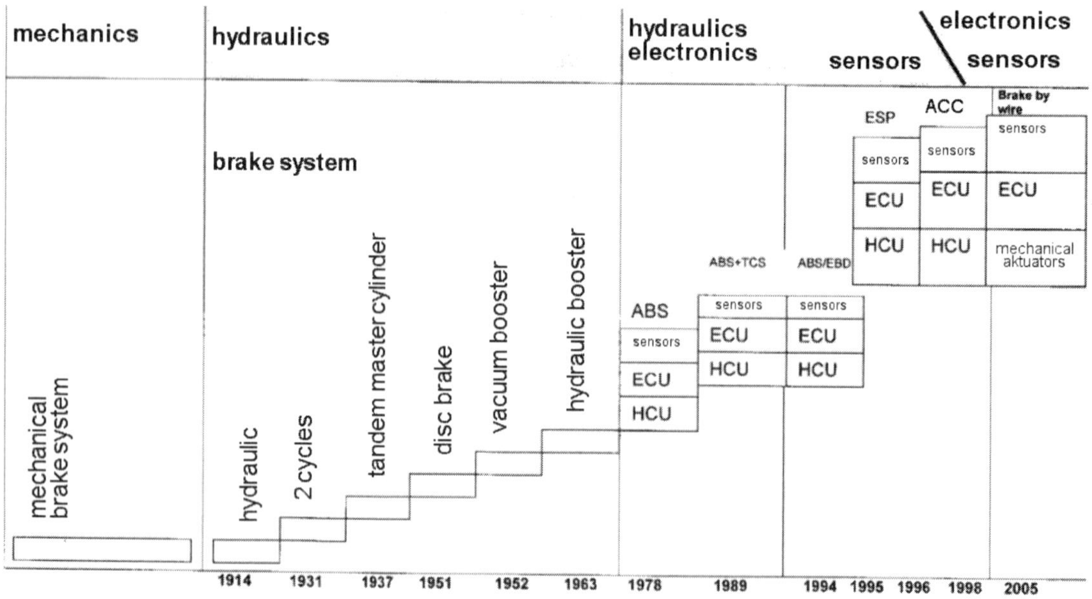

Fig. 19.13: Development of the braking systems [19-7]

It shows that use of the fully electro-mechanical brake will be possible by 2005. For two reasons, it is anticipated that the time frame will be greater; both cost and complexity appear higher than with the electro-hydraulic brake and the answer to the question as to which manufacturer will be the first to use the electro-mechanical brake in series produced vehicles and carry the risk in consideration of the Product Liability Law.

For future braking systems too, the brake lining / brake disc pairing must match. In addition to the real requirements such as temperature resistance, low wearing and constant friction, low noise, no "low speed" squeal and no "juddering" can be considered "soft factors". However, despite every development effort and the most up-to-date test bed technology, one of the parameters in the series still fails to conform.

The positive effect of the improved controls for chassis systems is shown by the new all-wheel systems from Volkswagen, the essence of which is the Haldex clutch, **Fig. 19.14**.

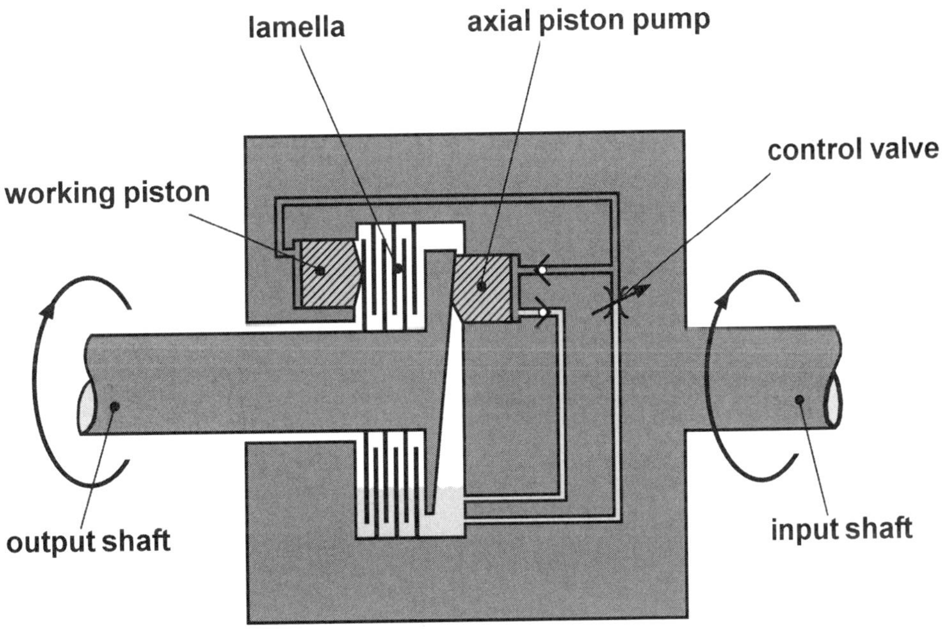

Fig. 19.14: Function of the Haldex clutch [19-9]

This third-generation all-wheel system is a combination of electronics and clutch and differential system control.

Generally speaking, the level of use of the all-wheel drive system will increase only slightly.

19.6 Comfort

Reference has already been made to the required increase in the level of comfort. Low tire noise, low impact force with a high level of driving safety, i.e. low wheel load fluctuations characterise development. The suspension and shock absorber system of the new S-class [19-10] **Fig. 19.15**, which uses pneumatic suspension with adaptive shock absorbers.

This solution enables an optimum suspension system to be used for improved comfort and in the event of a change being required suddenly, such as an evasive manoeuvre, driving stability is maintained.

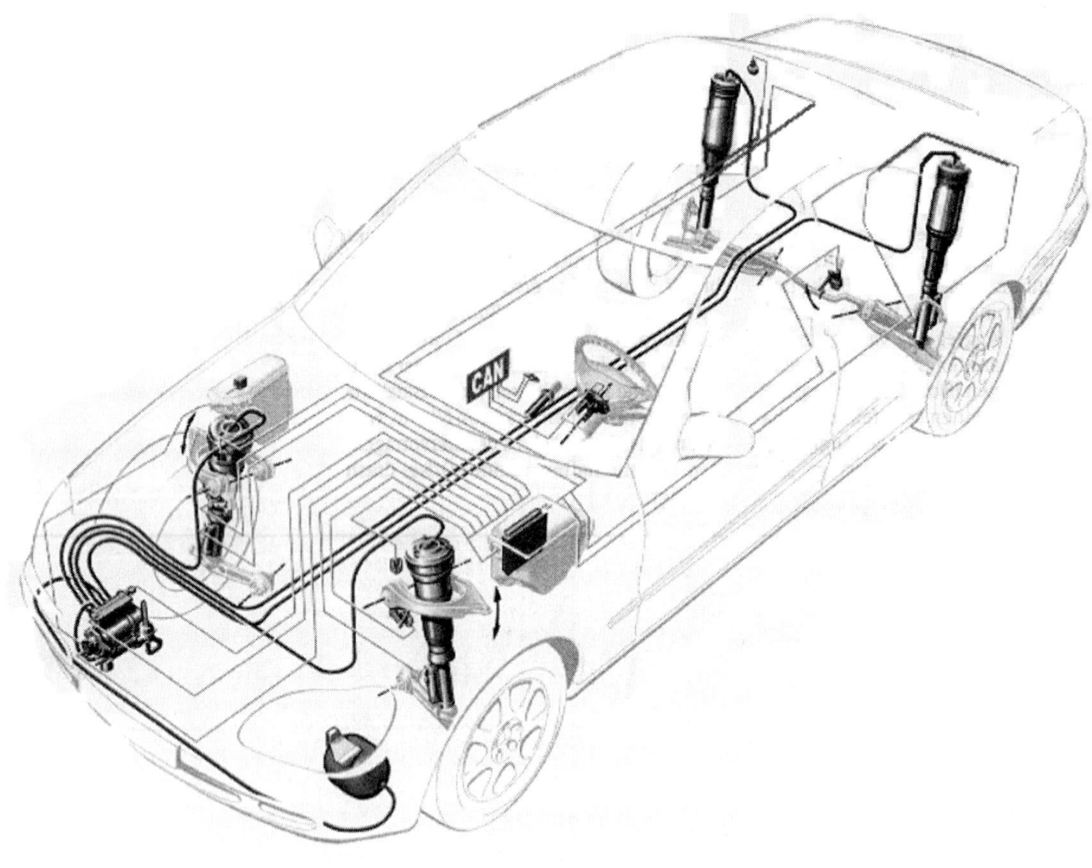

Fig. 19.15: The suspension and shock absorber system of the new S-class [19-10]

19.7 External assistant systems

As realised in the new Daimler/Chrysler S Class with Distronic, the first proximity systems have been brought onto the market. In conjunction with the speed regulator, the engine, braking systems and gears are controlled in dependence of the logistics of the proximity control. **Fig. 19.16** shows the external sensors and the possibility, in addition to the warning functions, of increased intervention with steering and braking systems.

Automatic driving can be illustrated using other devices such as cameras and communication facilities. Within the period up to 2010 being considered however, combined use in road traffic of vehicles which drive automatically on the same road as conventional vehicles that have no control facility, is barely feasible.

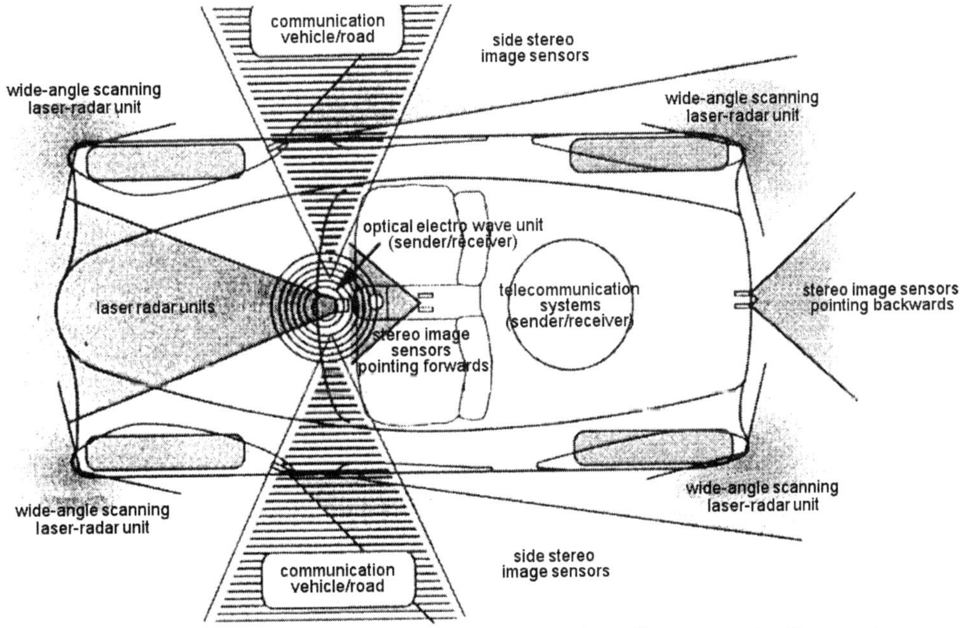

Automatic accident prevention - Multiy-eye and multi-communication system facilitates automatic driving and braking or steering in case of driver error, via computer

Fig. 19.16: External sensors for accident prevention

19.8 References

[19-1] M. Winterkorn, et.al. "The Volkswagen Lupo", ATZ, 100th Year, No. 11

[19-2] Klein, H., et.al. "Forged aluminium wheels as a contribution to the light build", Special Edition, ATZ/MTZ materials in automobile construction

[19-3] Hutmann, P., et.al. "Material competition in automobile construction", Special Edition, ATZ/MTZ materials in automobile construction

[19-4] ATZ/MTZ Special issue "The New Ford Focus", Wiesbaden, Januar 1999

[19-5] Konik, D., et.al. "Electronic Braking Management as the first step towards an integrated Chassis Management", ATZ 100th Year 1999, No. 4, ISSN 0001-2785, Wiesbaden

[19-6] Volkswagen-Press File, Wolfsburg

[19-7] Belschner, R., et.al. "Dry Brake-by-wire with fault-tolerant TTP/C-Communication System", VDI-Report No. 1415, 1999

[19-8] Huinink, H. "Tyres", Continental AG, 1999

[19-9] Volkswagen Workshop 1998, Wolfsburg

[19-10] Römer, M., et.al. "The Suspension and Shock Absorber System" Special Edition "The New S Class", ATZ/MTZ

20 Total Chassis Management – Heading for the Intelligent Chassis

H. A. Beller, Dr. P. E. Rieth, ContinentalTeves AG & Co. oHG

Frankfurt, Germany

Summary

The demand for more traffic and driving safety as well as riding comfort calls for innovative system solutions. In this connection, the chassis, which more than any other part determines the driving safety and riding comfort of a vehicle, is the focal point of interest. To date, conventional chassis designs at best strike a balance between these conflicting requirements. In order to cover different marketing and purchasing strategies, chassis subsystems, i.e. brake, spring and damper, suspension mounts, steering and tires, are currently designed as self-contained (stand alone) units. Standardized interfaces between the subsystems are rare. Based on the change in technology, system architecture and component topology in the vehicle and based on concrete concepts for a compound (global) control of the various chassis functions, this paper describes an integral and, at the same time, modular chassis control approach in the sense of Total Chassis Management and outlines the path leading to the intelligent chassis of the future. Viewed as a whole, the measures described reveal a high potential for both increased functionality and an improved cost-benefit ratio. Their implementation is expected to result in a substantial improvement of driving and traffic safety.

20.1 Introduction

Growing wealth and a rising standard of living go hand in hand with growing demands for more comfort on the one hand and more protection and safety on the other. These demands and needs extend to all areas of everyday life. Road traffic as an indispensable prerequisite for individual mobility and prime mover of our economy ranks high on this list because it represents the highest peril for life and health compared with other means of transportation.

Here, in turn, passenger cars lead the statistics. Of all persons killed on the road in Germany in 1997, 60% died in passenger cars. Passenger cars are involved in about 80% of all heavy accidents. In spite of a decline in absolute numbers, their percentage is rising continuously (**Fig. 20.1**).

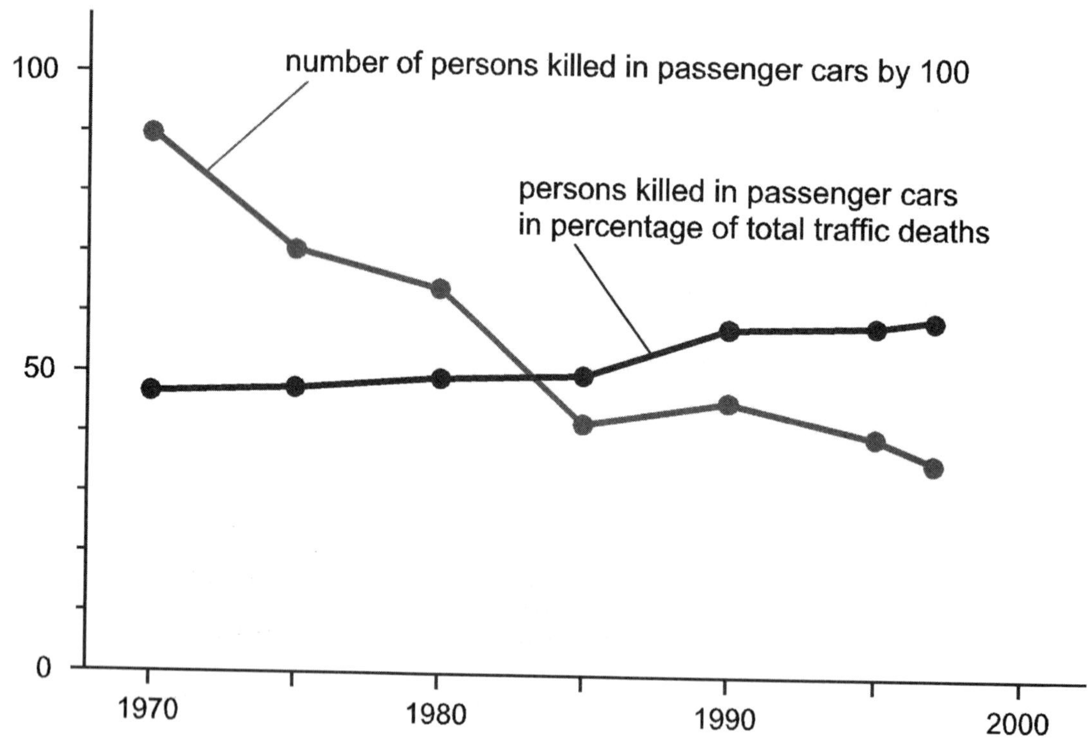

Fig. 20.1: Fatal accidents of passenger cars in Germany West [20-1]

In the past, efforts to improve traffic safety were mainly focused on measures aimed at reducing the severity of accidents and their consequences (passive safety). With the meanwhile comprehensive use of safety and restraint systems (air bags for driver and passenger, side air bags etc.) across all vehicle lines, a certain saturation can now be discerned in this field (**Fig. 20.2**). A reorientation is taking place which puts greater emphasis on active accident avoidance (active safety). This trend will decisively influence the development of vehicles in the future.

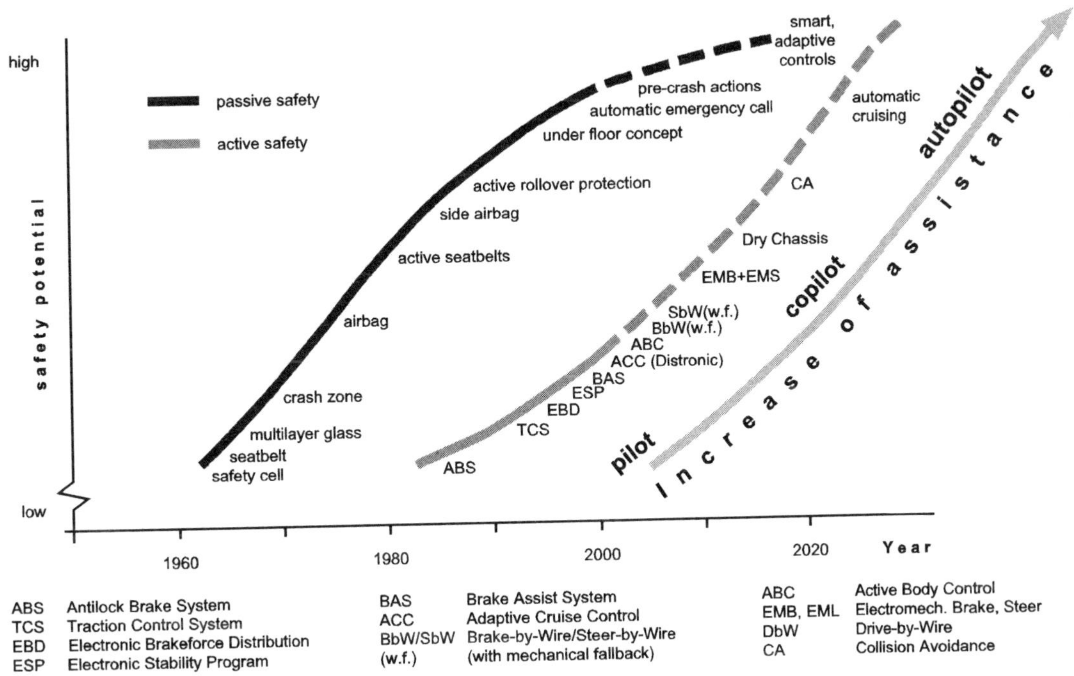

Fig. 20.2: Routemap scenario of automotive safety [20-2]

Transportation in terms of passenger kilometers has more than doubled in the last 20 years and will rise even more in the years to come. In Europe, millions of people are traveling by car every day: 150 million to their place of work (and back), 100 million on business trips, 90 million shopping. Even today, every German motorist on average spends more than half an hour in his or her car per day. Understandably, this creates a growing demand for comfortable, stressfree and relaxed driving, a demand which in view of our limited and largely overcrowded traffic space calls for innovative solutions, meaning electronically controlled modern comfort, safety and driver assist systems (see **Fig. 20.2**). The most important single aspect in connection with the implementation of these systems is the chassis, which, like no other part, determines the riding comfort and driving safety of the vehicle.

Looking at the so-called "Comfort-Safety Conflict Diagram" (**Fig. 20.3**), we see that the two criteria cannot always be met optimally at one and the same time.

Consequently, current chassis designs invariably represent a compromise between the two extremes: sporty (firm) and comfortable (soft). The chassis of the future, obviously, must be one which adapts itself to varying driving situations and is still fun to drive. To achieve this, the individual chassis functions (subsystems brake, spring/damper, chassis mounts, steering and tires), based on a sound axle design, will have to be networked to form a compound control system known as "Total Chassis Management". On the hardware side, this calls for adaptive or, better still, controllable active or autonomous subsystems. Up to a point, this approach will also permit the complexity demands (self-steering, elastokinematics) of current axle designs to be reduced.

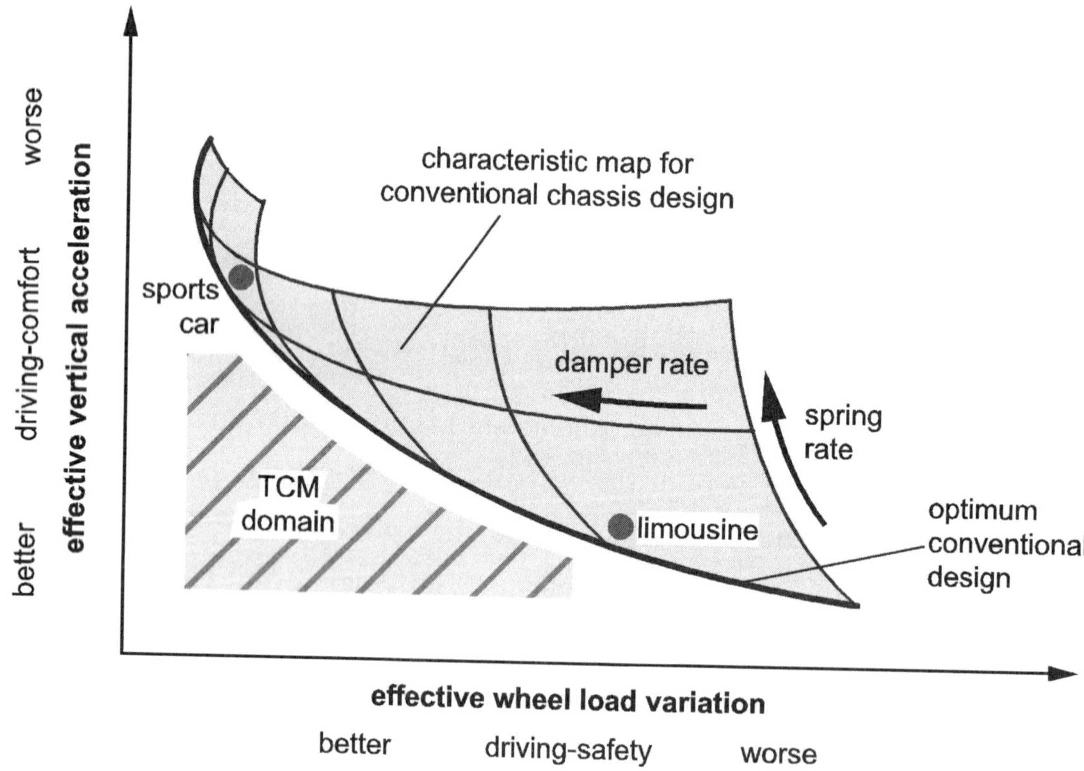

Fig. 20.3: Comfort/Safety conflict diagram [20-3]

20.2 Chassis systems - state of the art and development

The development of brakes has reached a satisfactory state. Modern computer-controlled braking systems can be applied automatically both centrally (for autonomous comfort or assisting emergency braking) and wheel-individually (for optimum acceleration, deceleration and stabilizaion) with satisfactory comfort (**Fig. 20.4**). They feature interfaces connecting them to powertrain controller. These 'upgraded' electronic braking systems are the result of consistent further development of the classical ABS and TCS technology. All these are power-assisted braking systems (servo brakes). Such systems operate on the basic principle of an energetic (hydraulic-mechanical) connection between the driver and the wheel brakes. The auxiliary power, provided by a vacuum brake booster, reduces the physical effort required to apply the brakes. However, control interventions of any kind will be noticed by the driver as brake pedal movement or pedal vibration caused by the hydraulic-mechanical coupling.

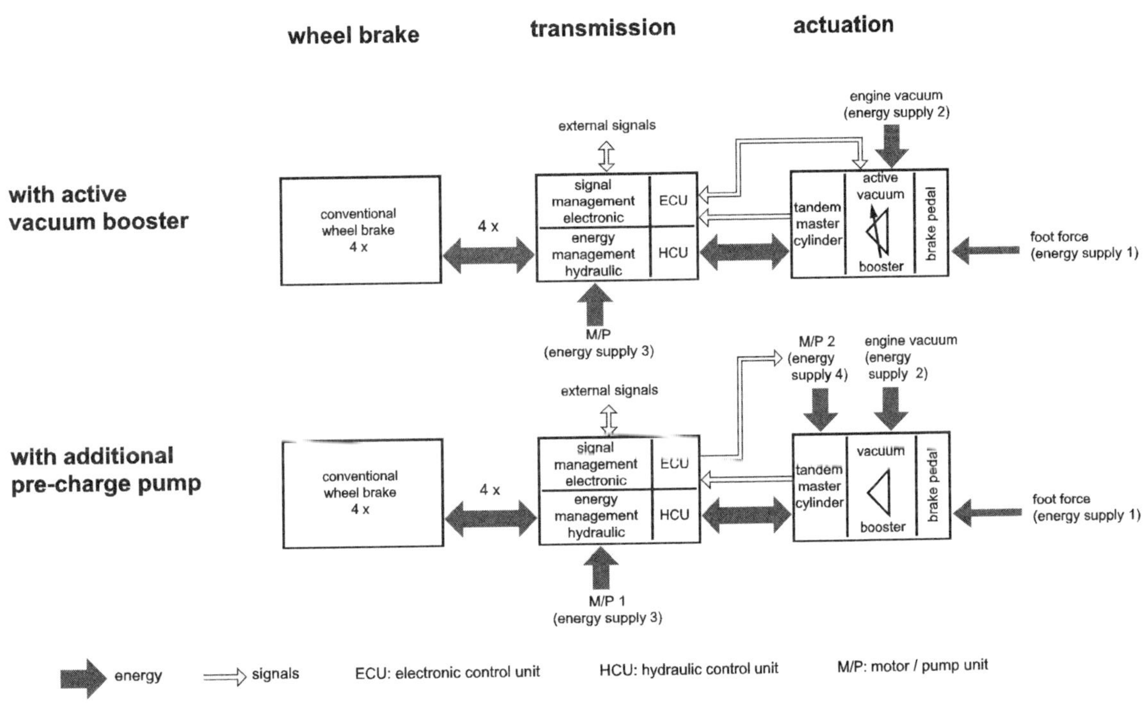

Fig. 20.4: Autonomous servobrake systems. Conceptual layout of systems on the market [20-4]

Fig. 20.5 illustrates the dramatic development which these systems have experienced since the early 80s. The steep slope of the curve for the computing power, but also the number of energy sources (**Fig. 20.4**) and the number of valves within the hydraulic control unit (HCU), 12 meanwhile for an ESP system, indicate the complexity which such power-assisted braking systems have reached.

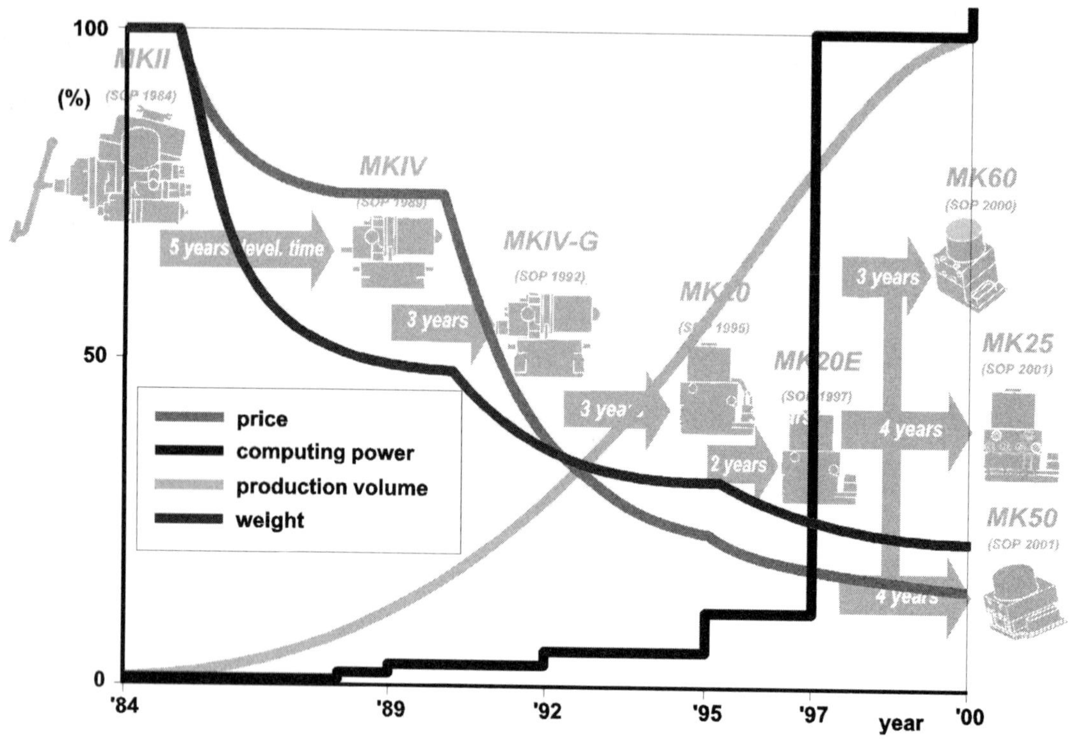

Fig. 20.5: Teves ABS technology in change. 1984 to 2001 [20-5]

It has been found that the approach used to date, namely upgrading existing or adding new components to the classical assemblies of braking systems, i.e. the actuating device, the transmission device and the wheel brakes, has resulted in a degree of complexity which is now difficult to handle in the increasingly more cramped spaces available for installation. Moreover, this add-on approach leads to an increasingly less attractive cost-benefit ratio.

These problems are expected to be solved by so called full power braking systems, such as brake-by-wire systems, which offer the driver an optimum pedal feel, reduced installation and space requirements for the vehicle manufacturer and combine all functional requirements, such as pedal force boost, brake force proportioning, ABS, traction control, stability control etc. in one unit. They also lend themselves ideally to networking with higher ranking vehicle and traffic guidance systems. These full power braking systems are being developed or about to be launched on the market (see section 20.4).

In the field of suspension/damping, a first (albeit still very expensive) milestone has been set with the introduction of the hydraulic active body control, ABC (**Fig. 20.6**). A hydraulic lift cylinder connected in series with the spring of the classical wheel suspension with a (stiffer) steel spring and a (softer) passive shock absorber actively controls body movements up to 5 Hz. This affords a significant improvement, i.e. reduction, of the rolling and pitching motions of the vehicle body without adverse effects on riding comfort at higher-frequency road excitations.

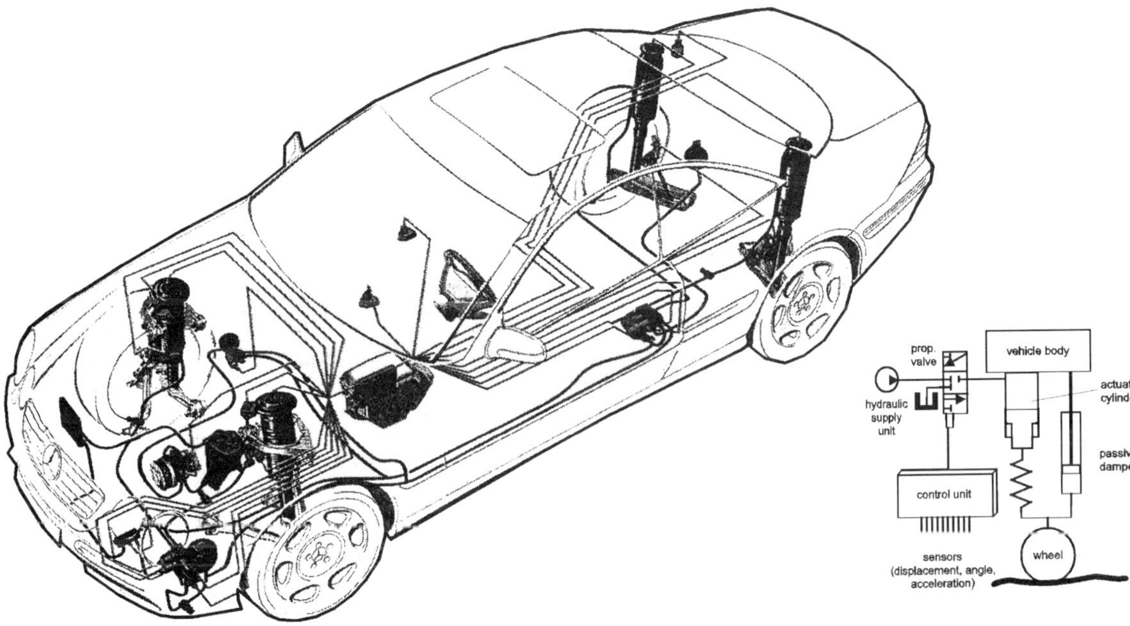

Fig. 20.6: Active Body Control (ABC) on DaimlerChrysler S-Class [20-3]

In addition to these most advanced chassis control systems, numerous other spring/damper adjusting systems are known which, as a rule, are offered as options. These are almost exclusively adaptive systems with steel springs or air springs acting as support and suspension between body and wheel. Not only the spring rate, but also the vehicle level is influenced hydropneumatically (combined steel and gas spring) or purely pneumatically (air spring). An energy supply and accumulator unit (hydraulic pump / accumulator, compressor / pressure vessel) is required in both cases. The damper rate is influenced almost exclusively by switching between different damper settings (such as soft, medium, firm). Different chassis control strategies can be implemented, depending on the sensor configuration used.

Progress in the development of bellows, which meanwhile permits wall thicknesses of less than 2 mm, has improved air springs in terms of responsiveness and hysteresis (so-called harshness) to a point where these now also meet exacting demands for comfort. In fact, air springs are now superior to hydropneumatic solutions in terms of comfort. Their ease of adjustment permits them to be used for level control (of the rear axle or the front axle and the rear axle). This means that the vehicle level can be adjusted largely independently of the load or adapted to different conditions of use (off-road, motor highway) (**Fig. 20.7**).

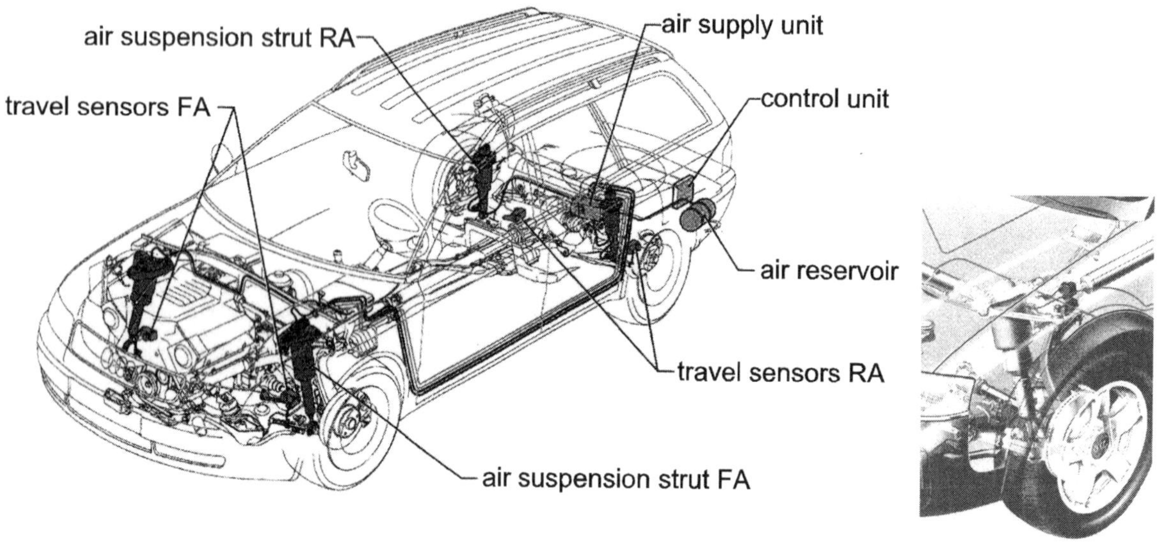

Fig. 20.7: Active 4-corner airspring of ContinentalTeves on the Audi Allroad Quattro [20-6]

In the field of steering, only power-assisted steering systems (servosteering) are available on the market at present. High-grade systems, so-called controlled power-assisted steering systems, permit the steering boost to be adapted to different driving situations (high: parking, low: fast driving). Regardless of their design, whether electrohydraulic or electromechanical, they cannot be externally controlled in the sense that a computer-demanded steering angle overrides the steering angle selected by the driver. This type of overriding affords a wealth of potential improvements relieving the driver and increasing steering comfort (see section 20.3). Such steering systems are being developed. **Fig. 20.8** and **Fig. 20.9** show two systems which are expected to be launched on the market within the next few years.

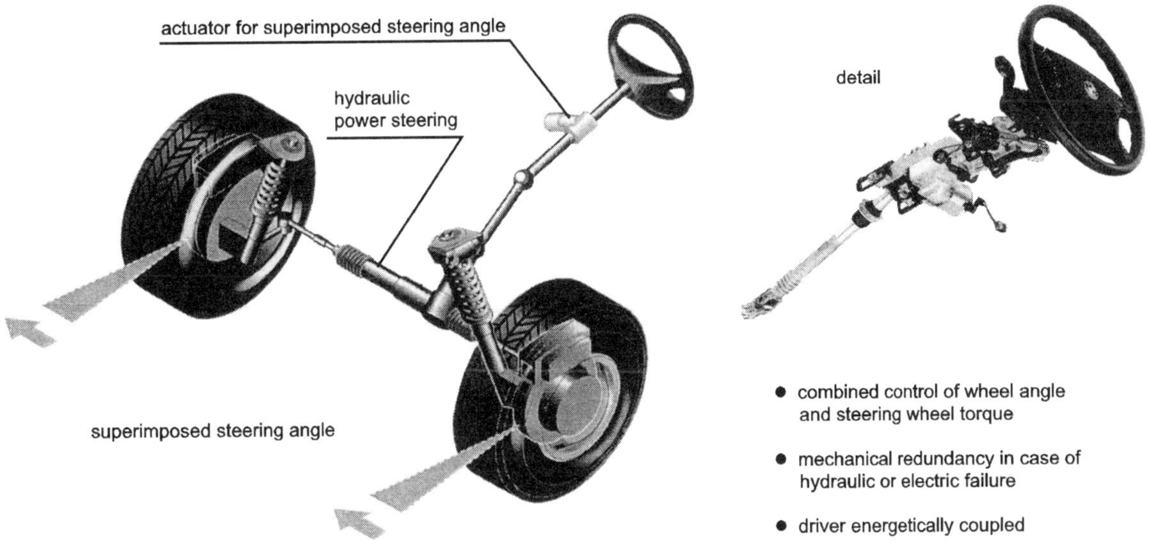

Fig. 20.8: Active front axle steering with manual steering effort [20-7]

Fig. 20.8 shows a solution for which a conventional controlled power-assisted steering is extended by the possibility of superimposing an additional steering angle. This calls for an additional actuator, sensors and compound control of the steering power assist and the additional steering angle. This compound control is necessary in order to keep the irritating effects of the steering wheel torque feedback to the driver caused by the superimposed steering angle as low as possible.

Fig. 20.9 illustrates the principle of a full power steer-by-wire steering system with manual fallback. In normal operation, the driver's steering wheel is energetically uncoupled from the actuator steering the wheels. The driver steers the vehicle purely electrically 'by wire'. In the event of an energy failure, the manual fallback feature permits the vehicle to be steered by muscular power, albeit without the functionalities available in normal operation. Like the system mentioned above, this so-called electrohydraulic steering system EHS, too, is based on the classical steering hydraulics technology.

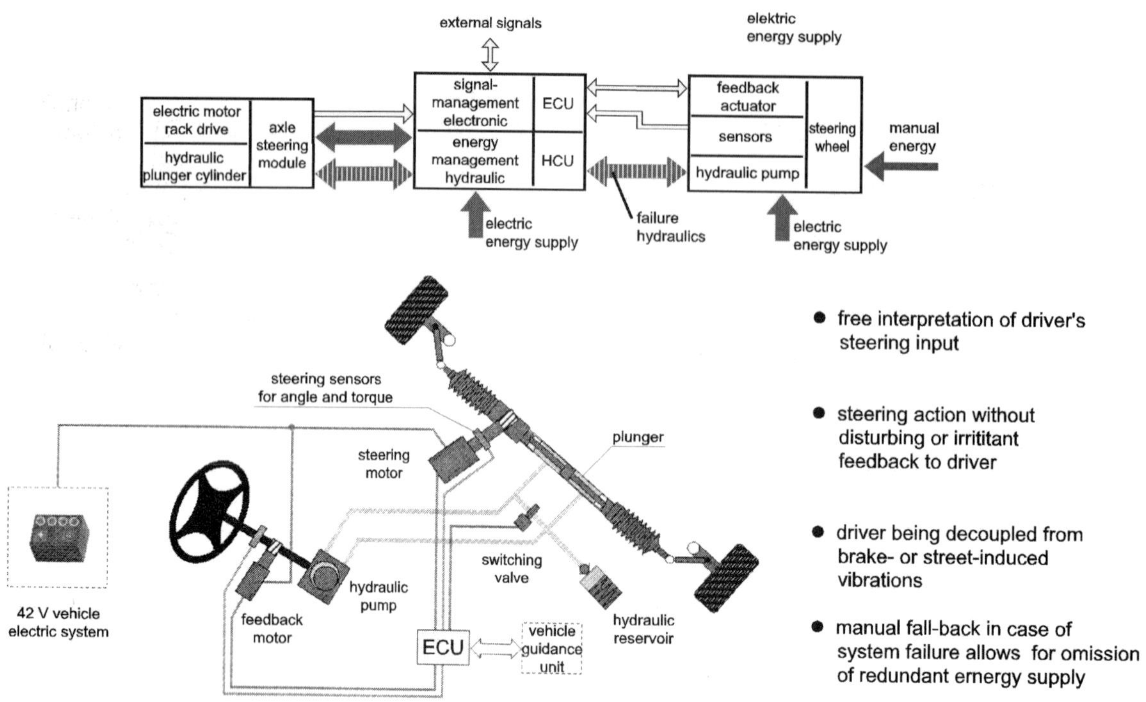

Fig. 20.9: Conceptual representation of an active (steer-by-wire) steering system with manual fallback solution [20-8]

Rear axle steering, which is about to be introduced in the North American market for particularly long, so-called full size light trucks, potentially also permits external control interventions. However, initially this system will only be used as a rear axle steering system electrically fed by signals from the (classical) front axle steering system, primarily with a view to improving the maneuverability of these vehicles.

For chassis suspensions, unlike engine suspensions (**Fig. 20.10**), switchable or even continuously adjustable bearings and mounts are not yet in use. However, bearings which can be adjusted to improve comfort or safety, depending on the driving situation, are being developed.

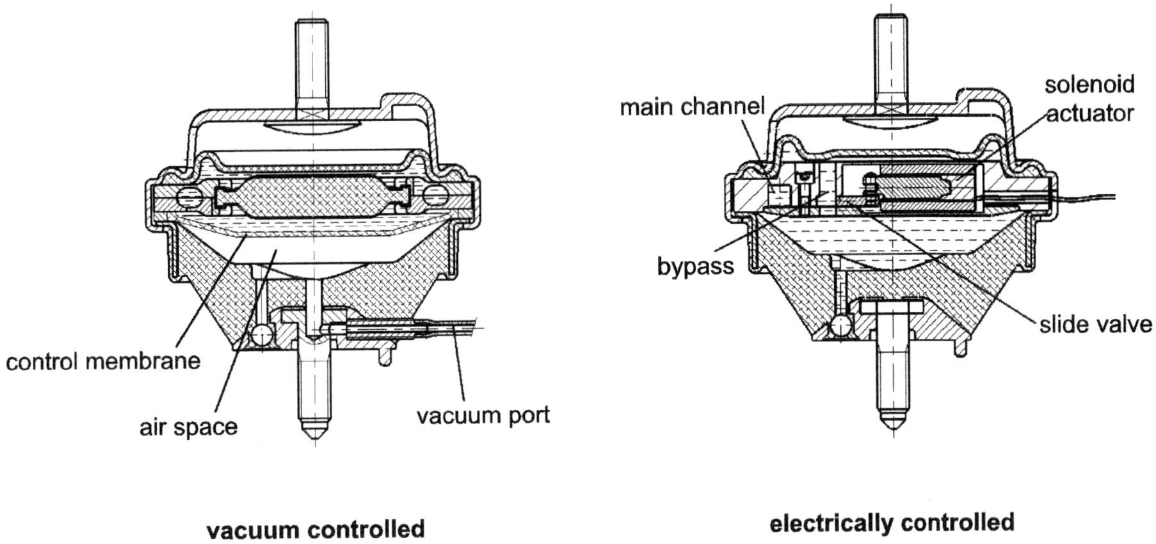

Fig. 20.10: Design variants of switchable hydromounts, shown by the example of adaptive engine mounts

In the field of tires, enormous progress has been made in the last few years. The so-called cobweb diagram (**Fig. 20.11**) illustrates the improvements achieved in terms of mileage, weight, rolling resistance, maximum permissible speed, braking on wet and dry roads. **Fig. 20.12** shows the influence of mixing technology on ABS performance on wet roads and rolling resistance. The change from soot to silica as filler and consistent further development of the silica polymer compounds with respect to structure and network density have improved braking efficiency on wet roads by 20% and simultaneously reduced rolling resistance. Our most advanced tire, the Continental Premium Kontakt, combines all the above-mentioned quality criteria (**Fig. 20.13**). Its bionic contour enables it to adapt its performance to changing operating conditions, following the example set by nature.

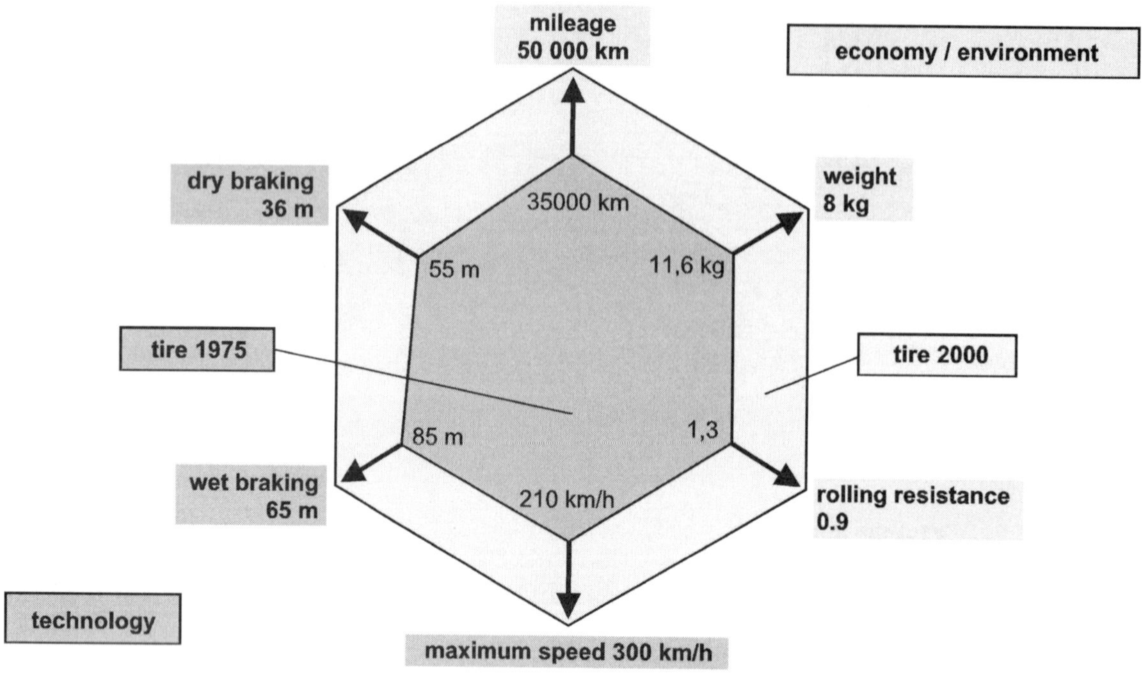

Fig. 20.11: Development of tire properties from year 1975 until today [20-9]

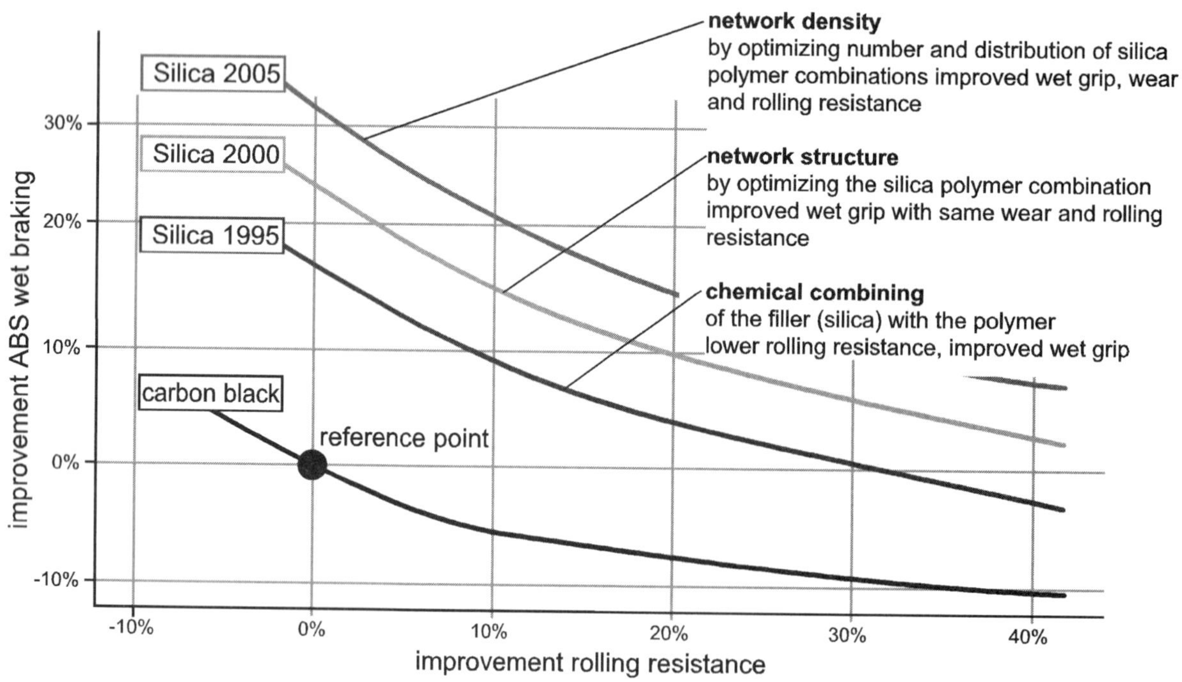

Fig. 20.12: Influence of the compounding technology on ABS wet performance and tire rolling resistance [20-10]

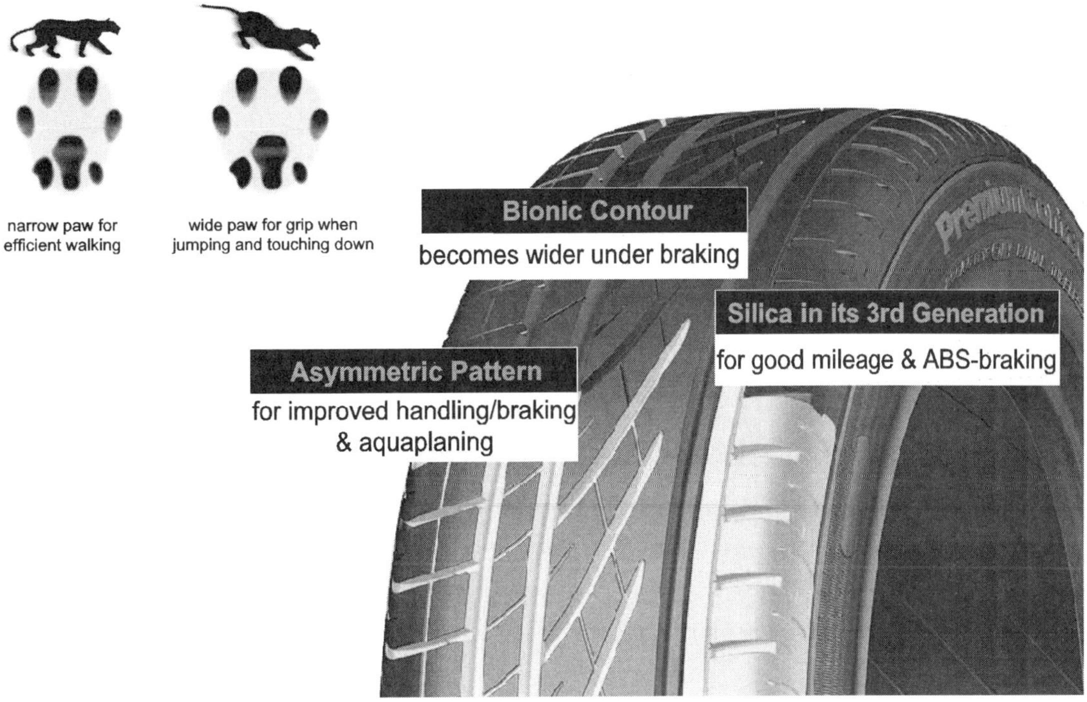

Fig. 20.13: Continental Premium Contact, state of the art tire [20-10]

The recently rekindled discussion concerning the tendency of certain vehicles to rollover in the event of a fast lane change has also demonstrated that the chassis of a number of vehicles with higher center of gravity have not been able to keep up with the latest achievements in tire development. In conjunction with an obsolete axle design, higher lateral stability achieved by the tire increases the risk to roll over during severe obstacle avoidance maneuvers.

An analysis of a total of about 2,500 passenger car accidents made by GDV, the Central Association of the German Insurance Industry [20-11], shows that is werving in the pre-crash phase' is much more frequent than has been assumed up to now. Approx. 25% of all passenger car accidents with severe occupant injuries are caused by collisions after swerving. Three quarters of these accidents are caused by inattentive drivers or driving errors, about one quarter by obstacle avoidance maneuvers in an emergency situation. Most of these swerving situations end in side collisions. Electronic systems controlling the dynamics of vehicle movement, such as ESP, can significantly assist the driver attempting to stabilize the vehicle and are expected to reduce the number of accidents caused by swerving substantially.

However, the advantages of vehicle stability control systems are not limited to accident avoidance, i.e. active safety, but also improve occupant protection and, thus, passive safety. Even if a collision cannot be avoided, ESP increases the probability of a collision with the vehicle front because it helps to maintain directional stability so that swerving is at least reduced. This could significantly reduce the high share of the extremely dangerous side collisions. The GDV therefore demands that systems controlling the dynamics of vehicle movement should be offered as standard for all vehicle classes.

Electronic Brake Systems available today exclusively use wheel speeds (ABS, TCS) and additional system- and vehicle dynamics sensors (ESP) to interpret the vehicle movements describing driving stability.

Here, the yaw rate and the sideslip angle, which cannot be measured directly, are the most important factors. The yaw rate can be measured directly with the ESP sensors available today. As far as the sideslip angle is concerned, only its time rate of change, i.e. the slip angle velocity, can be calculated with sufficiently reliable stability. For this reason, only the yaw rate has been controlled up until now with the slip angle velocity serving as an auxiliary variable.

The quality of current ESP control systems could be further improved if it were possible to also control the sideslip angle itself. Various methods are available for an indirect determination of the sideslip angle [20-12]. However, most of these methods use models and produce comparatively instable results owing to the non-linear tire characteristic and the influence of inclined road surfaces.

If the tire forces were known, then these models could be supported so that it would be possible to generate usable sideslip angle information. Ideally, the tire should provide information not only on its speed, but also its longitudinal and lateral adhesion forces.

Such a tire is taking shape at Continental (**Fig. 20.14**). Integration of a sensor system known as Sidewall Torsion (SWT) sensor makes this tire a source of information for the brake and stability control systems. With this sensor system, the side wall of the tire, which is 'soft against torsion' and incorporates sections of alternately positively and negatively magnetized ferrite powder, acts as magnetic pole wheel. Two magnetoresistive sensors arranged in 12 o'clock position on the chassis, one near the rim, the other near the outside diameter of the tire, supply sinusoidal signals whose phase shift and frequency is a measure of the twisting of the side wall of the tire and, thus, the longitudinal tire force. Their amplitudes are a measure of the lateral force since the air gap between the side wall of the tire and the sensor heads changes with the lateral force. Air gap and signal amplitudes are interdependent with this measuring principle.

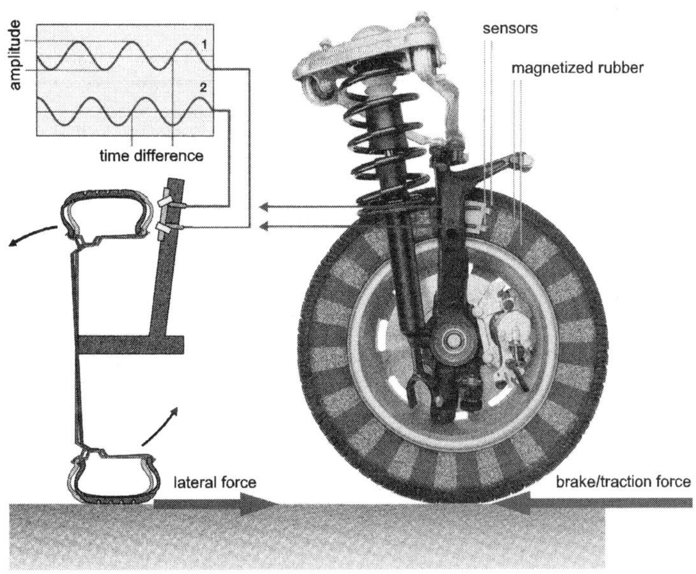

Fig. 20.14. Side Wall Torsion Sensor [20-9]

A further tire sensor which is primarily intended to provide information on the instantaneous adhesion potential between tire and road surface, is still in the research stage (**Fig. 20.15**). Vulcanized into the tire tread, a sensor chip operating on the wavegraph principle records the local deformation of the tread while passing through the area in contact with the road surface (tire print). The special advantage of this principle lies in its passive operation. Like the SWT sensor, it does not require energy to be transmitted to the tire or an energy source to be carried in the tire. A telemetrically coupled signal transmitting / receiving and processing unit, installed in the wheel house, interprets the pattern of the measuring signal as the 'fingerprint' of instantaneous adhesion. This local information permits conclusions to be drawn with respect to global performance, namely instantaneous adhesion utilization, maximum coefficient of friction (adhesion potential) and the risk of aquaplaning. This information can be used as valuable additional information for the driver, the system controlling the dynamics of vehicle movement or the headway distance control system, e.g. for determining the safe distance required to follow a preceding vehicle.

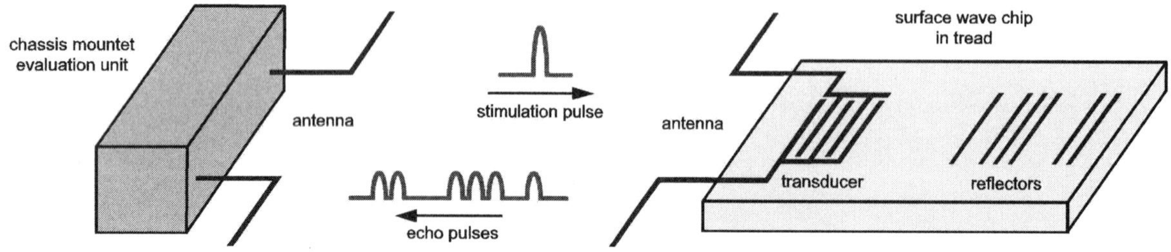

Surface wave chips are embedded in the tread of a tire. A 433 MHz high frequency-wave sent out from a chassis- mounted evaluation unit is converted into a structural wave. Reflectors on the chip generate echos on this wave, whose delays depend on the structural stress of the chip and thereby depend on the force transmitted by the tread. Echos are received for being evaluated.

Fig. 20.15: Local tire sensor based on surface wave technology, function principle and chararacteristics

In order to cover different marketing and purchasing strategies, all of the above-mentioned systems and components are currently still designed as self-contained (stand alone) units. Where systems are combined, networking is primarily enforced by the need to avoid critical driving situations and less by the desire to achieve synergetic effects (e.g. by sensor fusion) or to save costs (e.g. by sensor omission). Standardized subsystem interfaces are defined by way of exception only. A comprehensive and nevertheless modular chassis control system meeting the demand for Total Chassis Management, ideally located in a central chassis control unit, is not available today.

Thus, in the Audi Allroad Quattro, for which ContinentalTeves supplies the adjustable air suspension and another supplier the ESP system, the ESP is prevented from being switched off in off-road use. Special precautions in the chassis control unit of the air suspension make sure that after a change from cross-country to motor road use any errors occurring during the time which it takes to adjust the vehicle level will not lead to safety-critical driving situations.

Frequently the networking of systems is rendered extremely difficult or completely defeated by a lack of system compatibility (on the part of both technology and suppliers). The integration of the adaptive cruise control ACC function with ESP and possibly BA (Brake Assist) into the control electronics of braking system and engine/transmission control is a typical example. In cooperation with our joint venture partner Automotive Distance Control GmbH/Lindau who, incidentally, supplies the distance radar for the DaimlerChrysler S-class, we are in a position to supply such a complex system as a single contractor.

20.3 Concepts for a compound (global) chassis control

20.3.1 Tire-force-based upgraded ESP control

A first step towards compound chassis control has been made with the introduction of the tire-force-based vehicle dynamics control system. For this purpose, the conventional ESP was upgraded by adding SWT tire sensors (**Fig. 20.16**).

	average rating (0 10)
1. **"low-cost" ESP without yaw rate sensor** yaw rate sensor signal replaced with signal generated by observer	6
2. **standard ESP MK 60,** (SOP 09/2000)	8
3. **"high performance" ESP** **with yaw rate sensor and side slip angle control** side slip angle estimated by observer, combined yaw rate control and side slip angle control	9

evaluation criteria
stability, steerability, efficiency, noise
approximately 100 testpersons (wintertest, Arvidsjaur 1999/2000)

Fig. 20.16: Comparison of SWT-information supported ESP-systems (1.,3.) with the system in series (2.)

This addition permits two system configurations. In the first configuration, the SWT sensors are substituted for the comparatively expensive yaw rate sensor. Compared with the current production system, the cost savings which may be expected from this 'low-cost' ESP are, however, offset by a slightly poorer performance.

If the SWT sensors are added to the current ESP sensors, then the combined yaw and sideslip angle control described in section 20.2 can be implemented (**Fig. 20.17**). This permits a more sensitive and, for the average driver, even more efficient stability control. Slideslip angles, which are difficult to control and a challenge for the driver, are avoided. The steering effort is substantially reduced.

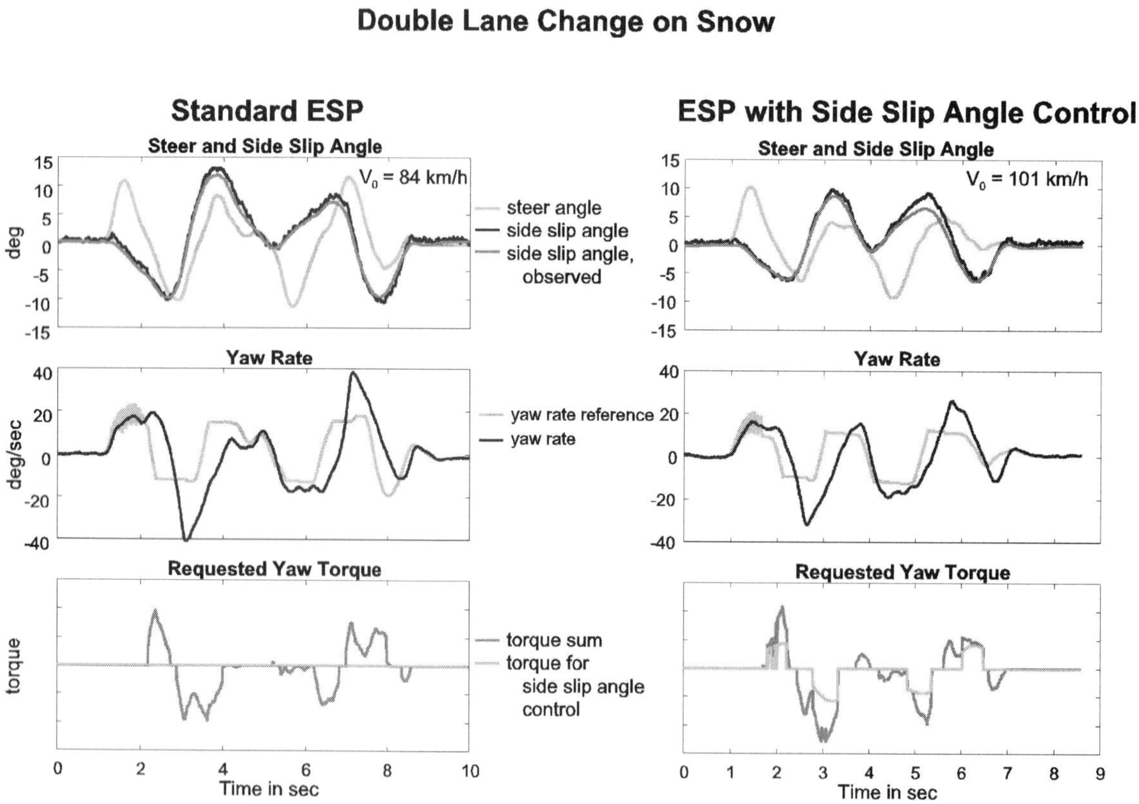

Fig. 20.17: Comparison of SWT information supported high performance ESP system (right) with the system in series (left) by the example of a double lane change

20.3.2 Compound control of wheel load and braking intervention

Chassis control systems should provide a high degree of riding comfort without necessitating the usual sacrifices in terms of driving safety. They influence primarily the vertical dynamics of vehicle movement. In the event of disturbances acting on the chassis from outside (side winds, rough roads etc.), caused by the load or initiated by the driver (steering, braking, acceleration), they should keep the kinematic variables of the chassis (track, toe-in, camber, ...) in design position and minimize wheel load fluctuations as far as possible (see **Fig. 20.3**). This generally provides a good basis for the systems controlling the horizontal dynamics of vehicle movement, i.e. ABS, TCS, ESP. For further improvements, such as improvements of the lateral dynamics of vehicle movement, stabilizing yawing moments during cornering can be generated by a selective combination of wheel-individual braking intervention and active wheel load shift (**Fig. 20.18**). Braking intervention generally produces higher yawing moments than a wheel load intervention. A disadvantage of braking intervention is that this type of control directly affects the longitudinal dynamics of vehicle movement. Major braking interventions lead to severe longitudinal decelerations and pitching motions of the vehicle which, in turn, disturb the harmony of vehicle handling. Especially in the event of a severe intervention, these disturbances will be clearly noticed by the driver. This type of intervention thus may result in a loss of comfort and is therefore used in critical driving situations only, as far as this is possible.

		increase of longitudinal force	loss of lateral force	wheel load shift to front
front	outside	turn out	turn out	turn in
front	inside	turn in	turn out	turn in
rear	outside	turn out	turn in	turn in
rear	inside	turn in	turn in	turn in

Influence of wheel- selective brake intervention on yaw torque and maximum achievable yaw torque over lateral acceleration

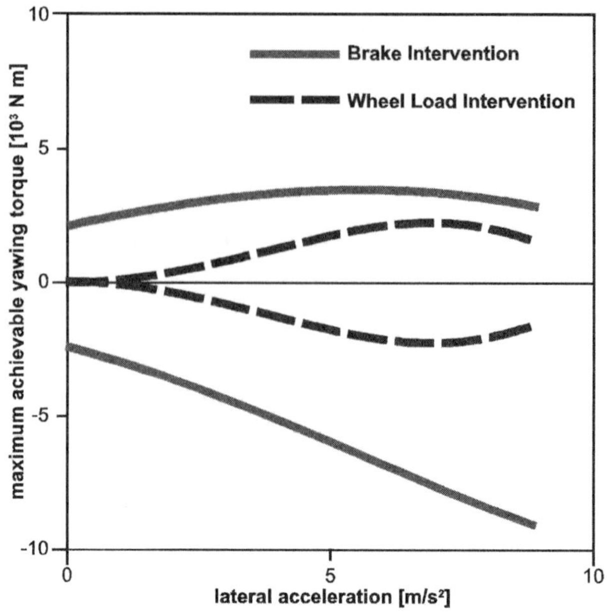

Fig. 20.18: Influence of wheel selective brake intervention on yaw torque (left) and maximum achievable yaw torque over lateral acceleration (right) [20-13]

Fig. 20.19 demonstrates the potential of a compound wheel load and braking intervention control system. According to Smakman [20-13], vehicle response to steering inputs can be improved by up to 10% by means of a compound control system. The overall effect is that of a greater harmony of vehicle handling.

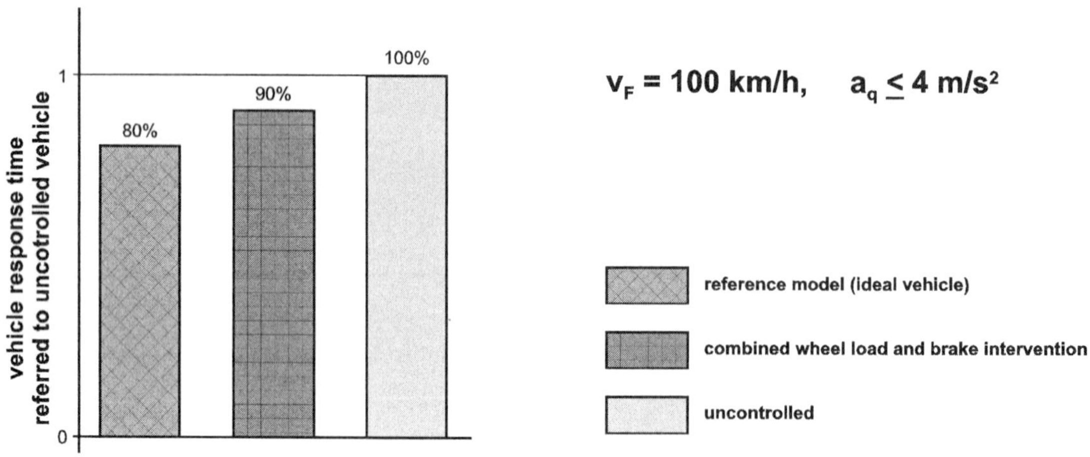

Fig. 20.19: Comparison of vehicle response time for stepped steering input with lateral accelerations up to $4 m/s^2$ [20-13]

20.3.3 Compound control of steering and braking intervention

Comfort, agility and stability can be increased with steering systems permitting a steering intervention which overrides the driver's input. In special situations, a system which networks this intervention with the braking system is particularly advantageous in terms of the resulting sensor synergy effect and especially in terms of the resulting improvement of the dynamics of vehicle movement.

Both systems influence the yawing moment which determines the dynamics of vehicle movement, the steering system via the lateral tire forces, the braking system via the longitudinal tire forces, the lateral forces with a longer lever arm than the longitudinal forces. Unlike braking intervention, which results in a longitudinal deceleration of the vehicle noticed by the driver, steering intervention remains unnoticed, at least if the overriding angle is small. Combining the two systems permits the yawing moment requested by the driving stability control system to be introduced less abruptly. The stability limit of the dynamics of vehicle movement can then be spread over a wider range and predicted more reliably by the driver.

Disturbances such as side wind and road surface inclination can be controlled in cooperation with the sensors of the ESP system monitoring the dynamics of vehicle movement. **Fig. 20.20** demonstrates the situation created by a vehicle equipped with overriding steering control moving past a side wind installation. Side deviation is significantly reduced by the quick steering intervention. Unlike an uncontrolled vehicle, the controlled vehicle leaves the side wind area with the heading angle unchanged.

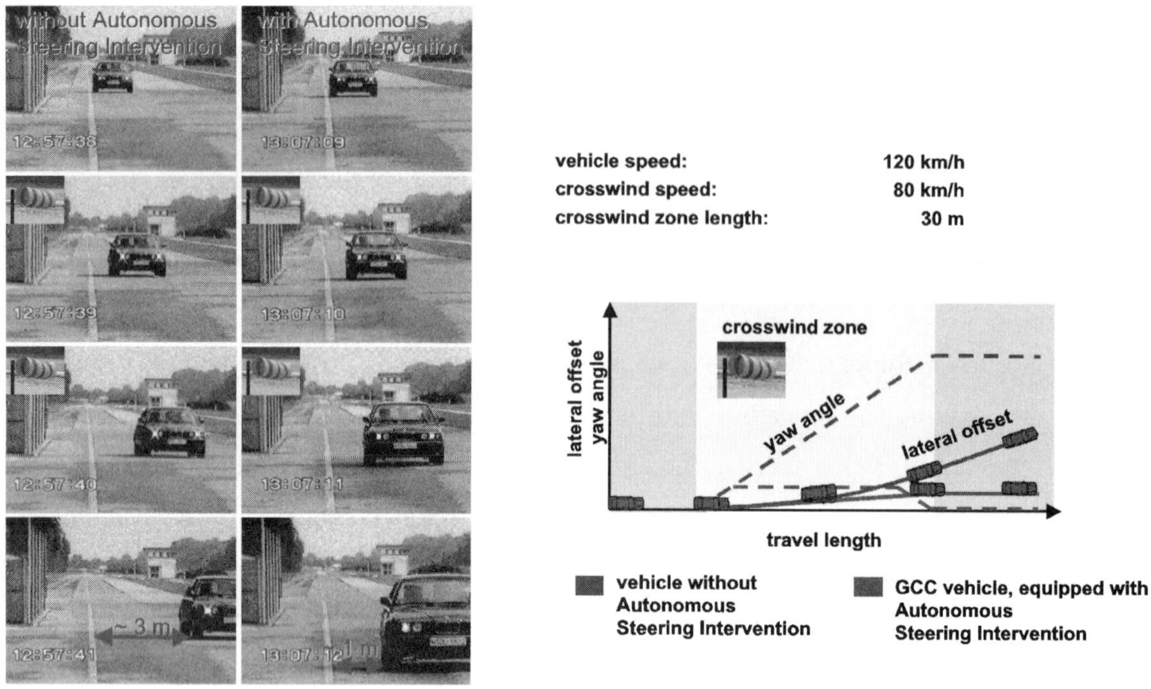

Fig. 20.20: Comparison of sidewind sensitivity without (left picture sequence) and with autonomous steering intervention (right picture sequence) [20-14]

During emergency braking on roads with large differences in the coefficients of friction on the right and left, a combined braking and steering intervention produces a safety advantage. The stand-alone braking system underbrakes the rear axle for stability reasons by adjusting both brakes to the lower pressure level of that axle ('select low'). Furthermore, the 'pressure gap' at the front axle is not built up to the full extent possible and only very slowly in order to keep the need for countersteering by the driver within realistic limits. Both factors increase the length of the stopping distance. These precautions are not required if a networked system of steering and braking control is used. Instead, the brakes can be applied with greater force. The resulting yawing moment (in the direction of the high-friction side) is immediately compensated for by an autonomous steering intervention (in the direction of the low-friction side). The stopping distance is reduced significantly (**Fig. 20.21**).

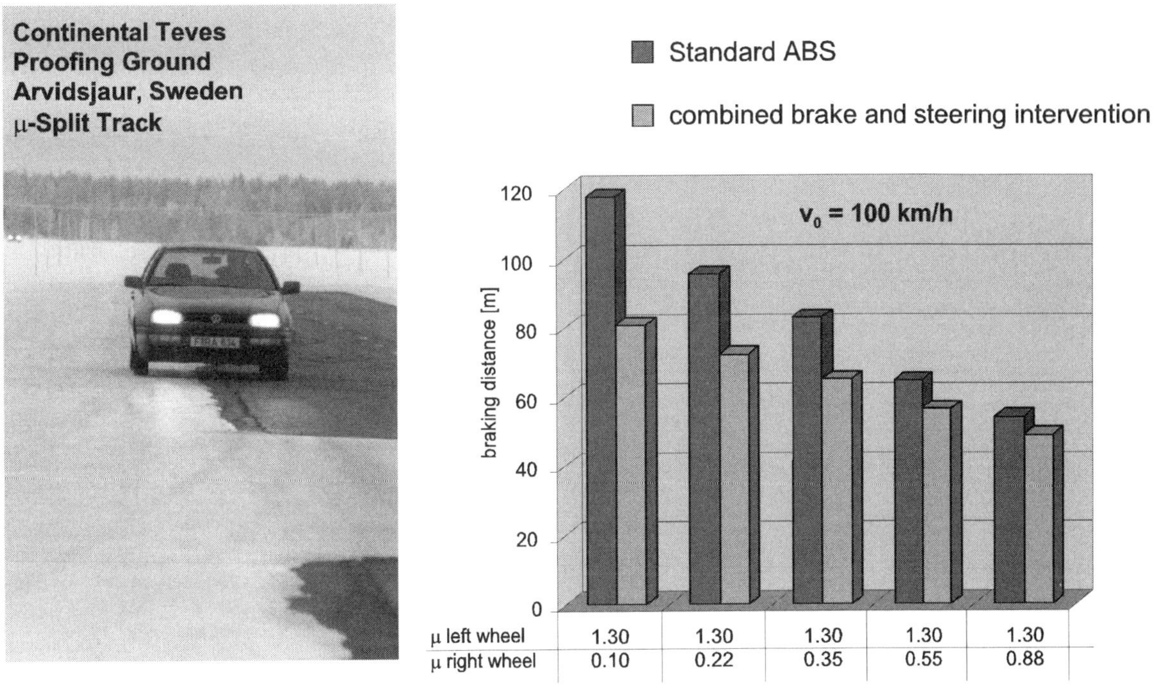

Fig. 20.21: Comparison of μ-Splitt ABS braking without and with combined brake- and steering intervention

20.4 Heading for the intelligent chassis

When it comes to taking advantage of the potentials described in the preceding section, ContinentalTeves is ideally equipped to put these into practice. The product range includes assemblies and subsystems which meet the functional requirements.

In the field of brakes, the electrohydraulic brake (EHB) will shortly be introduced into regular production as the most innovative single aspect in this domain (**Fig. 20.22**). It represents a decisive step on the path to Drive by Wire. Similar to E-Gas, the driver's wish to brake is transmitted to the hydraulic / electrical control unit by purely electrical means. With EHB, all wheel-individual braking interventions that can be imagined can be effected without the pedal feedback to the driver still produced by the current autonomous power-assisted braking systems. The dynamics of the system are generally better, especially at low temperatures. Braking pressure control has been analogized and is therefore more sensitive and comfortable even during ABS, TCS and ESP control cycles. EHB still uses the hydraulic wheel brakes.

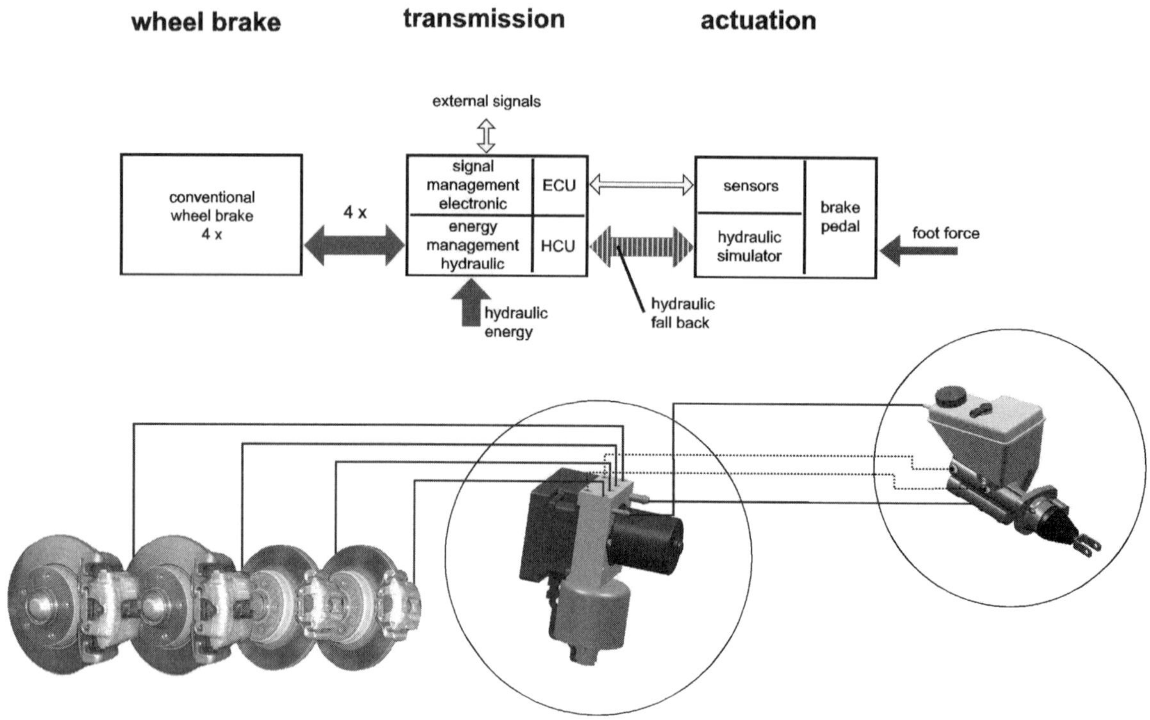

Fig. 20.22: Basic layout and components of Continental Teves' electrohydraulic brake system EHB [20-4]

New high-performance wheel brakes, such as the Teves FNR caliper (**Fig. 20.23**), which permits a larger brake disc to be installed for a given rim thanks to an integrated fist / frame design of the caliper housing, and the ceramic brake disc (**Fig. 20.24**) guarantee that the growing demands for higher performance can also be met in the future, even though the space available for installation may reach an upper limit.

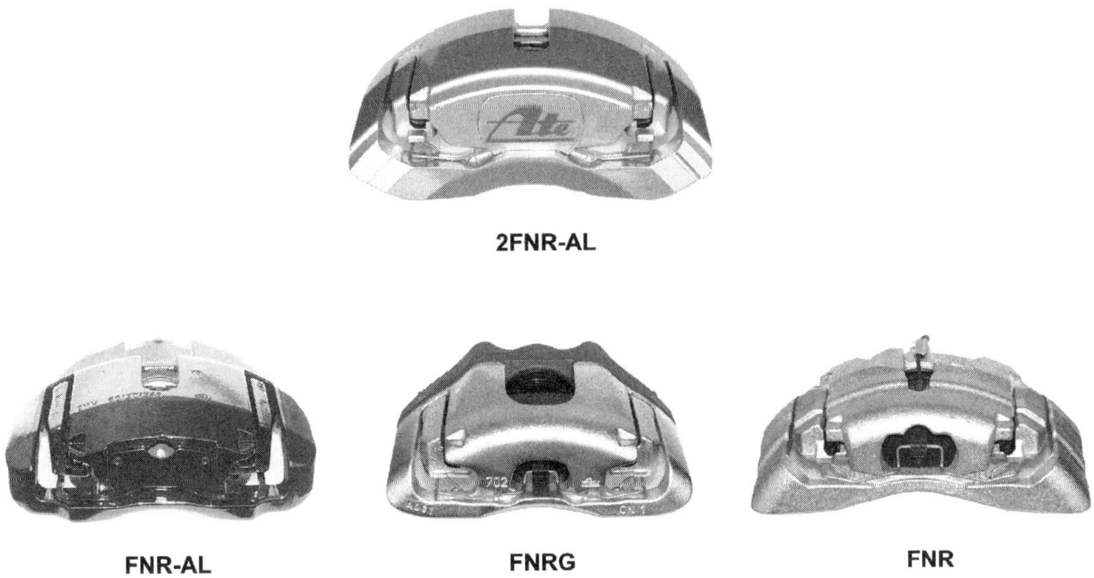

Fig. 20.23: FNR-type high performance discbrake family [20-15]

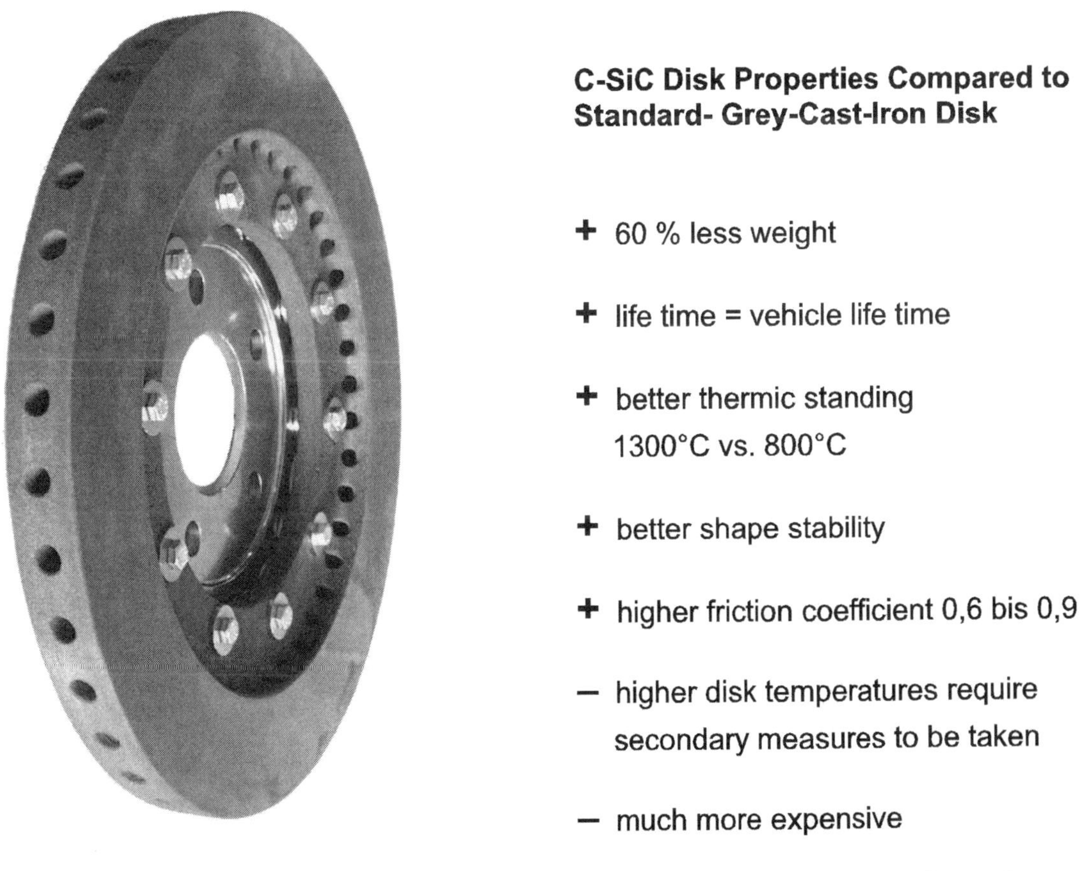

Fig. 20.24: Prototype and properties of new C-SiC ceramic brakedisc [20-15]

The electromechanical brake (EMB) (**Fig. 20.25, Fig. 20.26**), which has reached an advanced stage of development at our company, will afford further advantages, including:

- a pedal module combining E-Gas and E-Brake Pedal in one unit, ergonomically optimized and adjustable thanks to the omission of a manual fallback feature
- number of parts reduced to a minimum
- packaging and installation advantages
- brake-fluid-free, 'plug & play' instead of 'fill & bleed'
- easy to network with traffic guidance systems
- optimum cost-performance potential

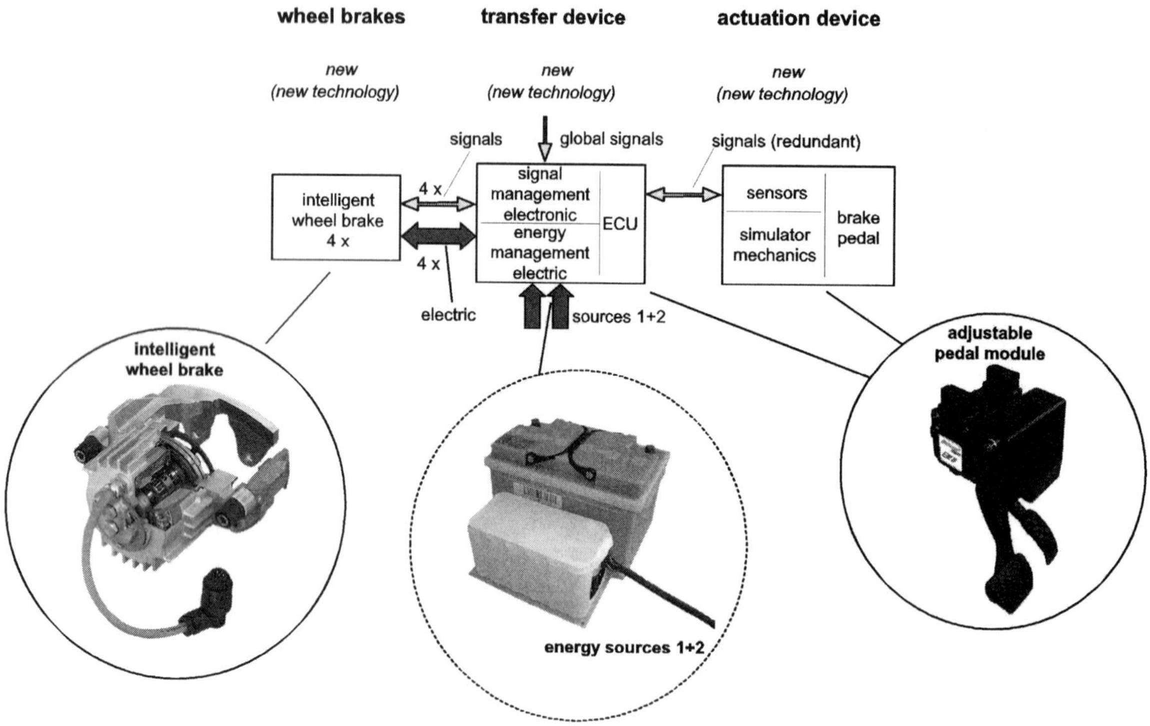

Fig. 20.25: Basic layout and components of Continental Teves' electromechanical brake system EMB [20-4]

Fig. 20.26: Prototyp of the elektromechanical wheel brake EMB [20-4]

Benefits expected from further developments of the controlled air suspension chassis include active roll and pitch compensation, higher control speeds and lower energy consumption.

Development of the SWT tire for series production has reached an advanced stage. **Fig. 20.27** shows the magnetizing equipment.

Fig. 20.27: Magnetization machine for SWT-sensor tire

20.5 Outlook

With the integration of Teves, Continental has become a supplier of comprehensive chassis control systems. Already, the Automotive Systems Division boasts a higher sales volume than any other division of the group. The goal which we have set ourselves is to create products and services at the highest technological level in order to make driving even safer and also more comfortable than it is today. With the recently started project initiative Global Chassis Control (GCC) we propose to make an active contribution to the design of the chassis for the next decade (**Fig. 20.28**).

Motivation

Continental chassis competence augmentation using the approach of an entire reflection on

- function potential (Global Chassis Control)
- devices topology and E/E architecture (Drive-by-Wire)

Targets

- Presentation of a global (dry) chassis concept with active brake (EMB), SWT sensor tire, adaptive air spring, controllable damper, adaptive bearings and active steering (EMS)
- Demonstration perceptible consumer benefits (safety, comfort and fun)

for vehicle user

- broader and easier controllabe vehicle stability range
- enhanced ESP functionality
- shortest brake- and stopping distances
- minimial vehicle body motion
- minimal handling effort while braking and steering

for vehicle manufacturer

- modules architecture (Plug and Play)
  - intelligent corners, smart actuators
  - safe and reliable E/E- architecture
- less constraints in chassis and suspension tuning
- less production costs

Fig. 20.28: Motivations of Continental's global chassis control (GCC) initiative

This drive- by-wire chassis will be made up of individual corner modules which, in turn, break down into intelligent submodules for steering, braking, spring/damping (**Fig. 20.29**). The steering and braking systems will be designed as mechatronic (electromechanical) modules. These intelligent modules will be networked with each other and with the operator control module. The electronic chassis control unit will be located in the latter.

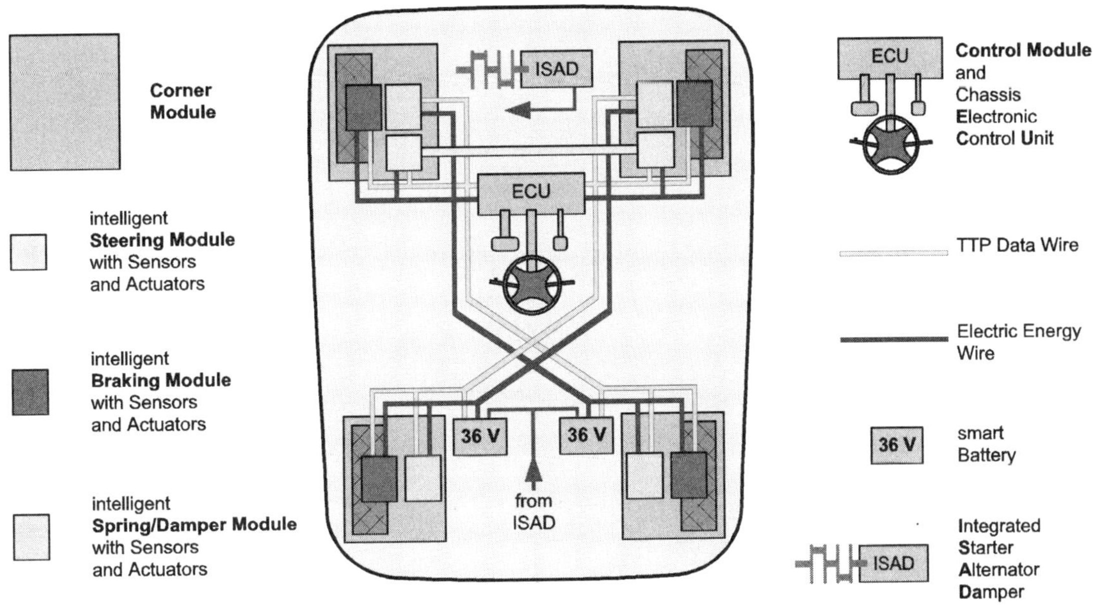

Fig. 20.29: Layout of the intelligent chassis

Fig. 20.30 shows the basic hardware structure for the various modules and the modular structure of the software. The safety-relevant systems - steering and brake - will be consistently designed as two-circuit (redundant) systems. For safety reasons, a time-controlled communication path (BUS) will replace the event-controlled (CAN) BUS on the signal side. The actuators will operate on 42 volts, which permits a better power density and a more advantageous overall cost structure. The energy will preferably be supplied by the Integrated Starter-Alternator-Damper (ISAD) available from Continental ISAD. Two so-called smart batteries, which know their state of charge and state of health and report critical conditions, will serve as energy accumulators for peak loads and as energy buffers in the event of a failure at the energy source. **Fig. 20.31** gives an impression of the distribution of the various components in the vehicle.

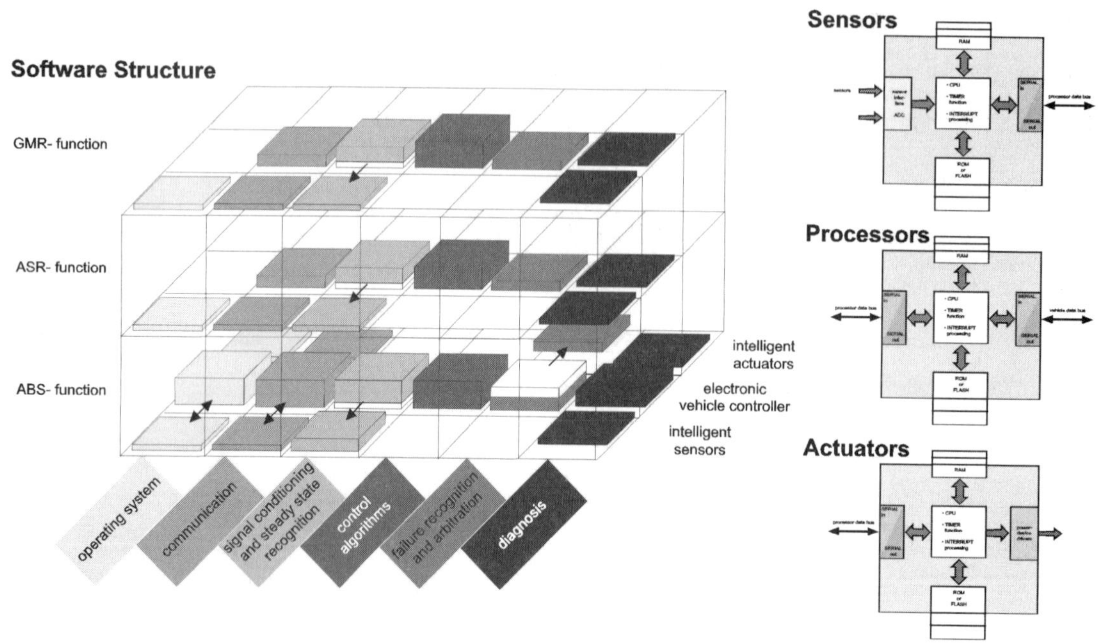

Fig. 20.30: Structuring of the hardware and the software within the intelligent chassis [20-16]

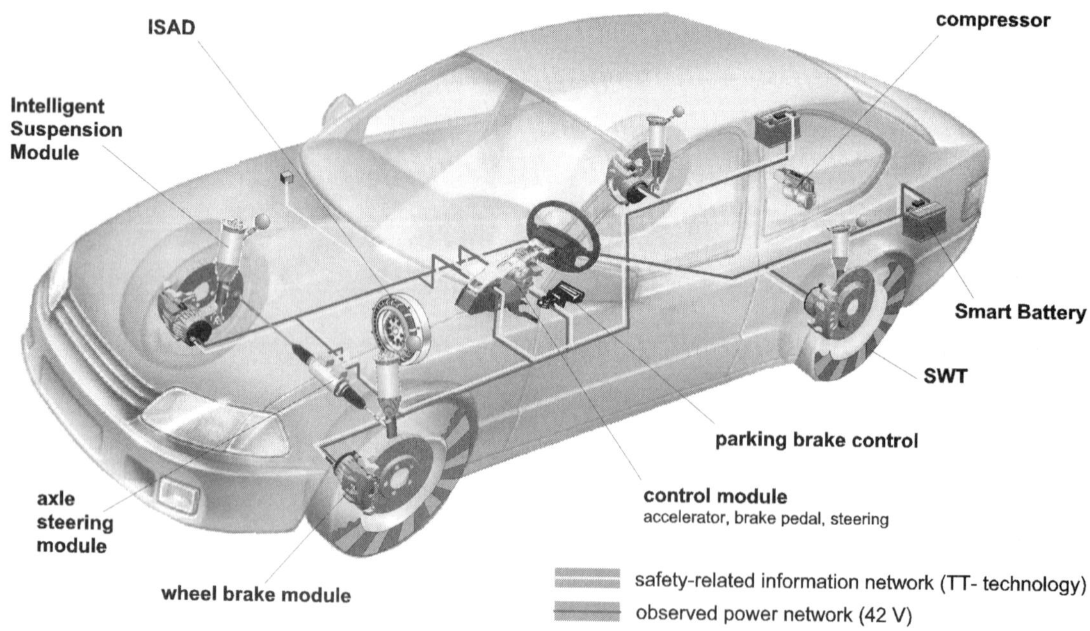

Fig. 20.31: Intelligent chassis, arrangement of components within the vehicle

Under the motto "30m Auto", a subproject of this group initiative is aimed at significantly reducing the stopping distance in emergency braking situations through a comprehensive study of all phases involved in the stopping process and all chassis components influencing the length of the stopping distance. In this connection, reducing the reaction distance is receiving the same attention as reducing the braking distance proper (**Fig. 20.32**). First results are promising. By rigorously combining all our technical potentials with all our expertise in the field of tires, brakes and suspension we expect to reduce both the reaction distance and the stopping distance in an emergency situation by as much as 15 to 20% compared with what is possible today.

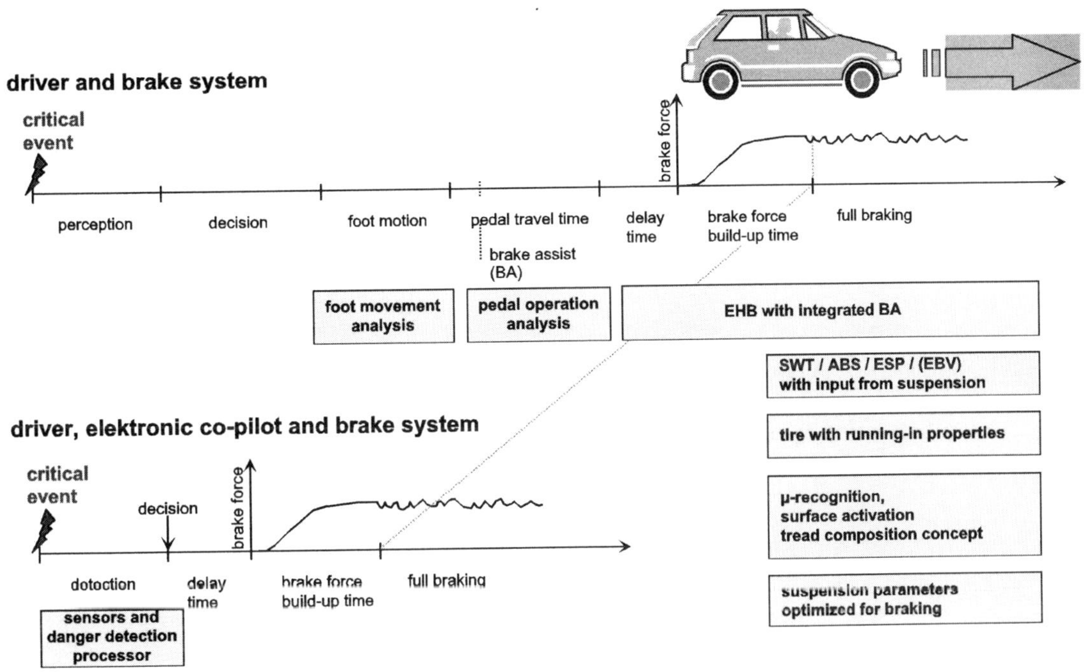

Fig. 20.32: Sequences of an emergency braking and measures for the reduction of the entire stopping distance

The ongoing efforts to put the intelligent GCC chassis into practice will contribute not only to expanding the driving dynamics potential, but also to improving the feed back by which the driver is informed about the stability limit of the vehicle. Naturally, this will go hand in hand with a special effort to improve comfort wherever this is possible (**Fig. 20.33**). All these activities will undoubtedly contribute to further improving the standards of safety on our roads.

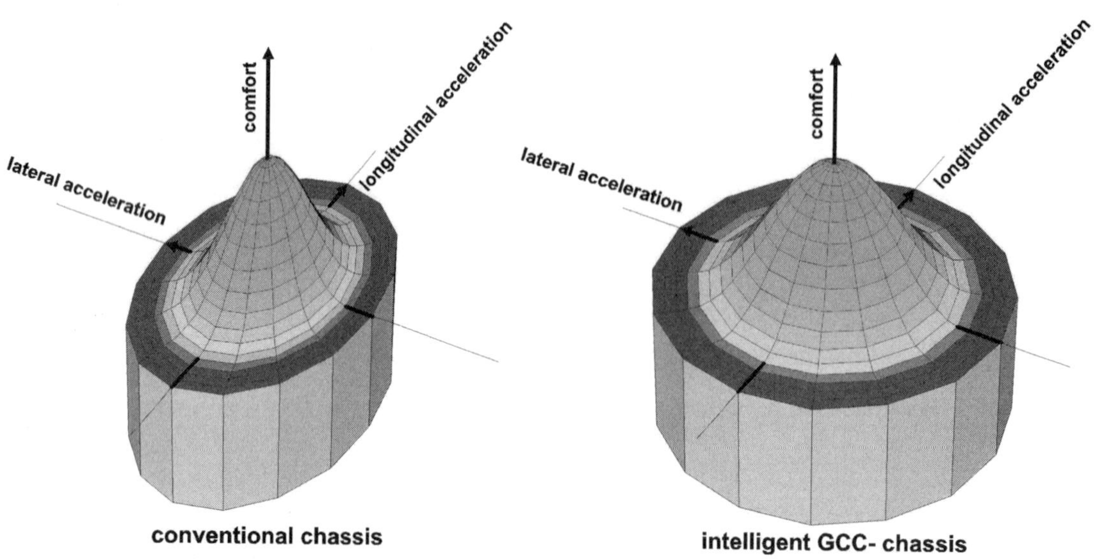

Fig. 20.33: Development of a larger dynamic and comfort potential with the intelligent chassis

20.6 References

[20-1] Rompe, K.: Die Notwendigkeit aktiver Sicherheitssysteme im Fahrzeug. EUROFORUM Konferenz Aktive Systeme zur Fahrzeugsicherheit. Köln, 3.3.1999.

[20-2] Rieth, P.: Technologie im Wandel X-by-Wire. IIR Konferenz Neue Elektronikkonzepte in der Automobilindustrie. Stuttgart, 13.,14.4.1999.

[20-3] Wolfsried, St., W. Schiffer: Active Body Control (ABC) - das neue aktive Federungs- und Dämpfungssystem des CL-Coupés von Daimler Chrysler. VDI Fachtagung Reifen Fahrwerk Fahrbahn. Hannover, 21., 22.10.1999.

[20-4] Rieth, P.: Autonome Bremssysteme. VDI Fachtagung Reifen Fahrwerk Fahrbahn. Hannover, 21.,22.10.1999.

[20-5] Rieth, P.: Bremsentechnologie im Wandel. Fahrzeug- und Motortechnisches Seminar der TU Darmstadt. Darmstadt, 11.2.1999.

[20-6] AUDI-Presseinformation.

[20-7] Pauly, A., H. Köhnle: Active Front Steering. IIR-Fachkonferenz Innovative Sicherheitssysteme im KFZ. Stuttgart, 5.,6.7.2000.

[20-8] Kilgenstein, P.: Die Ausfallsicherheitsanforderungen bei elektrischen Lenksystemen. VDA Technischer Kongress. Frankfurt, 20.,21.9.1999.

[20-9] Huinink, H.: The tire as an integrated element within the electronic chassis. TOPTEC Symposium Active Safety in Vehicle Traction & Stability Systems. Vienna, 27.,28.9.1990.

[20-10] Becker, A.: Influence of Rubber Friction on Steady State Tire Properties. HdT Veranstaltung Fahrwerktechnik. München, 6.,7.6.2000.

[20-11] Langwieder, K.: Rahmenbedingungen und Potentiale für den Insassenschutz von morgen. IRR Fachkonferenz Innovative Sicherheitssysteme im Kfz. Stuttgart, 5.,6.6.2000.

[20-12] Fukada, Y.: Slip-angle estimation for vehicle stability control. Vehicle System Dynamics, 32 (1999), pp 375 ff.

[20-13] Smakman, H.: Entwurf von integrierten Fahrwerkregelsystemen unter Anwendung von objektiven Beurteilungskriterien. HdT Veranstaltung Fahrwerkstechnik. München, 6./7.6.2000.

[20-14] Ackermann, J., et al: Robuste Schleuder- und Kippvermeidung durch Lenkungseingriff. VDI/VDE - GMA. Frankfurt, 13.4.2000.

[20-15] Neitzel, B.: Moderne Scheibenbremsenentwicklung. HdT Veranstaltung Brems- und Regelsysteme. Essen, 9.,10.5.2000.

[20-16] Fennel, H.: Strukturen für den Fahrzeug Regelsystemverbund der Zukunft. ATZ/MTZ Sonderausgabe Automotive Electronics, Januar 2000.

21 Automotive Technology – Where are we going?

Dr. W. Steiger, Volkswagen AG

Wolfsburg, Germany

21.1 Introduction

Global environmental awareness is increasing as pressure from densely populated countries push for more environmentally friendly automobiles, **Fig. 21–1**. Noise, accidents, polluting discharges, and CO_2 emissions from fuel consumption are the essential concerns.

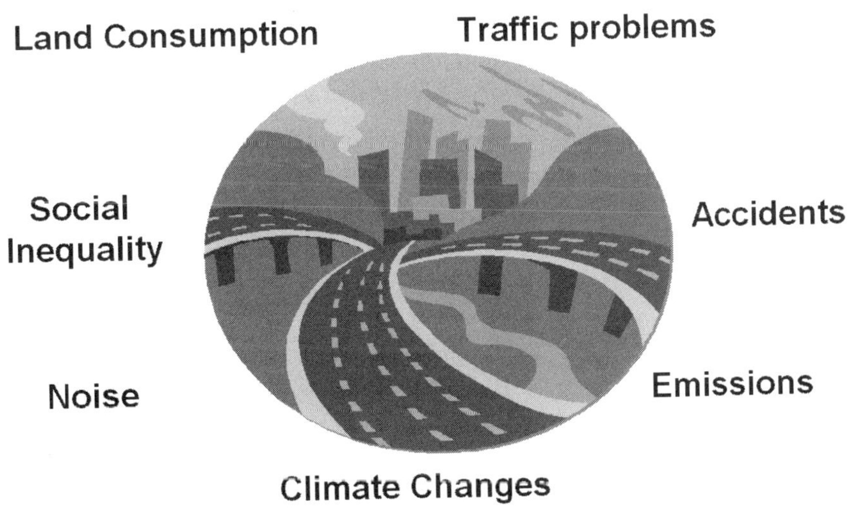

Fig. 21–1: The Global Challenge

The reduction of these environmental challenges as well as increasing comfort and driving performance is an intense challenge for the automobile industry. Safety issues present another important challenge.

The worldwide desire for increased mobility is evident, **Fig. 21–2**. Thus, it is expected that driving performance will be further enhanced. This expectation places a responsibility on automobile manufactures to provide the best possible technologies and innovative products, to protect the environment and enhance personal safety.

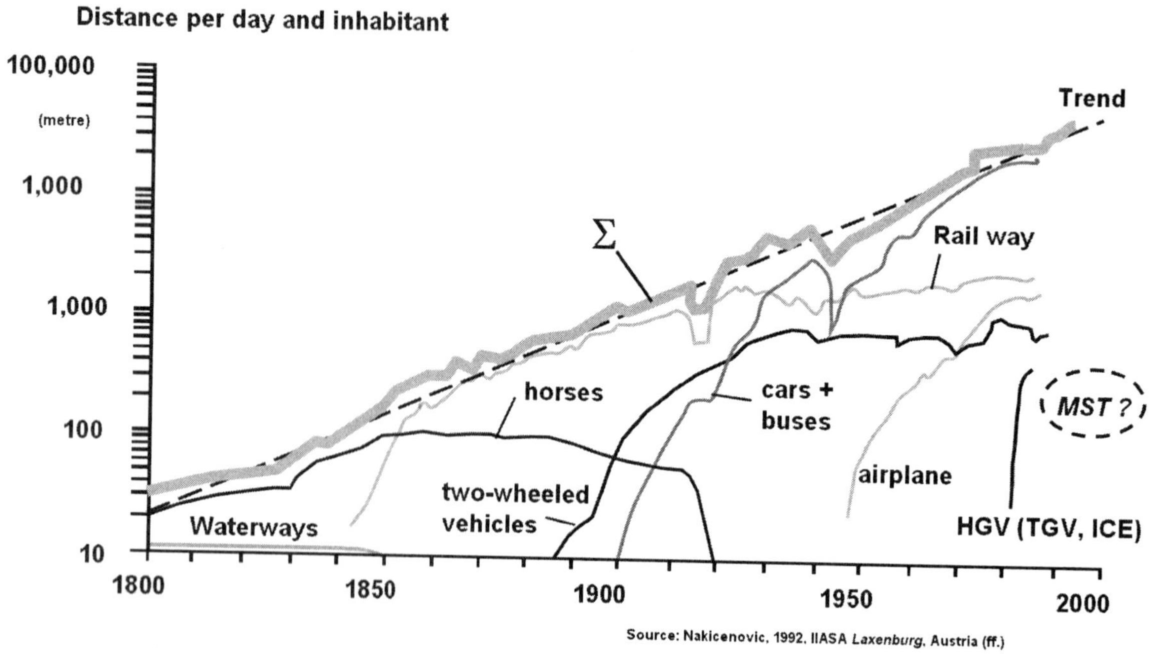

Fig. 21–2: Development of Individual Mobility

Among various modes of transportation (**Fig. 21–3**), vehicles are distinguished by their high flexibility, and the ability to travel any distances. Intelligent solutions that efficiently use the existing traffic infrastructures are needed, especially in urban areas. This has a far-reaching effect on information and communication technologies in vehicles. The design of future vehicles are required to fulfil these demands continuously.

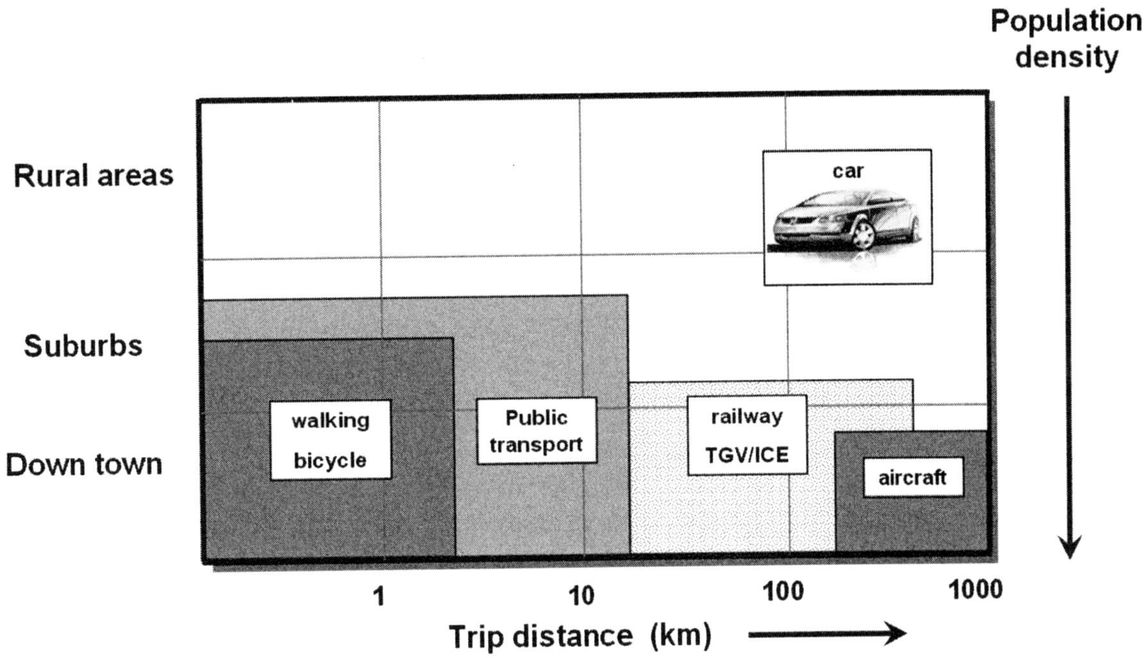

Fig. 21–3: Transportation Modes

21.2 Multimedia

Prior to the modern technology age, there was no rapid movement of data. Now the utilisation of comprehensive information is easily accessible through the internet and the data storage that can be accessed through it, is always in focus. Information processing has many different applications. Services like appointments and contact detail management, communications via e-mail, and online shopping are more and more accessible in mobile locations. People who use and depend on these services in their everyday life expect to have the technology available on their home PC, mobile telephone, their organisers or communicators and certainly in their vehicles as well.

As a result, the automobile industry has a new field to expand. Driving comfort, safety, and mobility are the traditional requirements of a vehicle, however, the need for information technology in a vehicle is emerging fast. A vehicle is no longer just a transporter, but increasingly serves as an information terminal and service provider. Nevertheless, the other functions of a vehicle should not be neglected, but is constantly improved further. Access of information technology services is an additional feature of a modern vehicle. The requirements of the next generation of vehicles are decided by a combination of traditional and innovative customer groups, **Fig. 21–4**.

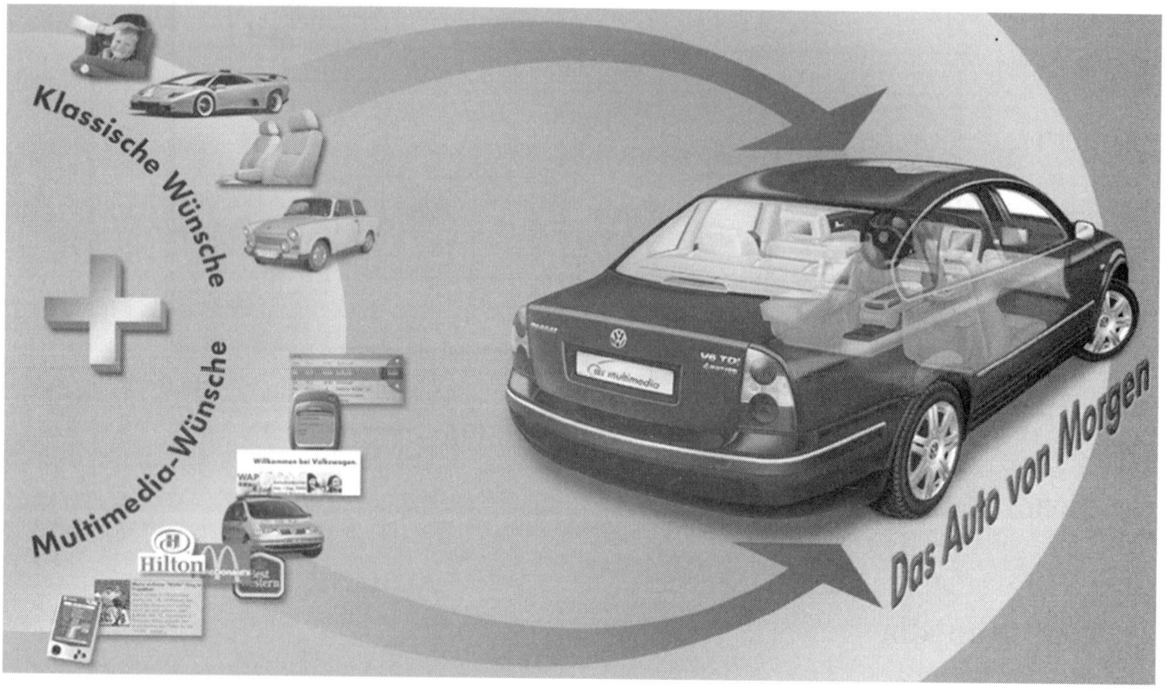

Fig. 21–4: The car of tomorrow

The application of multimedia in vehicles started with the usage of radio systems. In the beginning, the objective was simply to receive radio programs. Today, it is common to find a combination of radio, CD player, and navigation system in a vehicle. In the future, more appliances such as the DVD player, TV, radio and the internet will be included in a vehicle multimedia package, **Fig. 21–5**.

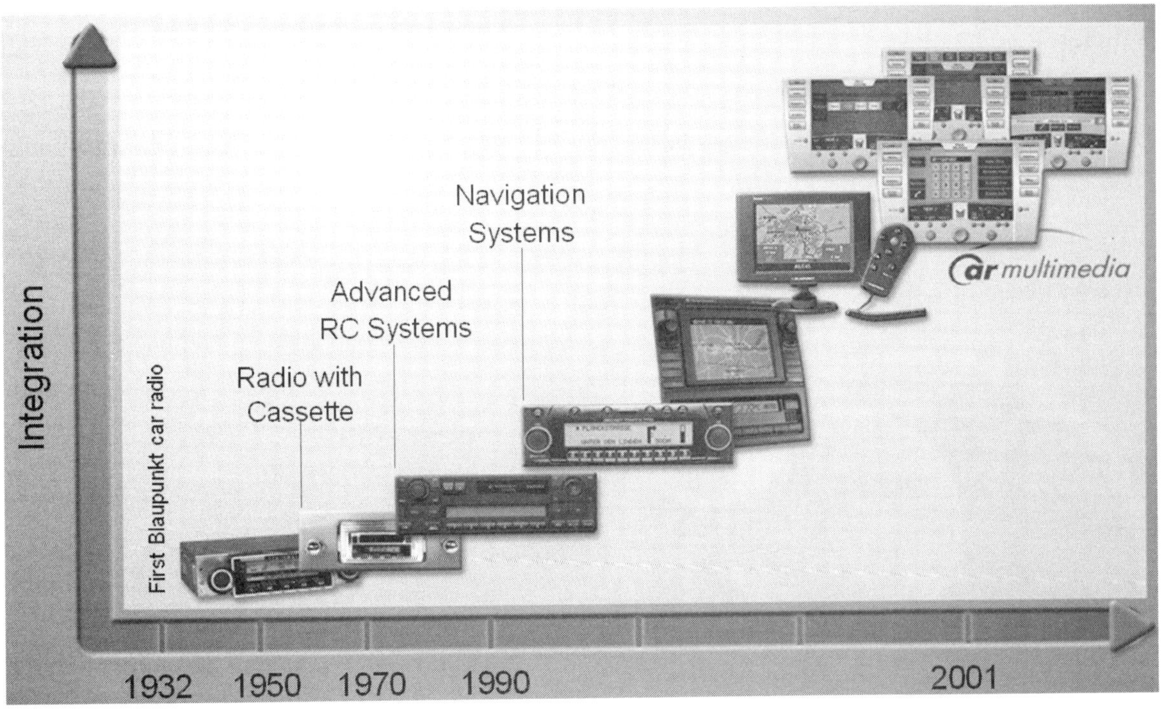

Fig. 21–5: From Radio to Service Platform

To accommodate these new features, new operation systems need to be developed. In the future of automobiles, the multimedia controls will be different and more comprehensive. While developing the more comprehensive operating systems, safety needs to be taken into consideration. The new system should not be a hazard to the driver and motorists on the roads. Multimedia controls in a vehicle should be able to handle more complex tasks with no additional complications to the users. The new concepts and methods include simplifications like: centralised menus, voice recognition, self-explanatory display and operations. For instance, in the past, adjusting radio and climate controllers are two unrelated actions. Automobile manufacturers need to integrate these two features. A core competence of the future will be a universal, intuitive, completely integrated display and control design of multimedia vehicle systems.

The new generation of services requires a great understanding of the internet. Often, contents of some web sites are vulgar and offer no profitable information. For example, there are internet sites that only inform users that a Cinema X exists, **Fig. 21–6**. This gives very little useful information to the user. User requirements are met when a chain of actions is an end-to-end solution and substantial results are generated. For instance, the contents of the Cinema X web site would be enhanced if film details and session times can be accessed. Users should be able to visit the box office interactively. Seating plans should be available, showing vacant seats, perhaps with some information about the person sitting next to the chosen seat as well. Prices could be calculated or even negotiated and delivered to the customer either electronically or in hard copies.

Fig. 21–6: Infotizing

A combination of digital wireless technologies and mobile phone services is an approach used to supply data for vehicles using hybrid transmission systems. Large data packages of general interest are transmitted economically and efficiently by wireless technology. Personal data such as e-mails are transmitted separately via mobile phones. Cost is indispensable for these services. As in the "Rich Media" concept, if music selections (audio on demand) or films (video on demand) are sent separately, the bandwidths even on UMTS is not sufficient. User costs will also be unacceptably high. A third transmission path could be the key to solve this problem. Volkswagen is doing experiments on short-range transmissions; for instance, petrol stations could become "DataKiosks". Besides filling up, customers could also get data of interest transmitted to their vehicles at the petrol station.

An advantage of this alternative concept is that in principle, the service is free. Since, the transmissions do not belong to any mobile telephone companies and the transmission equipments are provided by customers themselves or petrol stations.

The NetConnect Passat (**Fig. 21–7**) is a prototype that implements the short-range transmission concept, as well as the principle of action chains. Short-range transmission makes it possible for vehicles to access data at any intelligent "DataKiosks", such as petrol stations or homes, at almost zero cost. The service could be used for route planning of a holiday, getting information for the next business trip, or for those who wish to install audio or visual entertainment materials.

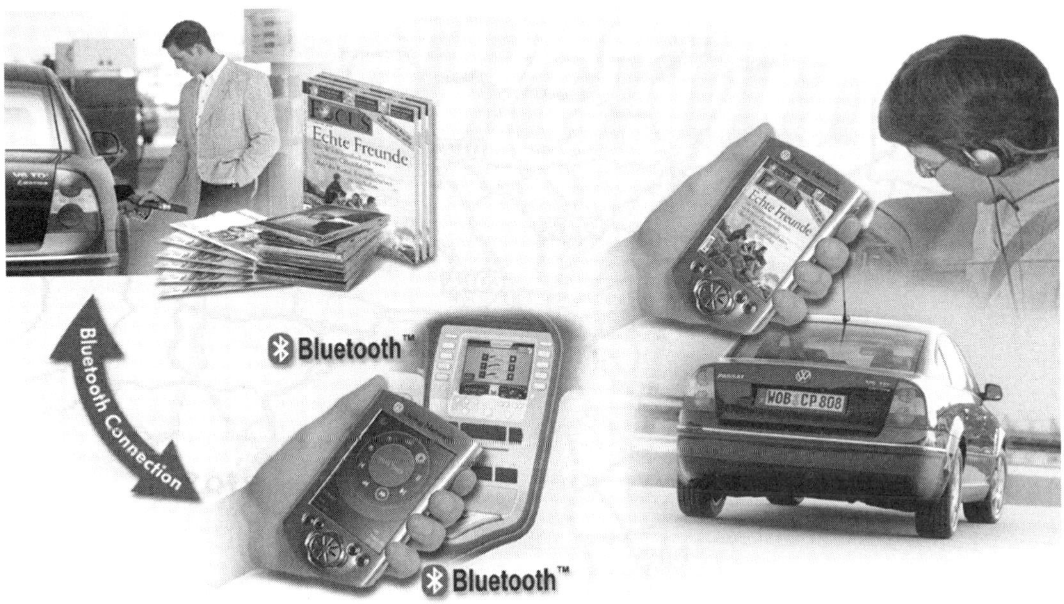

Fig. 21–7: Intelligent Region ... Service Station

Since the necessary transmission equipments either in a vehicle or at home belong to the customer, the cost of transmission is only the negligible charge of electricity that used to operate the data channels. The NetConnect Passat also offers remote control features on PDAs via Bluetooth technology or on mobile phones via SMS. The second generation of wireless technology is here in a vehicle, as a working chain action. By using the remote control features, doors can be locked or unlocked and seat positions can be adjusted remotely. Besides the research value of this technology, one particular remote control feature is very practical. The NetConnect Passat has an external lighting system, which can be turned on remotely via SMS or Bluetooth. In a dark underground car park, this feature helps to increase the safety around the vehicle and also the external lighting system can be used to help find vehicle in a car park.

21.3 Powertrain and Fuel Strategy

Throughout the progress of engine development, the pollution in exhaust emissions has been evidently reduced. With the implementation of the EURO4 standard, the percentage of pollutants is close to the detection limit, **Fig. 21–8**. To reduce carbon dioxide, the German automobile industry has pledged through its association, VDA, to reduce fuel consumption by 25% (based on 1990) by 2005.

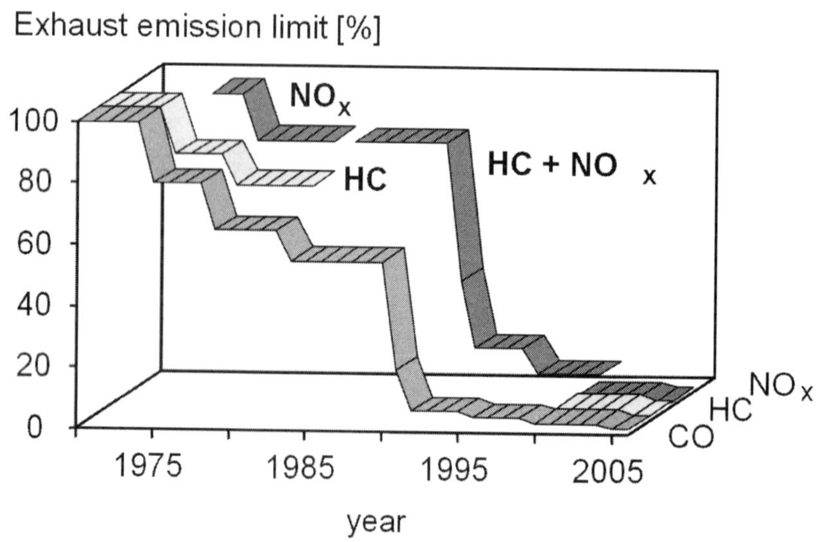

Fig. 21–8: Trends of exhaust emission limits in Europe

By using advanced combustion technologies like TDI and FSI, Volkswagen group's products have contributed significantly to the reduction of fuel consumptions and hence a reduction in CO_2 emissions. The average vehicle in the VDA fleet today emits approximately 32 g/km less than in 1990. In the year 2000, vehicles alone have saved 1,600,000 tonnes of CO_2 annually.

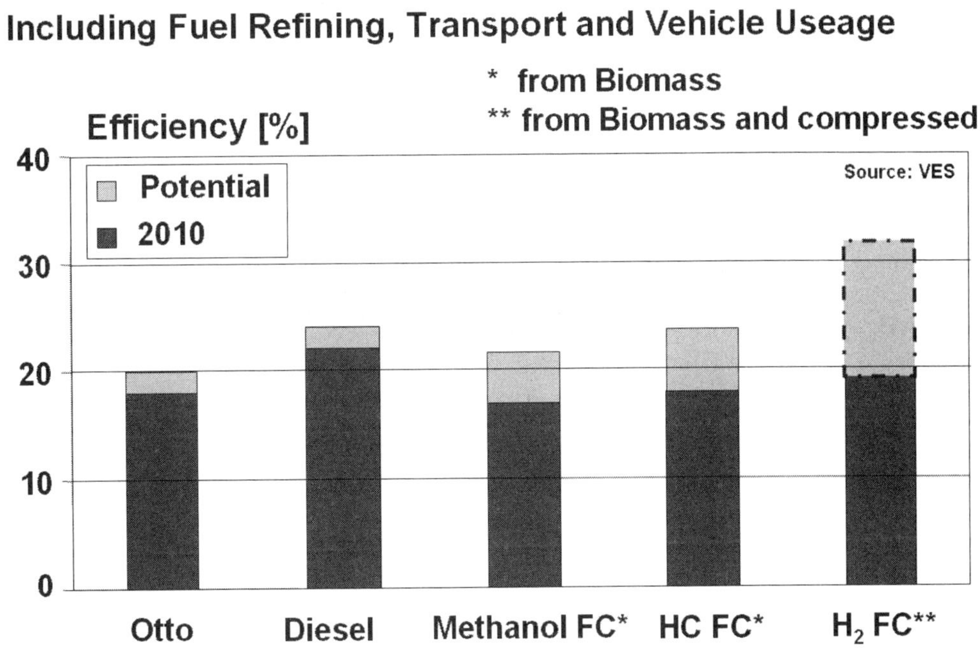

Fig. 21–9: Efficiency of modern Powertrain Concept

Hydrogen is considered by Volkswagen as a possible fuel in the future, in combination with fuel cell propulsion. Due to technological limitations in large-scale production of hydrogen, safety and storage of elemental hydrogen, Volkswagen projects that an appreciable number of vehicles using hydrogen (about 100 per day production) will not be in operation till 2015, **Fig. 21–10**. This would be the earliest time to expect to see CO_2 reduction from hydrogen fuel. Since this CO_2 reduction essentially comes from renewable energy, it is not coupled for the use of elemental hydrogen as an energy carrier. Due to this reason, Volkswagen is in favour with the synthetic designer fuel, "SunFuel", which comes from biomass, wood, or fast-growing plants farmed for this purpose, as fuel of the future. This follows the example of nature, to bind hydrogen to carbon in a highly calorific from, which creates a closed CO_2 cycle and is thus CO_2 neutral.

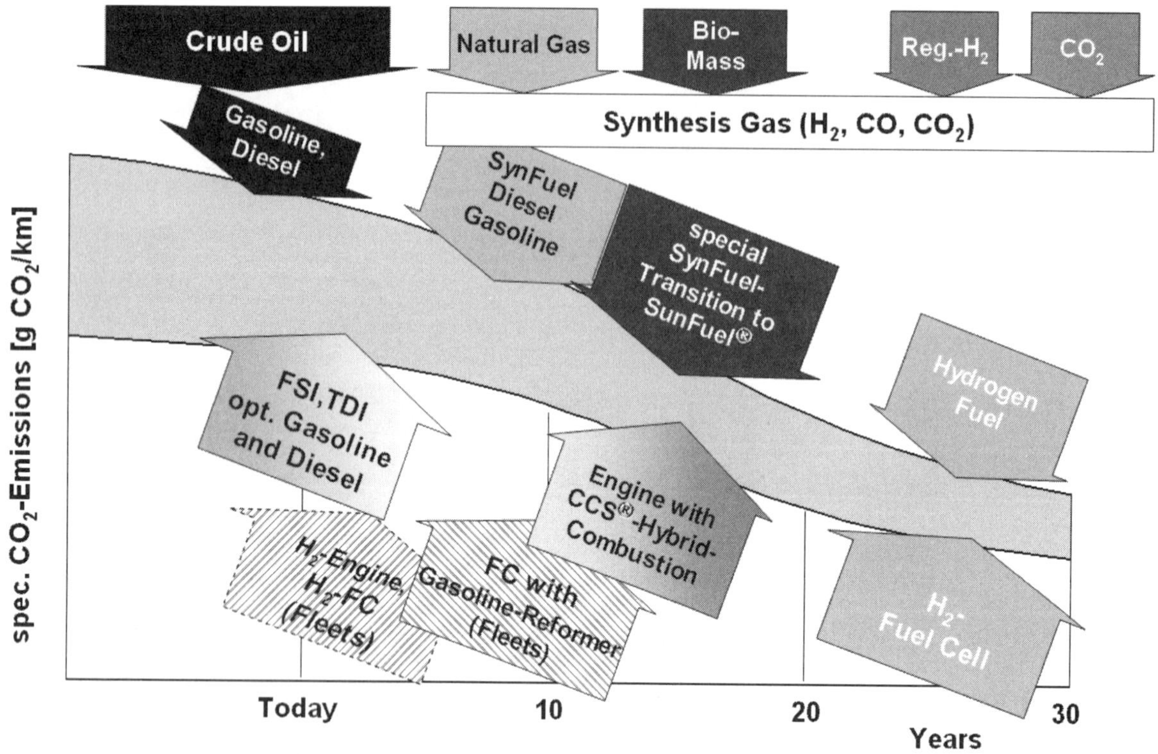

Fig. 21–10: Fuels and Powertrains

The designer fuel is very appealing as fuel for combustion motors that could be developed in closed loop, due to their pre-determinable characteristics. In the next generation of combustion systems and exhaust gas treatment, the conventional pollutants will also be barely detectable. This technology path will lead to noticeable environmental improvements within a short time. When combined with traditional fuel, it can also be applied to the existing fleet of vehicles.

Volkswagen is pursuing both approaches, Hydrogen and SunFuel. The predomination in the long run depends on the overall environmental and energetic balance, costs of vehicle and fuel, infrastructure, driving characteristics, and nevertheless, safety.

An example of a modern engine is the VW Fuel Stratified Injection (FSI) motor, **Fig. 21–11**. It runs on an adjusted, low-sulphur fuel. Direct injection will also become prevalent in spark injection engines in the future. It is the only most effective way to save on fuel. The FSI motor has high performance (1.4L has 63 or 77 kW; 1.6L has 81 kW) and is highly economical. In the frequent used partial load operations (city traffic or country roads), it can be driven in stratified charge mode and on average, burns only a lean fuel mix. Its exhaust purification is also a pioneering technology, which uses a $DeNO_x$ storage catalytic converter and a special NO_x sensor that monitors the build-ups in the catalytic converter.

Lupo FSI
- 1,4 l
- 77 kW (105 PS)
- 130 Nm @ 4250 min^{-1}
- Euro4
- 4,99 l/100 km

Polo FSI
- 1,4 l
- 63 kW (86 PS)
- 130 Nm @ 3500 min^{-1}
- Euro4
- 5,9 l/100 km

Golf FSI
- 1,6 l
- 81 kW (110 PS)
- 155 Nm bei @ min^{-1}
- Euro4
- 6,2 l/100 km

Fig. 21–11: VW FSI-Engines (Fuel Stratified Injection)

21.4 Minimal consumption vehicles

Comfort, driving performance and safety are the most important customer driven criteria for automotive research and development. With every new generation of vehicles, the requirements are perceptibly raised. This results in increased vehicle weight and more motorisation, retaining the existing expectations. To retain the driving performance, it is necessary to upgrade the undercarriage, and it is the same principle for tank volume and chassis stiffness. Thus resulting in higher weight and requires higher motorisation. This weight spiral (**Fig. 21–12**) can only be countered by a new light construction concept.

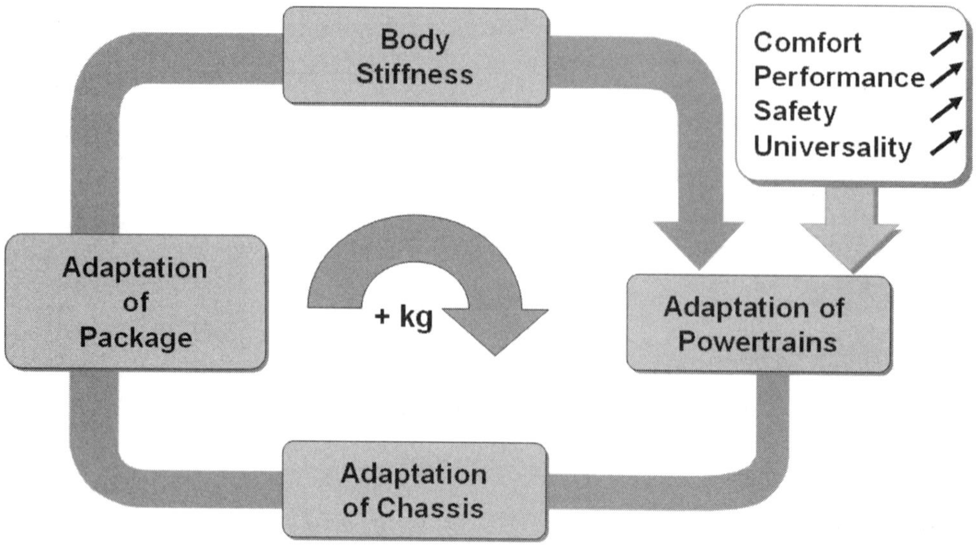

Fig. 21–12: Mass spiral

The 3 litre Lupo exemplifies a number of solutions, besides the 150 kg weight reduction compared to the standard Lupo, a number of other improvements have been implemented into the vehicle, **Fig. 21–13**.

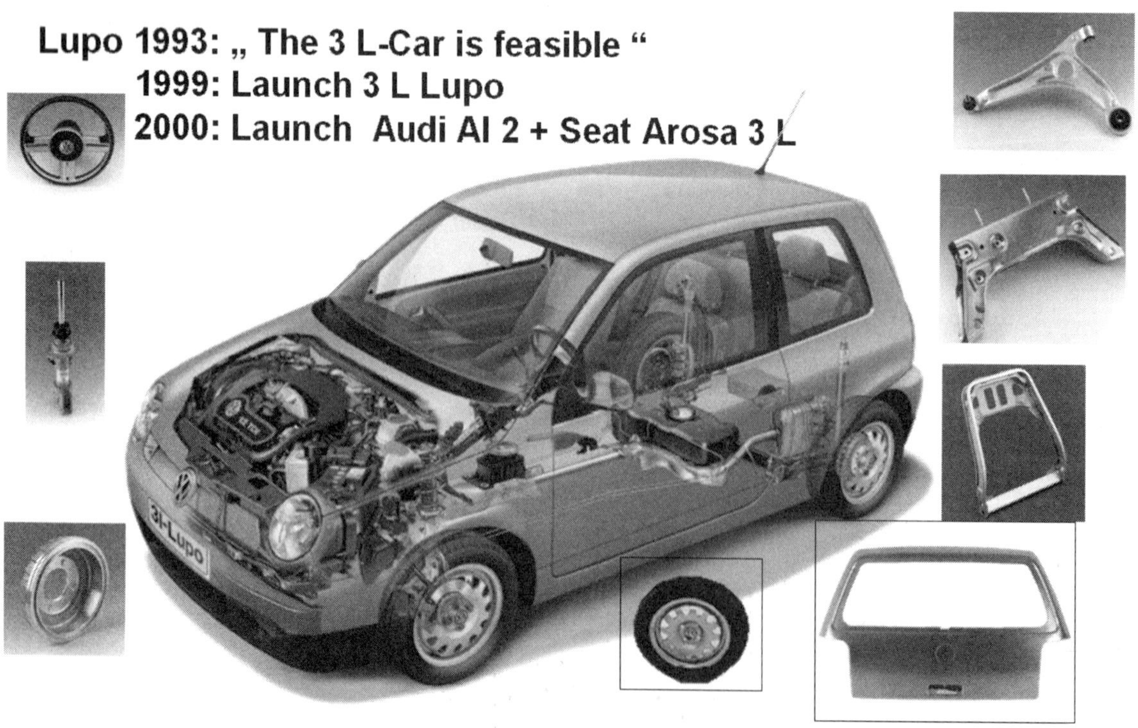

Fig. 21–13: The 3L - Lupo: -150 kg

Light construction materials are used for its tailgate. The standard tailgate uses conventional steel construction and it weights 10.5 kg. The newly designed tailgate for the 3 litre Lupo is made of aluminium-magnesium construction. It satisfies the additional requirements, with an integrated spoiler to reduce air resistance and which weighs only 5.4 kg. This is a 48% reduction in weight. This innovation has then moved extremely quickly from research into production.

It was a significant contribution of Volkswagen to the reduction of fuel consumption and emissions. The Lupo 3L TDI has the ability to compete with the best public transport systems in terms of energy consumption. As a vehicle, it is advantageous to combine optimal mobility in downtown areas and long distance driveability. Since the summer of 1999, the 3 litre Lupo has offered the opportunity for customers to experience mobility with future technologies and extremely low CO_2 emissions. From the well-known tests, the 3L series yields fuel consumption of 2.99L per 100km and was only attainable as a cumulative effect of improved vehicle detailing. The notable results are, the air resistance coefficient, exceptional for this class is at $c_v = 0.29$; the very efficient 3 cylinder TDI engine with unit injector technology; and the automated 5 gear direct shift transmission with Tiptronic mode. Overall, 80% of its components are newly developed.

Volkswagen has confirmed its technological leadership in the reduction of fuel consumption filed, with the 1 litre vehicle. It is a vehicle that uses 0.99L of fuel per 100km and it weighs 290kg in total. The weight reduction is achieved by the use of a combination of a magnesium spaceframe and an exterior shell of carbon fibre composite material. The vehicle has an extremely low air resistance coefficient of 0.159, and surface area of 1.0 m^2. Based on these boundary conditions, a maximum speed of 120 km/h can be attained with a one–cylinder, 6.3 kW motor.

With the demonstration of this achievement, it has proved that a fully functional vehicle for 2 people using 1 litre of fuel per 100km is not an unfeasible dream, **Fig. 21–14**.

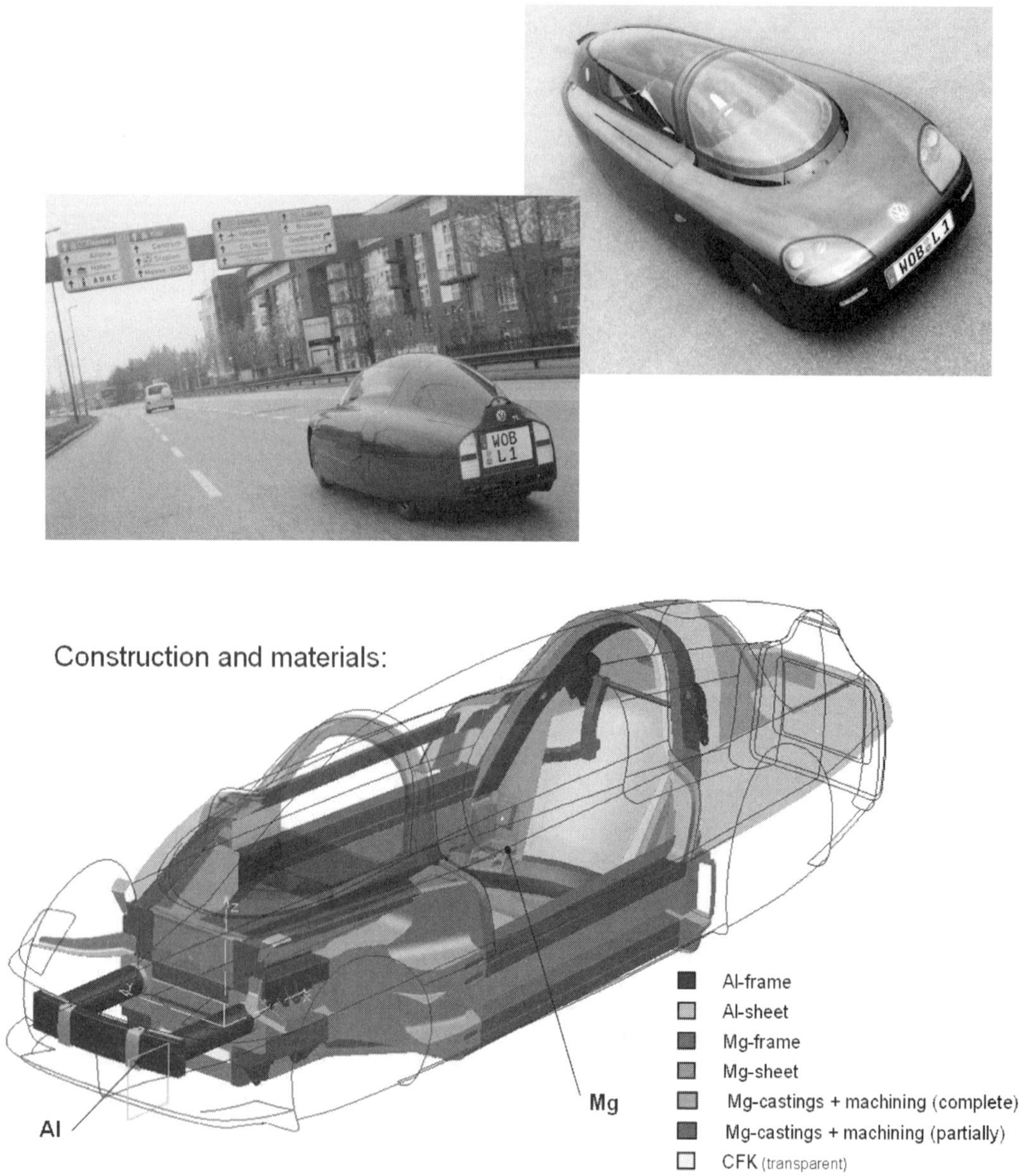

Fig. 21–14: The 1-Litre car: Demonstration of Feasibility

21.5 Vehicle safety

Despite increasing mobility and the consequent higher traffic loads, the number of people involved in traffic accidents has remained constant. The number of fatalities has rather decreased significantly. This inclination is a benefit from the introduction of various measures for active and passive safety, **Fig. 21–15**.

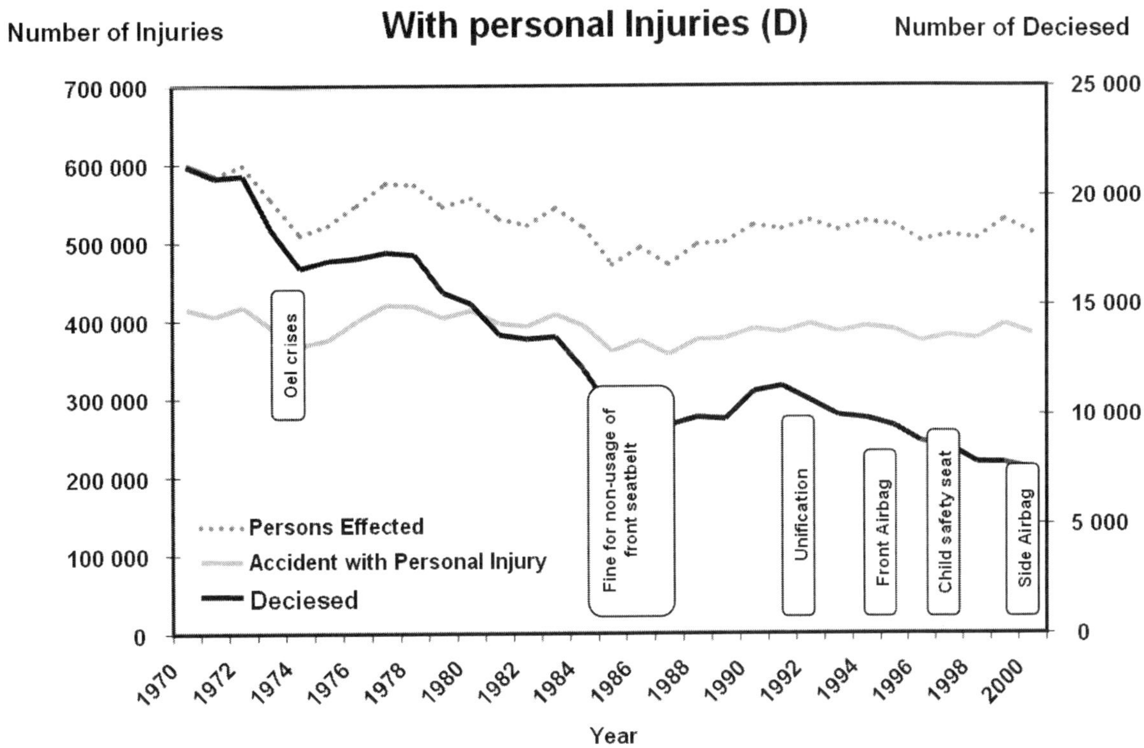

Fig. 21–15: Accident Data Analysis

New electronic options allow a new quality of accident prevention, especially through active safety. Pre-crash systems detect critical traffic situations earlier, and activate crash-active measures to avoid the accident entirely or at least reduce its severity. Typically, radar and video sensors are used for observations, **Fig. 21–16**.

A special pre-crash safety system, the automatic emergency break (ANB), is in the research stage. The system would automatically brake the vehicle with maximum deceleration if an accident is inevitable. The optimal triggering time is the critical criterion, since usually a collision is avoided by steering rather than braking. Thus, the system can brake only when the driver has no chance of avoiding the collision by steering. The swerving radii that determines this point in time depends on the relative collision velocity, road conditions, and obviously the geometry of the other vehicle. These data must be detected and processed by the system.

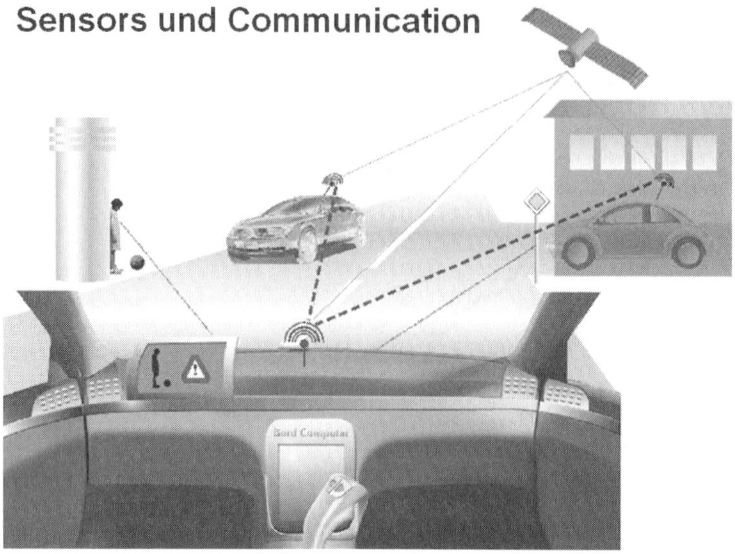

Fig. 21–16: Collision Avoidance - the Future?

As a step on the way to the development of an electronic co-driver, research has been conducted under the leadership of Volkswagen with supports from the government of Lower Saxony in the "Autonomous Driving" joint project, **Fig. 21–17**. The goal of the project is to continuously carry out automotive stress testing on vehicles in production on a special track in the VW Ehra-Lessien proving grounds. Where, up to 10 vehicles are able to drive the "Intensified Course" simultaneously, using precision locating, picture processing, distance sensing, and driving robots. A modular system is required for fast equipping of the vehicles in production. All components are so universal that they can be installed within a few hours. Almost every vehicle in production, even those with manual transmission can be operated this way.

The driving robot "Klaus", named after his human "father", puts the control commands of the trajectory-planning and the precision position location from a computer into action. He drives the course the same way as a human driver would do. He operates pedals, gear shifting (manual or automatic) and the steering wheel simultaneously. There are also servomotors for operating windshield wipers and blinkers. The dynamics of the actions are set up, so that a professional driver can be simulated (fast shifting speed) and critical manoeuvres carried out (like swerving).

Fig. 21–17: Autonomous Driving

The results gain in the "Autonomous Driving" joint project and other projects make major contributions to the development of an "electronic co-driver". The co-driver will support the driver in critical situations. The calculated steering and braking actions of the driver will contribute to the reduction in the consequences of accidents. At the end of the development, there will be reliable collision avoidance as results of information fusion from human driver, vehicle and surroundings. The driver always remains in control and being responsible, electronic systems are there to assist the driver.

In the field of active and passive safety, continuous vehicle improvements will be seen leading towards further reductions in accident statistics and serious injuries.

The need for individual mobility is essential, it could be seen as one of the basic needs in modern civilisation. By using the solutions and systems described earlier, the side effects of the increment of vehicles will be further reduced in the future. At the same time, transport performance, driving experience, and customer comforts behind the wheels will be further improved in the future. This ensures in the long run, vehicles will keep playing the dominant role in mobility of everyday life.